Theory and Methods of Metallurgical Process Integration

Theory and Methods of Metallurgical Process Integration

Ruiyu Yin
Central Iron and Steel Research Institute, Beijing, China

ELSEVIER

AMSTERDAM • BOSTON • HEIDELBERG • LONDON
NEW YORK • OXFORD • PARIS • SAN DIEGO
SAN FRANCISCO • SINGAPORE • SYDNEY • TOKYO
Academic Press is an imprint of Elsevier

Academic Press is an imprint of Elsevier
125 London Wall, London EC2Y 5AS, UK
525 B Street, Suite 1800, San Diego, CA 92101-4495, USA
50 Hampshire Street, 5th Floor, Cambridge, MA 02139, USA
The Boulevard, Langford Lane, Kidlington, Oxford OX5 1GB, UK

British Library Cataloguing-in-Publication Data
A catalogue record for this book is available from the British Library.

Library of Congress Cataloging-in-Publication Data
A catalog record for this book is available from the Library of Congress.

ISBN: 978-0-12-809568-3

Publisher: Matthew Deans
Acquisition Editor: Simon Tian
Editorial Project Manager: Naomi Robertson
Production Project Manager: Lisa Jones
Designer: Mark Rogers

Typeset by MPS Limited, Chennai, India

Contents

Author Biography

Ruiyu Yin was born in July of 1935 at Suzhou City of Jiangsu Province, China. He is a metallurgist with very high reputation in China, and is an Academician of the Chinese Academy of Engineering.

Ruiyu Yin has focused on the scientific and development strategic research of Chinese steel industry, especially on the strategic selection and continuous popularization of the technological progress in Chinese steel industry since the 1990s. He has paid his full effort to the breakthrough and popularization of several key technologies of steel industry. He studied theoretical problems of the physical essence of the dynamic operation and the operation dynamics of the steel manufacturing process, discussed the optimization of the operation mode of steel plants, and furthermore, he proposed the theoretical frame of the new generation steel manufacturing process. He prospectively proposed the new philosophy that the steel plant should have three functions, that is, steel product manufacturing, energy conversion, and waste treatment and recycling, which provide the support for the top-level design and key technologies for the establishment of the new generation steel plants. He also first put forward the theoretical frame and methods of the tailored dynamic design of steel plants, studied and described the concept of the high-efficiency, low-cost clean steel production platform, and its connotation and dynamic operation rules. He investigated and proposed the energy flow behavior of the steel manufacturing process and the theory and the design principle for the establishment of the energy flow network. He studied issues of the engineering practice of the cleaner production and circular economy of steel plants. He has published several books, including *Metallurgical Process Engineering* (both in Chinese and English), *Theory and Methods of Metallurgical Process Integration* (both in Chinese and English), *Philosophy of Engineering* (in Chinese), and *Theory of Engineering Evolution* (in Chinese), etc.

Foreword

Ruiyu Yin, an academician of the Chinese Academy of Engineering, is a successful scholar in the field of ferrous metallurgy in China, with strong academic background and well experienced in the organization and management of steel production. Ruiyu Yin worked as the vice manager and chief engineer of Tangshan Iron and Steel Corporation, the director of the Bureau of Metallurgy of Hebei Province, the chief engineer and vice minister of the Ministry of Metallurgical Industry of China, and the president of Central Iron and Steel Research Institute, China. Ruiyu Yin always devotes himself to improving technical progresses and equipment modernization of steel industries. When he was the vice minister of scientific technology and production for the Ministry of Metallurgical Industry of China in 1990s, he paid his extreme effort to the application and spreading of continuous casting in Chinese steel industries. As a result, within 10 years, 100% continuous casting was achieved in many large and medium key steel plants in China, and the national continuous casting ratio of Chinese steel industries quickly grew from 25% to over 87%, which narrowed the gap of steel industries between China and developing countries. Furthermore, the technologies and manufacturing process of steel production achieved a revolutionary promotion in China. This has been considered as one of the outstanding contributions of Ruiyu Yin to Chinese steel industries.

It is well-known that the steel industry is a typical process industry, in which numerous interconnected links of production and various smelting and processing equipment are involved during the conversion from main raw materials, such as iron ore, coking coal, and limestone, to steel products. For former metallurgical engineering, the manufacturing process was divided into individual reactors of ironmaking, steelmaking, ingot/continuous casting, rolling, and heat treatment according to the point view of reductionism, and the scientific investigation and description for these batch reactors are isolated or considered as closed systems. Undoubtedly, the classical thermodynamics is the basis for the chemical reactions and physical phase transitions during steel manufacturing process in these individual reactors or steps, such as the reduction of iron ore, decarburization during steelmaking, phase transition during solidification, deformation during rolling and forging, and phase transition during cooling. However, this micro and static way of thinking is hardly adapted to the process analysis and the integrated management of modern steel enterprises and leads to the fact that each process and reactor works in its own way without any coordination. To solve this problem, buffer reactors, such as torpedo tank for the mixing of hot metals, heating furnace for molten steel, and soaking pit before rolling, had to set up among continuous manufacturing processes (such as blast furnace, continuous casting, and tandem rolling mill) and intermittent production reactors (such as steelmaking furnace and cogging mill). More seriously, when disorder and contradiction by the production discordance at each branch plant, workshop, and process step occur, human orders of intervention and arbitration have to be employed. However, from the point view of the macroscopic physical nature and laws of dynamic operation of steel manufacturing process, mass flow, energy flow, and information flow go through the entire production process. The integration, configuration, and interaction of the three "flows" not only satisfy the steel production, but also have an influence on smooth running of the entire manufacturing process such as the production pace and labor productivity, on the product quality, and the environment, and meanwhile, they reflect the utilization efficiency of various factors, such as material consumption, energy consumption, and land and capital investment. Based on the deep study of the dynamic intersection and reasonable transformation of the three "flows," Ruiyu Yin has put forward the three major functions of the new generation of iron and steel manufacturing process, that is, clean steel production platform with low cost and high efficiency, clean and efficient energy conversion, and nonhazardous treatment process dealing with large amount of social wastes. This new concept of resource-conserving and environment-friendly iron and steel production has been realized during the design, construction, and operation of Shougang Jingtang Iron and Steel United Co. Ltd at Caofeidian region in China, which stopped formerly designing steel plants in China just by copying foreign industries, and stopped designing steel plants just by uniting static pieces of individual reactors without smooth system for the production.

Therefore, many issues during steel manufacturing process can be summarized as and upgraded into the rational cooperation and integration of mass flow, energy flow, and information flow. This new logic for the way of engineering thinking can be applied to the optimization control of process, the planning of metallurgical engineering, the design of iron and steel plants, the dynamic operation and management of steel production, also used as the assessment of the life circle of steel industries, accordingly, the effect of which on natural and social environment would be evaluated in a more scientific and reasonable way.

Based on the concepts and spirits above, Ruiyu Yin has performed prospective and rigorous thinking and studies on the integration theory and its analysis method of metallurgical process during the past 10 years, and he proposed the following theories: concepts and fundamentals of dynamic operation of the process, fundamentals and feature analysis of dynamic operation of steel manufacturing process, theories of dynamic tailored design, and integration of steel plants. In addition, a new generation of iron and steel manufacturing process is proposed, combined with several practical case studies. All the results mentioned above are introduced in this book.

Through the efforts done by generations of metallurgists in the past 60 years, China has turned into a leading country of steel production in the world with over 800 million tons per year and 46% of the world's steel production capacity from an iron-poor country of only 190 thousand tons and less than 0.1% of the world's steel production capacity in 1949. However, Chinese steel industries have been the target of public criticism due to their high consumption of resources and energy and their serious pollution and emission problems. In order to change the current view of our society to steel industries, it is essential to establish a resource-conserving and environment-friendly steel industry system, which is hardly achieved simply by starting up several waste water treatment plants, off-gas purification apparatus, or solid waste landfill sites. It has to start from the beginning and from the new manufacturing process and transform the ironmaking and steelmaking plant into a new manufacturing one with three functions of quality steel production platform, efficient energy conversion, and digester of social wastes, only by which the sustainable development of steel industries can achieved.

The current book "Theory and Methods of Metallurgical Process Integration" by Ruiyu Yin provides fundamental theories and guidance ideology for the transformation, and this book is his new and important contribution to the development of Chinese steel industries in the 21st century.

This book can be used as a text and reference book for faculties and students in the field of ferrous metallurgical engineering in colleges and universities, as well as an advanced reference book for senior engineers in iron and steel research institute and managers in steel industries.

Kuangdi Xu
Beijing
February 2013

Preface

I have worked in the field of metallurgical engineering for almost 60 years. From 1953 to 1957, I received professional education in the former Beijing Steel and Iron Institute, currently known as University of Science and Technology Beijing. Thereafter, I was engaged in production, technology, and research works in Tangsteel for 26 years. Afterwards, I dedicated myself to professional works, such as development strategy study, technological progress study, and industry management, in Hebei provincial department of metallurgy and ministry of metallurgical industry. After that, I worked in Central Iron and Steel Research Institute focusing on studies and explorations of process engineering and engineering philosophy. I deeply realized that work and study at different positions have benefited me with different visions and thinking methods, the combination of which was philosophical.

There are two reasons for me to write this book. First, there is still some further work to be done for the book *Metallurgical Process Engineering*; it took me 10 years to write the manuscript of almost 400,000 words under the advice and encouragement of Changxu Shi in 1994. It has been revised twice, printed three times, and published both in English and Japanese versions. Second, the advisory panel of total strategy study for the national long- and medium-term scientific and technological development was founded by the State Council in 2003 and it was my honor to be one of its members. Meanwhile, the committee of experts was established, consisting of 20 experts. The president of the Chinese Academy of Engineering, academician Kuangdi Xu, was concurrently the leader of the manufacturing industry group, which includes two subgroups, ie, process manufacturing subgroup and equipment manufacturing subgroup. Academician Shourong Zhang and I were in charge of drafting and writing the science and technology development program of process manufacturing. Academician Kuangdi Xu led the proposal of proposition, "the new generation of steel manufacturing process" and its concepts and connotations, which were established as a major research topic by the government. Meanwhile, the strategic relocation of Shougang started. Therefore, the proposition of the new generation of the steel manufacturing process was combined with the design, construction, and operation of Shougang Jingtang Steel. The concept study as well as the theory and method of the top-level design and dynamic-accurate design during process design was included. It is hard to break through these problems using traditional concepts and methods.

In the view of the theory and methods for metallurgy and metallurgical engineering, the methodology of the reductionism played an important role in contemporary science and the development of the modern science. In the traditional knowledge system, it is true that the steel manufacturing process has been divided into single production procedures or devices and then these procedures have been further divided into chemistry reactions and transformation processes of motion, heat, and mass. However, time and space are not included in the study of the steel manufacturing process. This relationship and some interface technologies among procedures for the dynamic operation have been ignored.

In the steel production practice, the traditional organization mode is that every procedure or devices operates, respectively, upstream procedures and upstream procedures wait for each other, then, the procedures combine randomly. The entire manufacturing process is in a state of disorder or chaos. For example, there are many phenomena of deadlock, waiting, and randomly combining in the entire process during blast furnace tapping→mixer→open hearth furnace→ingot casting→blooming/cogging→rolling. It is concluded that the production efficiency is low, the consumption is high, and the quality is unstable.

In the original research and development, the traditional method is to analyze and segment an independent procedure. This way of thinking has been limited by the idea of the "isolated system." The open and dynamic ideas are lacking, and the study of dynamic-orderly structure and coordinated operation in the process has been ignored.

At the same time, in the original process of manufacturing and engineering design, the production scale of an enterprise is calculated habitually by simply selecting and combining in series or in parallel from the appearance of the local operation of process/device. The study on the physical essence of the manufacturing dynamic operation and the innovation of the theory and method for dynamic-accurate design has been overlooked.

In order to design, build, and operate a new generation steel manufacturing process, some traditional ways of thinking and design methods had to be replaced, and the new integration theory and method needed to be proposed and created. Hence, the author has written 27 academic papers, research reports, and consulting reports, etc. for different demands and cases. During the writing process, the way of thinking and direction of the author has been expanded and deepened, and the new research works include the following.

The physical essences and running rules of the metallurgical manufacturing process dynamic operation have been studied. The basis of the theory has not only included Newtonian dynamics, classic thermodynamics, and chemistry thermodynamics but also paid attention to the application, transplant, and conversion of Prigogine's Dissipative structure theory. Based on all of the above, the physical essence of the metallurgical manufacturing process has been concluded and some new concepts such as flow, flow network, and operation program have been proposed. According to this theory, the metallurgical manufacturing process is developed to the direction of dynamic-orderly and coordinating-continuous operation. The minimum consumption of the manufacturing process will be achieved. From the viewpoint of research, all of the above is the concept exploration for a new generation of the steel manufacturing process.

On the basis of an understanding of the physical essences and operational rules of the metallurgical manufacturing process, the function of the steel manufacturing process has been expanded to three functions, including steel product manufacturing, energy transformation, and waste elimination (or recycling). Expanding the functions of the steel manufacturing process is helpful in order to increase the competitiveness and sustainable development ability. Furthermore, some new concepts, such as high-efficiency and low-cost clean steel production platform, behavior of energy flow, and energy flow network, have been proposed.

The theoretical research has been applied to the field of engineering design and metallurgical engineering discipline development besides manufacturing enterprises and research institutes. Actually, marketing competitions include not only product competition but also design competition, especially the competition of new concept research and top-level design. In order to achieve an optimized top-level design for the steel manufacturing process, a new design theory and method system should be studied and established. Procedure and equipment adoptions should be optimized according to the new top-level design. Hence, the perfect top-level design should include optimization adoption of procedure/device, expansion or reasonable arrangement of manufacturing process functions and high efficiency of manufacturing process operation. From another point of view, the innovation of engineering design philosophy, integration theory, and method should be promoted.

The dynamic-accurate design theory and method has been further studied, including concepts and connotations such as flow, flow network, and operation program. This theory includes the compiling rules and methods of dynamic operation GANTT and the innovation and design of interface technologies. A perfect engineering design should provide not only the engineering drawings but also the operation physical system frame for the manufacturing process.

The manufacturing process operation dynamics for a new steel plant or an existing steel plant has been studied. Under the principle of dynamic-orderly and coordinating-continuous operation, six rules of dynamic operation have been summarized. These rules have referential value for engineering design, production technology, production management, and technology management, etc. In other words, it means transformation and innovation of engineering thinking, ie, finding new ideas away from the deficiencies of reductionist methodology.

Seven chapters are included in this book, ie, introduction on overview and thinking method, concepts and theories of dynamic operation of the manufacturing process, basic elements in dynamic operation of the steel manufacturing process, characteristics and analysis of the dynamic operation of steel manufacturing process, dynamic design and integration theory of steel plants, case studies, engineering thinking and new-generation steel manufacturing process. The characteristics of the dynamic operation of the steel manufacturing process have been studied in the current book. It is helpful for readers to concentrate on the dynamic integration of mass–energy–time–space–information in order to realize further promotion of the steel manufacturing process and optimization of steel plant structure. The study of "space of point," ie, metallurgical reaction or physical phase change, and "space of field," ie, process/device, is not enough. Only the study of "space of flow," ie, the steel manufacturing process, can reveal new understandings, new concepts, and new rules of dynamic operation.

In conclusion, this book is the extension and furtherance of "*Metallurgical Process Engineering*." They can be considered as companions to each other.

Ruiyu Yin
Beijing
February 28, 2013

Acknowledgments

The author would like to mention: I have received a lot of support and help from many authorities and experts of metallurgy during my research, writing, and the summarizing process of this book. Academician Kuangdi Xu has always been concerned, supportive, and advocatory about the establishment and concept study of the new generation of the steel manufacturing process. He organized seminars in Shougang Jingtang Steel himself, to find solutions for production issues. He also wrote the foreword for the current book despite his busy schedule. Academician Shourong Zhang has participated in the discussion of concept, connotations, propositions, and engineering design plan. So to speak, he understands and supports the theoretical viewpoints in this book. Academician Yong Gan has put great effort into the organization and implementation of "the new generation of recycling steel manufacturing process." Academician Zhongwu Lu supports the theoretical study of energy flow behavior and energy flow network. The 356th Xiangshan scientific conference was held in 2009 under his suggestion. Chairman of the board Jimin Zhu, general manager Tianyi Wang, and Yi Wang (former vice general manager) from Shougang Jiangtang Steel have played important organizational and leading roles during the program establishment, design, construction, and manufacturing processes. They adopted and supported theoretical perspectives, technical measures, and design philosophies of the new generation of the steel manufacturing process proactively. The author would like to make grateful acknowledgment to them all.

My former college teacher, Professor Ying Qu, has participated in the whole process of writing and summarizing of the book. He proofread and revised the book chapter by chapter, section by section. A lot of helpful suggestions were provided by him. My young partners Dr Chunxia Zhang, Dr Anjun Xu, Dr Xiuping Li, Dr Dongfeng He, Master Fangqin Shangguan, Master Xuxiao Zhang, and PhD student Jicheng Zhou, Heng Yu, Weigang Han, Haifeng Wang, etc. have put a lot of effort and intelligence into this book. I would like to acknowledge them with best thanks.

I received help and support from experts in metallurgy during the writing of this book, especially in the section of case analysis. I would like to thank Professor Xinhua Wang from University of Science and Technology Beijing, vice manager Chunzheng Yang from Shougang Jiangtang Steel, factory manager Jinbao Chang from the second steelmaking and Rolling plants of

Tangsteel, general manager Wei He, vice manager Fuming Zhang, Professor senior engineer Shichong Qian, and division chief Jianxin Jie from Beijing Shougang International Engineering Technology, vice manager Jianfu Liu, Yixin Shi, and factory manager Weidong Wang from Shagang Group, for the data they have provided, which enrich the book.

Professor Naiyuan Tian is my lifelong companion, who has always been supportive of my work and theoretical exploration. She has promoted the teaching and popularization of metallurgical process engineering, as well as supported me writing this book, during her education career. Her concerns and support are essential to this book.

I have to mention the concerns and support from my senior Changxu Shi, whose reminders and encouragements are unforgettable. We met several times every year, and he asked about my work and research every time. He encouraged me to do theoretical studies and explorations, and emphasized the importance of combining theory with practice to promote the development of steel industries. His constant reminders, encouragements, and edifications will always be remembered and put into practice. With this book almost complete, I give my deep gratitude and respects to Mr Shi.

Writing and publication of cases in this book are helped and funded by IIP (Institute for Industrial Productivity), whom I would like to thank.

At last, I have to say, due to the limited theoretical level and knowledge scope, there may be many inappropriate places in this book. Any criticisms and corrections from readers and professionals are welcomed.

I love metallurgy. I love steel. I dedicate this book to domestic and overseas peers in metallurgy.

Ruiyu Yin
Beijing
February 28, 2013

Chapter 1

Introduction

Chapter Outline

In the 21st century, steel industries are facing a new comprehensive challenge, including complete optimization targeting operation cost, product quality and performance, production efficiency, reasonable scale, material consumption, energy consumption, process emission, ecological environment, etc. During the development of modern steel enterprises, engineering science propositions for solving comprehensive and integrated problems on the basis of fundamental science and technological science are put forward with the purpose of solving complex, systemic, and integrated problems on a larger scale and higher level. Therefore, metallurgical process engineering at the macroscopic level of engineering science is drawing more and more attention for study.

For metallurgical and metal materials science, if studies are only limited in the field of fundamental science or the level of technological science, it will be difficult to synergistically solve complex problems occurring in the production or engineering design of steel plants, and it will have a detrimental effect on the effective application of information technology to the throughout manufacturing process of steel plant.

Process engineering is a very wide discipline that is expanding its borders. The traditional discipline of metallurgy is being supplemented by or interacting with some new knowledge disciplines. Nowadays, isolated individual studies on the scientific problems of process industries, including the steel industry, are out-of-date. We are facing an era of engineering science with interacting disciplines and multiple levels.

In the old days, we learned to analyze problems using abstract and static viewpoints and isolated methods. However, in modern society, dynamic viewpoints have been employed in almost all fields of science. Systematic, open, complex, irreversible, and probabilistic objects exist everywhere, which can

Theory and Methods of Metallurgical Process Integration.
DOI: http://dx.doi.org/10.1016/B978-0-12-809568-3.00011-5

hardly be solved by pure and classical thermodynamics methods. The thermodynamics of an equilibrium system is hardly suitable for the open, complex, irreversible, and probabilistic propositions at a process engineering level.

The steel manufacturing process is a nonequilibrium dynamic open system, which is constructed by the integration of a series of related, different functional and heterogeneous units (or procedures) with random fluctuations. A certain self-organization is formed by the fluctuations and nonlinear interactions among units or unit-processes. Hetero-organization measures are needed in order to improve the self-organization level and generate the dynamic-orderly and synergistic-continuous dissipative structure of steel manufacturing processes. These measures include the design of optimized and simple process networks, the compilation of dynamic-orderly running programs, and the dynamic-reasonable adjustment of mass flow and energy flow intensity (throughput) by information technology, etc. It is aimed at controlling the unit (procedure) under dynamic operation within a reasonable fluctuation range and creating a dynamic-stable nonlinear coupling relationship among units so that the self-organization level of the manufacturing process is improved and the desired dissipative structure is created. This dissipative structure reflects the synergy, continuity, compactness, and rhythmicity of the entire dynamic operation process, and eventually achieves the minimal amount of energy dissipation, materials consumption, and process emission.

In the dynamic open systems of the steel manufacturing process, deliberately and unilaterally pursuing local equilibrium may even cause chaos-disorder of the entire system, and lead to an increase of dissipation during the operation process.

After the establishment of the static frame of the steel manufacturing process, for example, the general layout, the basic parameters that keep the "synergistic-orderly, continuous-compact" running of the process system are "mass," "time," and "temperature." This means that the "mass," "time," and "temperature" are basic parameters throughout the entire process of steel manufacturing and all the heterogeneous procedure units.

The dynamic operation of steel manufacturing processes is hardly independent of spatiotemporal parameters. The concepts of "process," "dynamic-orderly," and "continuity-compactness" are all directly related to spatiotemporal parameters. In particular, the irreversibility of the time arrow requires great attention. The formation of the dissipative structure and the amount of process dissipation are affected basically by the time factor. The issues of time order, time point, time interval, time location, time cycle, time rhythm, temporal sequence, spatiotemporal structure, etc., are rarely considered in classical dynamics. However, they are commonly and inevitably relevant during the practical manufacturing processes in steel plants due to the simple fact that the scientific propositions in current process engineering are hardly a single and static target problem in an abstract and isolated system but a comprehensive optimization of dynamic problems with multitarget and multiscale.

As an open nonequilibrium system-oriented self-organization theory, the dissipative structure theory (Prigogine and Nicolis, 1989) reveals features and rules of a dynamic open system on the basis of general physical science and it solves the theoretical issues of fundamental science. However, long-term collaboration and the efforts of experts and scholars in related fields are required to transform the general theory of fundamental science to engineering science, and then to the theories and methods of engineering design, to the controlling and adjustment technologies and organization methods of industrial operation, and furthermore to lead to the development of new technologies, especially generic and key technologies. We may be now in the subsequent era of utilizing the theories and methods of dissipative structure and synergetics (Haken, 1983; Holland, 1995) to promote the development of process engineering, supervise the engineering design, production operation, and investment direction of companies, and the industrial upgradation of process manufacturing.

As the sister book of *Metallurgical Process Engineering*, the current book contains the fundamental theories of open and nonequilibrium thermodynamics, which have the features of uninterrupted exchange of materials, energy, and information between process systems and environment. In other words, the system structure coexists with environment. Negentropy constantly inputs into the process system to compensate for the generation of entropy during the operation of the process system. An open system is always dynamic. Besides the meaning of material and energy, dynamic operation also has the meanings of time and space, and of course, the meaning of relevant information. It is meaningless and impossible to discuss the dynamic operation without the spatiotemporal concept. The problems encountered in steel plants can be divided into three categories. The first problem is the process of the relation between molecules and molecules, atoms and atoms, such as the reduction process and the oxidation process. The second problem is the operational process of reactors or procedures, such as the operational process of the blast furnace, basic oxygen furnace (BOF), and continuous caster. The third problem is the process of the entire operation of the workshop or plant. The three categories have hierarchical and structural relationships and are expressed as a hierarchy, structure, and integrality. Therefore, in order to achieve the multiobjective optimization of the dynamic operation of the entire process, the minimization of the process dissipation must be the foundation. To achieve the minimization of the process dissipation, a proper dissipative structure is required, since a bad structure increases dissipation of the process and vice versa. So, what is structure? Structure consists of the integrity of five elements: material, energy, time, space, and information. Factors of nodes and elements and the interactive relationship and hierarchy between factors are definitely involved in the structure. Especially, the dynamic structure always includes the spatiotemporal concept. For the static structure, the concept of time is fuzzy, but it has the concept of space and space factors are usually solidified by the fixed general layout. Under certain conditions, certain space factors during the dynamic operation of the process can be indirectly converted to the effect and value of time. In short, the structure of the production process is the interaction relationship of various procedures or nodes, as well as the program for dynamic running.

The steel manufacturing process is an integrative dynamic operation system consisting of several interrelated procedures with different running patterns. It includes both continuous and intermittent operation modes. In order to organize these different procedures/devices into a coordinated running movement, one of the most important parameters of movement, namely time, must be analyzed as an objective function. This enables the transformation of unpredictable problems into dynamic-orderly movement under the guidance of the theory of dissipative structure and self-organization, that is, to organize the problems into dynamic-orderly, synergistic-continuous "flow." Since the movement of "flow" has features of dynamics, directivity, and process in the spatiotemporal concept, its output and input are characterized by the feature of dynamic vector. "Process network" is actually the spatiotemporal framework of "flow" during its operational process. Therefore, the "process network" has to be simple, compact, and optimized in order to minimize the process dissipation, otherwise, disorder and chaos will be caused during operational process of the "flow." "Process network," for example, the general layout, directly displays space concepts and space factors, while actually it indirectly or directly expresses the factor of time, and expresses certain procedural rules of time as well. This is one of the important principles for the new design or technological reformation of steel plants.

The dynamic process nature of "flow" definitely implicates the "running program." To a great extent, the "running program" of the process is the information command by "hetero-organizations." A proper "running program" promotes the "flow" to a dynamic-orderly and steady state. Otherwise, the "flow" enters a chaotic and unsteady state.

In order to realize the optimization of the steady input/output process of the "flow" in an open system, a proper "process network" and optimized "running program" should be built up so that a functional, spatial, timing self-organizing system, that is, a dynamic-orderly operation system, can be generated by the nonlinear interaction between "forces" and "flows" in the open system. In other words, nonlinear interacted fields (a dynamic structure) can be achieved by building a proper and optimized "process network," such as the general layout, with the movement fluctuations of "nodes" and the coherent and collaborative relationship among the fluctuations of the "nodes." A nonlinear coupling among each operation process steps/reactors in an open process network can be achieved by compiling a self-organizing and hetero-organizing program which reflects the physical natures of the dynamic-orderly process operation. A proper "dissipative structure" of the open system can, therefore, be generated, minimizing the dissipation of "flow" under the dynamic-orderly running process.

Further exploration of theories and methodologies for the design and production dynamic operation of steel manufacturing processes can be performed based on holism, hierarchy theory, and dissipative theory. The following theories and methods are involved.

1.1 DYNAMISM

According to the observation and analysis of the operation mode of different procedures/devices during the steel manufacturing process, the essentials of the operating processes and the actual operating modes of various procedures/devices are different. For example, the operation process of a sintering machine and blast furnace are essentially continuous, while the operation process of a BOF and electric furnace are essentially intermittent. The operation of hot metal pretreatment and secondary refining processes are also essentially intermittent, while the operation process of caster is continuous or quasicontinuous. The operation process of the reheating furnace is continuous or quasicontinuous, while its mode of delivering billets or slabs is intermittent. The rolling process is continuous, while the input/output of rolling pieces is intermittent. The semiendless rolling and endless rolling favor the improvement of the continuous operation of rolling mills. Through further overall observation and studies on the coordinative operation process of the entire steel manufacturing process, the roles of macroscopic dynamics of different process steps/reactors in the coordinative operation process can be revealed. From the perspective of running time of ferrous material flow in steel plants, different procedures/devices play the roles of "pushing force," "buffer," and "pulling force" in order to achieve a smooth, harmonious, and continuous timing progress of the steel manufacturing process.

Generally, the steel manufacturing process can be divided into two sections: the upstream section from ironmaking (including pelletizing or sintering) to continuous casting and the downstream section from continuous casting billet/slab to the end of hot rolling. The upstream section is mainly related to the chemical metallurgical and solidification process, that is, the liquid metal process, while the downstream section is the physical process for the transportation, storage, reheating, rolling, deformation, and phase transition of rolling pieces, namely, the solid process of high-temperature metals. Therefore, the analysis-integration of operation dynamics of the entire steel manufacturing process can be performed from the dynamic characteristics of the continuous operation of three continuous/quasicontinuous procedures of the blast furnace, continuous caster, and hot rolling mill. For the upstream section, the continuous operation of the continuous caster is the "pulling force," the continuous operation of the blast furnace is the "pushing force," and the intermediate procedures/devices (steelmaking furnace, hot metal pretreatment and secondary metallurgy devices) are the "buffer" during continuous operation of ferrous material flow. For the downstream section, knock-out of the billets/slabs from the continuous caster is the "pushing force," continuous operation of the rolling mill is the "pulling force," and intermediate process/devices (billet/slab stock, reheating furnace, and rollerbed) are the "buffer."

The operation rules of dynamism of continuous/quasicontinuous operation, that is, macroscopic coordinative operation rules of the process, are supposed to be as follows. The operation rules of intermittent running procedures/devices should be adapted and obey those of continuous running procedures/devices. The operation rules of continuous running procedures/devices should be the guidance and regulation of intermittent running procedures/devices. Continuous running of procedures/devices under low temperature should be adapted and obey those under high temperature. The mass flow among procedures/devices should keep in laminar type running when there are serial–parallel structures of procedures/devices. The interactive dependence compact layout between upper streams and down streams is the prerequisite of "laminar type running." In summary, the macrodynamic mode characterized by coordinating the continuous "pushing force," "buffer," and "pulling force" should be established.

1.2 STRUCTURITY

In order to achieve specific functions and excellent efficiency, the modern steel manufacturing process should have a reasonable and dynamic-orderly process structure established by the fundamentals of dynamism and synergism on the basis of holism, hierarchy theory, and dissipative structure theory, instead of simple stack, superposition, and assembly of elements (procedures/devices). The manufacturing process structure means the set of procedure units with different specific functions and nonlinear interaction relationships among procedures under certain conditions in the manufacturing process. The connotation of process structure does not merely mean a simple quantitative accumulation of and proportion of procedures in the manufacturing process. More importantly, it means the function set optimization and the relationship set adaptability of procedure units, the rationality of spatiotemporal relationships, and the dynamic synchronization of the whole manufacturing process. Therefore, the functions and parameters of procedure units in the manufacturing process should be analyzed and integrated under the requirement of optimization of the entire process. The objective is to optimize the entire dynamic operation, an objective of which is the optimization of procedure functions and parameters and relationship among procedures, including top-level design and hierarchical structure design. As a result, the optimization of dynamic operation of the manufacturing process can be promoted according to the coordination and integration among different levels. The detailed contents are as follows:

1. Optimization function of different procedures/devices are orderly and connectedly chosen, distributed and arranged, and coordinated properly so that a procedure function set of analysis-optimization can be established.
2. Interconnection and synergic relationships of different procedures/devices are established, distributed, and coordinated properly so that the synergetic-optimal relationship set among procedures can be established.
3. The procedures/devices of the new generation of manufacturing processes are integrated and evolved based on analysis-optimization of their function set and their relationship set. The reconstruction-optimization of procedures/devices and the emerging effect of overall operation of the manufacturing processes are achieved, for example, a clean steel manufacturing platform with high efficiency and low cost, high conversion efficiency and recovery energy flow network, so that the emergence of the new generation of manufacturing structure can be promoted.

1.3 CONTINUITY

Time is an indispensable factor for the study of dynamic operation of the open metallurgical processes. Variation of everything and evolution of various processes are spread out in the time coordinate. There is no motion and variation, even manufacturing process and procedure, if the time concept does not exist. Motion speed, variation rate, dynamic coupling, and coordinated operation are all expressed by time. For the need of coordination, continuity, and easy control of operation of the manufacturing process, the running pattern in the process of time should be studied. The form of time can not only be shown as time length of certain processes, but also time order, time point, time interval, time position, time rhythm, time cycle, etc. (Yin, 2009). In order to realize the continuation or quasicontinuation of the manufacturing process, the time factor is not only an important independent variable, but also an objective function to analyze its effects and meanings during the coordination and integration of procedures/devices. For process industries including ferrous metallurgical industries, the continuation degree of the manufacturing process is the symbol of technical progress, market competitiveness, and sustainable development potential, which also is becoming one of the economic targets of scientific-technological circles and business circles.

Based on the overall proposition of continuation/quasicontinuation and compactness of the manufacturing process, the operation target must focus on the dynamic-orderly and coordinating-continuous manufacturing, emphasizing the study on the integration of optimization of the running program and the network integration of dynamic running. Thus, an overall optimized-continuous (quasicontinuous)-compact manufacturing process and engineering system should be established by efficiently connecting, matching, coordinating, and interpenetrating the entire manufacturing process.

1.4 EMBEDDING

The so-called embedding means to properly and effectively embed low-level operating (moving) processes in a smaller spatiotemporal scale, into larger, more complex, higher level open systems. It is actually a multiscale integration theory and method, including "hierarchical embedding" requirements and "coordinating-continuous embedding" requirements, etc.

Combining the hierarchy concept of process engineering and the multi-scale "embedding theory," process operation in different levels can be brought into a unified and nested coordination system, which favors the network integration of the process system and the establishment of models for the dynamic running program.

Dynamic operation should be considered during design, production, and management. Dynamic operation must focus on the spatiotemporal concept, the hierarchical concept and cross-level coupling, that is, the choice and optimization of unit operation and elementary reaction must be connected to the dynamic running program of the procedure/device, and the dynamic running process of the procedure/device should be embedded orderly and efficiently into the dynamic running process from a higher level. This means that for the dynamic running of a manufacturing process, the running (moving) process from different levels should be able to be integrated together. And the principle of integration is that the process of molecular scale movement should be able to be embedded into the relevant dynamic operating process of the procedure/device scale, which should be able to be embedded into the dynamic running process of manufacturing process scale.

There are also "embeddability" issues for operating processes at the same level, for example, the dynamic operating process of procedures/devices because the synergistic-continuous running of the overall manufacturing process requires the "embeddability" of parameters such as the spatiotemporal-function set of the running of different procedures/devices. Sometimes the support of an optimized interface technique is also required to achieve the synergistic-continuous running between upstream and downstream procedures/devices. Whereafter, dynamic-orderly, continuous-compact, synergistic-steady running of the entire manufacturing process can be realized.

1.5 SYNERGISM

Synergetics is an academic discipline regarding an overall system consisting of plenty of subsystems. It studies how, under certain conditions, coordinating phenomena and coherent effects are generated through nonlinear interaction of subsystems so that the system obtains a self-organizing structure with certain functions; and then how function structure, time structure, space structure, or spatiotemporal structure at the macro level are generated and meanwhile a new, overall, and orderly state is reached. Therefore, synergism is a self-organization theory about how multicomponent systems lead to an orderly evolution of the overall structure through the coordinated action of subsystems.

In the dynamic running of a process system, as a theory and method, synergism helps to identify the nonlinear interaction caused among procedures/devices, and helps to achieve nonlinear interaction and dynamic coupling to stabilize the system. Specifically, according to synergetic theory and methods, the tool design, production operation, and management control can be implemented by compiling the Gantt charts for the dynamic operations of raw material yard → coking → sintering (pelletizing) → blast furnace process, blast furnace tapping → hot metal pretreatment → BOF → secondary refining → continuous casting process, slab transport → reheating furnace → rolling mill → coiler/cooling bed process. Furthermore, through synergetic theory and methods, new interface techniques between procedures/devices can be established, favoring the networking/programming integration in the manufacturing process and process engineering. This can lead to a new generation of manufacturing processes, and the new interface techniques within, to be established.

1.6 FUNCTIONALISM

For steel industries, the manufacturing process has a feature of extensive relevance and strong penetrability. The overall functions of the steel manufacturing process (manufacturing of steel products, energy conversion, and disposal and recycling of wastes) are reflected by the combination of functions of multiple processes, but it is definitely not a simple stack and superposition of functions of each unit procedure. Functions of each procedure must be coordinated and cooperate with each other to achieve a perfect function realization of manufacturing processes. The manufacturing process has great influence on the factors for market competitiveness of the corresponding enterprise (including comprehensive cost, material/energy consumption, product quality, product variety, production efficiency, investment benefit, etc.) and many aspects of sustainable development (such as the availability of resources and energy, the effect of emissions and their impacts on the environment during production process, the establishment of eco-industrial chains, disposal and recycling of social wastes and their recirculation in the circular economic society by ecological transformation). Therefore, the manufacturing process should not be limited only to the technological issues of metal production. The steel manufacturing process is directly related to the orientation, social role, and economic role of steel plants. Through the study of the physical essence of the dynamic running of steel manufacturing processes, it is revealed that modern and future steel plants should focus on the functions of steel production, energy conversion, and the disposal-digestion-recycling of wastes.

From the perspective of engineering philosophy, the engineering design and production operation nowadays focuses on local and analytical "substantiality," but often ignores the general, emerging, and integrated "flows." For future engineering design, production operation, and process management, both specific-local "substantiality" and generally integrated "flows" should be also emphasized. The effect of process and its dynamic running will be inefficient if the concept of "flows" are excluded. In essence, the design and operation of units are just illustrations of the appearance and local expression of the operation of overall "flows." Namely, the "substantiality" of procedures/devices is a part of the dynamic composition and operation of manufacturing process running. It is the "soul"—the objective of the production operation and engineering design of an enterprise—to achieve a dynamic-orderly, compact-continuous running of "flows," that is, to form a dissipative structure and process optimization. It is of importance to combine the "virtuality" with the "substantiality" for both the engineering design and production operation of a plant. The "virtuality" concept should be employed by establishing a dynamic-orderly, synergetic-steady, continuous-compact open system and an optimized dissipative structure. Then, the "substantiality" can be achieved by the application of these concepts through the engineering design by designing institutes and the dynamic production of enterprises, by which the minimum dissipation, the environment friendliness, and the multiobjective optimization of complex systems are realized. This is the meaning of dynamic-tailored design mentioned in the current book.

The production mode of a steel plant is definitely involved in its dynamic-tailored design. The optimization of the mode is closely related to the spatiotemporal process, dissipative structure, and dissipative mechanism, that is, related to the optimization of the process network structure and the dynamic operation program. Optimal parameters for engineering design and manufacturing operation can be further derived by the optimized mode of the steel plant. It is worth mentioning that a large production scale does not guarantee a better production mode. Instead, a more reasonable process time of ferrous material flow should be achieved by dynamic-orderly and coordinating-continuous operation. In a word, the theoretical core of dynamic-tailored design is that the dynamic-orderly and continuous-coordinating feature should be embodied during the engineering design, based on the perspective integrated optimization of element-structure-function-efficiency.

The general idea of this book is to highlight the theoretical system of the open system on the basis of the book of *Metallurgical Process Engineering*. The current book includes concepts and theories of dynamic operation of the manufacturing process, theories and methods of dynamic-tailored design, and the dynamic operation and control of the manufacturing process, as well as the description of the future functions of steel industries. The current book starts from systematic points of "element-structure-function-efficiency" and discusses the design idea, operation rule, and dynamic control of the dynamic operation of the manufacturing process. Four cases, including interface techniques and the clean steel platform, are analyzed based on practical industrial experience. The current book is the extension and spread of theories and their application in the field of engineering design and production operation of the book of *Metallurgical Process Engineering*.

REFERENCES

Haken, H., 1983. Advanced Synergetics: Instability Hierarchies of Self-Organizing Systems. Springer, Berlin.

Holland, J.H., 1995. Hidden Order: How Adaptation Builds Complexity. Addison-Wesley, Reading, MA.

Prigogine, I., Nicolis, G., 1989. Exploring Complexity: An Introduction. WH Freeman, New York, pp. 6–26.

Yin, Ruiyu, 2009. Metallurgical Process Engineering, second ed. Metallurgical Industry Press, Beijing, pp. 205–208. (in Chinese).

Chapter 2

Concept and Theory of Dynamic Operation of the Manufacturing Process

Chapter Outline

Theory and Methods of Metallurgical Process Integration.
DOI: http://dx.doi.org/10.1016/B978-0-12-809568-3.00012-7

The social and the economic development can hardly be isolated from the materiality engineering activities. Science, technology, and engineering interact closely and mutually promote each other. There is one kind of engineering that is developed under the guidance of the scientific research discovery. Namely, first the laws of science are revealed and the technology is therefore developed, then the engineering entity is formed, finally the new industry is established, such as the electronic information engineering and nuclear power engineering. However, there is another more popular engineering, namely, the exploration and discovery of scientific principles is provoked by technical innovation and invention based on the need of life and production and engineering practice. The development of science and technology is used to promote the progress of technology so that technology and engineering reach a new level. As a form of process engineering, metallurgical engineering belongs to this category.

Steel, as the fundamental material of the national economy, is regarded as the "material of choice." Because of its manufacturing process, the steel industry belongs to a process manufacturing industry. For the steel industry, manufacturing has a comprehensive influence on technical and economic target groups, such as production efficiency, product cost, product quality, product variety, resource consumption, and investment returns, and the sustainable development factors, such as emissions and the environmental ecology and availability of resources—energy.

Facing the tendency for economic globalization in the 21st century, the essential propositions of steel industry in the whole world are market competitiveness and sustainable development. The steel industry is requested to observe and solve problems regarding the overall strategy, instead of simply separating the significant propositions above into isolated issues of output, scale, quality, etc. Looking to solve individual issues on their own will only raise other issues, whereas it is more effective to consider the issues altogether. Nowadays, the contemporary proposition should be considered as a whole and be solved from an overall strategy.

In order to correctly solve this contemporary proposition, the process of "understanding, decision-making, and implementing" should be experienced. Understanding is to understand and consider the objective laws of things. Decision-making is to take the best, foster strengths and circumvent weaknesses, and make their own path on the basis of an in-depth understanding. The implementation involves transforming the strategies and decisions into the achievement of materials, thus validating the accuracy of the understanding and decision-making. The correct understanding is the starting point of all subsequent steps.

2.1 PROCESS SYSTEM AND BASIC CONCEPTS

2.1.1 Process Manufacturing Industry

Process manufacturing industry usually refers to an industry by which raw materials are transformed into products with special physical and chemical properties and special use through a series of processes. Sometimes it can be also called process industry to highlight the continuous processing, modification, and deformation features of the mass flow during the technological process. The production features of a process industry include: the "mass flow" consisting of various raw materials, the logistics, processing from heat transfer, mass transfer momentum transfer, and physical and chemical effects following special processes and with the motivation and interaction with the energy input, and conversion into anticipated products. In the production process of a process industry, their operation at each procedure (devices) are diversified, including chemical and physical conversion with continuous, quasicontinuous, and batch type operation modes.

Process manufacturing industry includes the chemical industry, metallurgical industry, petrochemical industry, building materials industry, papermaking industry, food industry, medical industry, etc. Specifically speaking, these process industries usually have the following features:

- Raw materials used are mainly from nature.
- Products are primarily used as raw materials for equipment industry; therefore, many categories of process industry have features of raw material industry. Some products of certain process industries can also directly be used for consumption.
- The production processes are mainly continuous, quasicontinuous, or are being developed into continuous, but some of them are batch.
- Raw materials are transformed into products or by-products through chemical–physical transformations in the form of mass flow and energy flow.
- The production processes are often accompanied by various emissions.

With respect to process manufacturing industry, the manufacturing process often contains the storage, transportation, and pretreatment of raw materials and energy, the reaction processes and the processing of reaction products, and it also involves the auxiliary materials and energy supply system connected with reaction processes for realizing the functions of the manufacturing process. The broadening connotation of the manufacturing process can be also understood generally as the process including the selection, storage, and transportation of materials and energy; the selection and design of products; the design and innovation of process structure; the control, utilization, and treatment of emissions and by-products; the treatment and elimination of toxic and harmful substances; and the discarding or recovery (recycling) of used products.

For process manufacturing industries, their manufacturing process is an engineering system with multifactor, multiscale, multilevel, multiprocedure, and multitarget, integrated with the control of mass flow, energy flow, and information flow. For example, the steel manufacturing process is an operation control system with multifactor, multiscale, multilevel, multiprocedure, multitarget that consists of matter state transformation, matter property control, and mass flow control with the coordination and control of process parameters of mass flux, temperature, time, and space (Fig. 2.1).

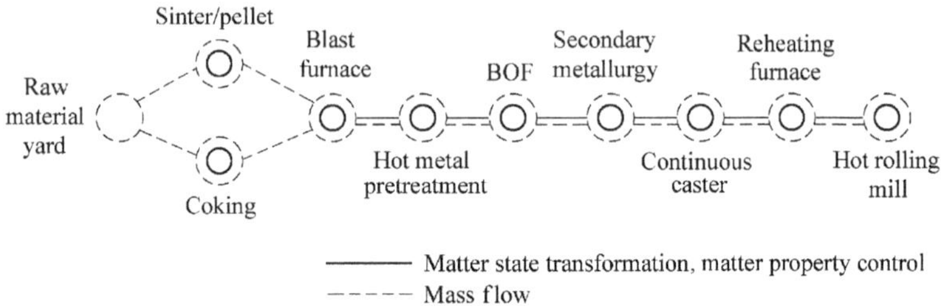

FIGURE 2.1 Schematic diagram of matter state–matter property–mass flow in a steel production process (Yin, R., 1997. The multi-dimensional mass-flow control system of steel plant process. Acta Metall. Sin. 33 (1), 29–38 (in Chinese) (Yin, 1997)).

An entire manufacturing process is composed of a number of heterogeneous units/procedures, which each have a structure and are closely correlated. The dynamic-orderly running of entire process requires an identical activity of motion for each unit procedure device in a quite long timescale. Because of the differences in the physical and chemical functions of each various unit (procedure), the relationships between different procedures are extremely complex and are affected by the external environment. It seems that the dynamic operation of a manufacturing process is an unpredictable complex problem and it is difficult to find its running rules. However, the theory of self-organization of the dissipative structure system founded by the school of Prigogine (Ilya Prigogine, 1917–2003) gives the possibility of studying the complex problem above. In order to know the dynamic-orderly operation rule of the entire manufacturing process, it is of importance to study the dissipative structure theory.

2.1.2 Spatiotemporal Scales of the Processes

From the viewpoint of the spatiotemporal scale, the manufacturing process is an integrated system of large-scale or larger scale units. With our current understanding about matter in nature, its structure is in different levels, and the movement and evolution of matter runs at different spatiotemporal scales, with great differences and different action laws. In the micro system, the elementary particle is of the size of 10^{-18} m (am) or even smaller. The atomic nuclei are of 10^{-15} m (fm), and the molecules, the atoms, and the ions are of 10^{-9} m (nm) order of magnitude. The polymer molecules are in micron (μm) order. But in the cosmoscopic view, the fixed stars, the Milky Way system, and so on take light-years as a spatial scale. The macrocosm between the two scales mentioned above is where humans live and undertake their production activities. The macrocosm is in the spatial scale from micrometer (10^{-6} m) to kilometer (10^3 m). From the view of engineering, a more thorough understanding of the different levels is required in order to know the different evolution rules, and a further spatiotemporal scale needs to be used, as shown in Fig. 2.2.

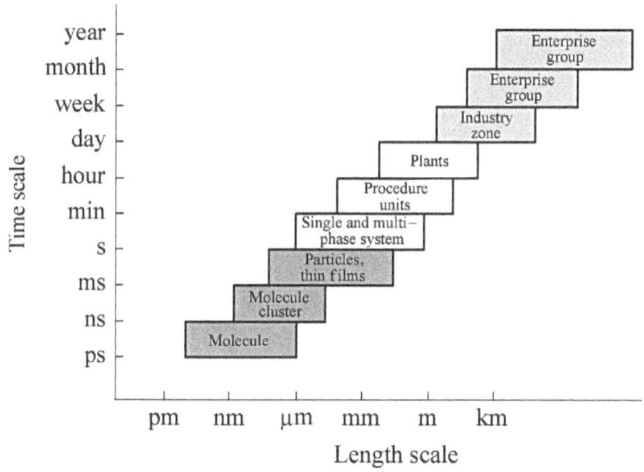

FIGURE 2.2 The space and time criterion level scheme of different processes (Grossmann, I.E.,2004. Challenges in the new millennium: product discovery and design, enterprise and supply chain optimization, global life cycle assessment. Comput. Chem. Eng. 29 (1), 29–39 (Grossmann, 2004)).

Metallurgical engineering can be taken as an example. Fundamental metallurgical reactions proceed among molecules, atoms, or ions at the scale of nm (or Å). The reaction is at high temperature with a fast rate and in a timescale of 10^{-6}–10^{-9} s. The phase transition and the deformation of solid metals occur in the nanometer scale, with a larger timescale than that of metallurgical reactions. The rules above are suitable for physico-chemistry metallurgy to study thermodynamics and kinetics, and belong to fundamental science in a micro scale. When the metallurgical processes proceed in the various reactors, due to the nonuniform distribution of concentration and temperature, the mass transfer and heat transfer as well as reactor geometrical shape (10^0–10^1 m) including the boundary layer (10^{-5}–10^{-6} m) needs to be considered; the timescale is 10^{-4}–10^4 s and the space scale is 10^{-6}–100 m. The rules of this scale are appropriate for the theory of transport phenomena and reaction engineering. The technological science of metallurgy is found at the mesoscopic scale. The entire manufacturing process involves a relative dependency operation with a long timescale ($>10^4$ s) and a large space scale ($>10^1$ m), and the study of which requires engineering science methods at the macro scale—process engineering.

2.1.3 Processes and Manufacturing Process

In the manufacturing process, the processes with various functions and performance are carried on. The two terms of *process* and *manufacturing process* are well related but their extension and intension should be differentiated and delimited.

Process is the incessant change of the state of a system in a period of time. *Process* concerns broad objectives and has a wide concept extension.

As large-scale as the evolution of heavenly bodies, the movement of stars in the universe, and as small-scale as reactions or the changes in the molecules/atoms, all may be called as a *process*. In different disciplines, according to the difference of the studied objects, the defined category and connotation of *process* must be standardized again.

The *Manufacturing process* particularly means the production process made up of different working procedures/devices under industrial production conditions. It is a complex process in an entirely integrated system. It consists of a process network, functions of procedures, the relationship among procedures, and the running program. It is a special process with several operation "flows" in it. The *Manufacturing process* is generally considered production at the plant scale or the workshop scale, with a larger spatiotemporal scale than that of unit operation and unit procedure. Through the course of modern study of steel manufacturing processes, the physical and chemical reactions in solid or liquid metals, such as smelting and refining, crystallization and segregation, crystal slip and dislocation, are first studied to solve problems at the scale of atoms/molecules. Then, the unit operation and energy/mass transfer in a device, such as fluid flow and mixing in a metal bath, are studied to solve problems at the scale of the procedure/device, far smaller than the spatiotemporal scale of the manufacturing process. Based on the success of these studies, the rules of the manufacturing process in an integral scale are integrated and studied, and then the problems of market competitiveness and sustainable development are also solved so that the *Process Engineering Science* is established.

The information above gives the distinctions between *process* and *manufacturing process* on the concept extension and the spatiotemporal scale.

Process can be divided into a reversible process and an irreversible process. A reversible process is not as simple as a process that occurs in its reverse direction, it particularly refers to a process in which each step can occur in its reverse direction without any impact on surroundings. For example, without the friction force, each steps of the movement process are in an equilibrium state, that is, the system makes an infinitely slow shift during an infinitely long time. Therefore it is an ideal process which can hardly be reached but can be infinitely approached. All actual processes are irreversible. In metallurgical manufacturing process, all actual processes are irreversible.

2.2 PROCESS ENGINEERING AND MANUFACTURING PROCESS ENGINEERING

2.2.1 Engineering and Engineering Science

2.2.1.1 Engineering

There are different approaches to know *engineering*, such as science ontology or engineering ontology. According to science ontology, *engineering* is generally defined as a kind of application based on the cognitive scientific and mathematical laws, through which natural substances and energy can be efficiently and reliably converted into useful things for human beings using various structures, equipment, systems, and processes.

According to engineering ontology, *engineering* is defined as an integrated system combining plenty of economic factors and techniques and aimed at real productivity under the specific natural and social conditions. *Engineering* includes a technology group (technology integrated system) of related necessary techniques through proper selecting, integrating, and coordinating. The technology integrated system embodies the features of specific structure and dynamic running, which means the dynamic-orderly integration of relevant heterogeneous techniques with different functions. The special technology integrated system must be cooperated with many natural and social economic factors (such as resources, land, capital, labor force, market, environment), and the prospective specific functions and values can be generated through structuring and operating. Technology integration is related to the constitution of elements and the relationships among them (Yin et al., 2011). Element integration is related to the engineering structure, while the structure is related to the function and efficiency of engineering.

Engineering is the integration of many relevant technologies with their own idiosyncrasies and functions. The heterogenic technologies are the elements of engineering. Only if the technology is efficiently embedded into an engineering system, can it efficiently play its correct role. If the technology is wrongly embedded into the engineering system rightly, it will not only reduce its own functions and efficiency, but also have a detrimental effect on the function, structure, and efficiency of the entire engineering system. *Engineering* should be an optimization-integration of related heterogenic and interacted technical modules, which leads to the formation of an optimized structure to realize the relevant function and to improve the efficiency of the system's operation.

2.2.1.2 Engineering Science

Engineering science starts from deep observation and thinking about objects and phenomena in the engineering practices. Through the analysis, study, and further exploration and induction of various objects, factors, technologies, and phenomena, the universal, simple truths of *engineering science* hidden in the internal engineering system can be revealed.

Solid and broad theoretical knowledge, enough practical experience, excellent imagination, and sharp judgment are needed to study *Engineering science*, and distinctive visions and methods are employed to know the essence and the motion law and to find out the complicated colorful inner truth and beauty hidden in an engineering system. This kind of beauty shows the identity of the inner rules hidden in the diversity of engineering phenomena, the relativistic invariance in some physical quantities and geometric dimension coming from the unceasing evolutionary process of objects, and the inner simplicity under the outer colorful phenomenon. Such beauty in science, including engineering science, is a rational beauty and is shown through the common structure (eg, the dissipative structure of an open system) and the operation laws (eg, the dynamic-orderly and dissipative process optimization) of universal objects. The beauty in art is expressed by the sensibility whereas it is expressed by rationality in science. Science takes the objective world as the object, and pays attention to the inherent laws of the relationship, the interaction, and the movement and change of objective things.

Generally, engineering science seeks the truth and beauty, including discovering the reality, the essence, and the motion rules of the phenomena in engineering systems; finding the inherent law from the simple to the complex and then from the complex to the simple of various phenomena; exploring the diversity and the internal simplicity of evolution; as well as discovering the continuity, coordination, and rhythmicity of different types of objects (Yin et al., 2007a).

The observation and research methods of engineering science must be different from that of fundamental science. The two have differences in spatiotemporal scales or the parametric dimensions in the studied area. For most basic sciences, analytical methodologies are used for their research, by which the objectives are divided, simplified into the elements of the matter and the matter structure, and the total pattern can be understood as long as elementary particles were clarified. Therefore methods of classification and simplification are used. However, for the research method of engineering science, it should start from the entire characteristics of the engineering, and then it is continuously improved and supplemented by repeated optimization procedures of analysis-integration and integration-analysis, and finally the deeper cognition of the nature of the integral feature and the law of motion are achieved. Scientists and engineers should learn from artists. Artists clearly know that the entirety is needed to be identified with the subsequent supplement of details to make a deeper description of the entirety. A painter can catch the vivid feature of the object using just a few strokes.

As engineering is hardly isolated from integration; engineering science must surpass the limit of reductionism and according to the method of analysis-integration, study the relevance or operation mechanism of different process units under different scales in the entire engineering system, and summarize the relationship between the microscopic-mechanism and macroscopic phenomenon and between integral structure and its function of different units. Correspondingly, the following contents can be investigated: the formation mechanism and evolution rules of the structure in a complex engineering system; the relationship between structure and behavior of the complex engineering system; and the occurrence of "mutation" and its adjustment of an engineering system; all of which are at the level of engineering science.

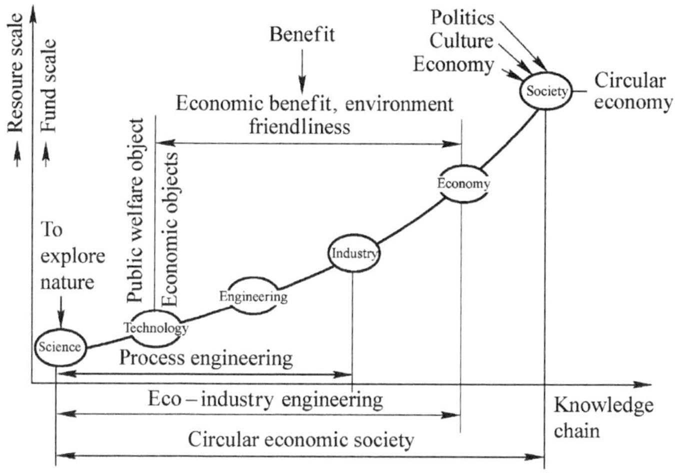

FIGURE 2.3 The relationship between expanding scale of resources and funds with knowledge chain (Yin, R., Wang, Y., Li, B. et al., 2007b. Philosophy of Engineering. Higher Education Press, Beijing, p. 7 (in Chinese) (Yin et al., 2007b)).

A good understanding and grasp of engineering and the engineering science category can be achieved through the knowledge chain, the time–space chain, and the constituting network of "nature–science–technology–engineering–industry–economy–society" (Fig. 2.3). For science and technology, engineering always plays the role of integration and "emergence." But for production and economy, engineering becomes the basic unit and component unit, and the industry is composed and integrated with identical and related types of engineering. Engineering science should pay attention to the physical essence and the inherent laws of the integration knowledge chains and the engineering constituting network. There are many themes and knowledge chains in engineering science and these chains and process networks are the key issues to explore and understand in engineering science. In the meantime, the interdependence among basic science, technological science, and engineering science in an engineering system can be gradually understood and the knowledge frame of the integration can be established.

2.2.2 Process Engineering

Process engineering was developed from chemical engineering. Chemical industry includes many sorts of chemical engineering plants so that it is impossible to establish a theory for each production process. Upon research by numerous scholars, it was realized that every kind of chemical production process includes physical processes and operations such as fluid flow, heat transfer, evaporation, desiccation, dissolution, distilment, crushing, classifying, filtration, sedimentation, mixing. Thus the common chemical engineering is based on unit operations (Coulson and Richardson, 1997; Richardson and Harker, 2002). In the 1950s, people knew that momentum, heat, and mass transfer are physical processes of the linear relationship of force and flow within analogue laws and analysis-methods, through which the theory of transport phenomena was established. The study of the combination of momentum, heat, and mass transfer with the rate theory of chemical kinetics led to the development of the discipline of chemical reaction engineering. Theories about transport phenomena and reaction engineering improve process optimization and reasonable design. The investigation and application of dispersion system (particles, drops, and bubbles) promote the development of multiphase fluid flow dynamics. With the development of computer technology and computational fluid dynamics, and also the extension to more fields (such as biochemical industry, metallurgy, ceramics), process engineering evolved gradually from chemical engineering.

2.2.3 Manufacturing Process Engineering

Manufacturing process engineering came from the technological development and profound change of the steel industry. Ferrous metallurgy is a process of iron-coal chemical engineering at high temperature. The manufacturing process consists of many procedures and consumes plenty of resources and energy. In some countries with high steel capacities, the energy consumption of the steel industry usually occupies 10% of total consumption or even more. The supply and prices of the resources and energy therefore have a significant influence on the steel industry. Due to the oil crisis in the 1970s, the technology of continuous casting (CC), instead of ingot teeming, in steelmaking workshops was developed. The connection of blast furnace ironmaking to continuous rolling process of steel via 100% CC process created the precondition for the quasicontinuous/continuous running of the manufacturing process of steel plants. The technology of near-net-shape casting further improved the establishment of continuous operating lines in steel plants. Since the 1990s, there had been six important key/common technologies widely applied in the Chinese steel industry: CC technology; blast furnace pulverized coal injection technology; blast furnace campaign elongating technology; the long-product continuous rolling technology; comprehensive energy conservation technology by means of process structure optimization; and the basic oxygen furnace (BOF) slag splashing technology. The successful application of these key/common technologies provided the physical foundations for the formation of metallurgical process engineering. For example, the quasicontinuous/continuous operation flow for a steelmaking workshop can be realized by coordination of the continuous solidification processing with the high-frequency high-speed periodically batch type converting. However, as the lining refractory life of BOF is short, the slag splashing protection technology has been applied to coordinate the maintenance cycle of BOF with caster, by which the dynamic-orderly operation flow of sequence CC for long time periods becomes realizable.

In a manufacturing process of ferrous metallurgy, the efficient operating technical conditions for various procedures are quite different. Seemingly, there are few internal coupling factors or linkage parameters among procedures/devices. To make a series of procedures/devices into a unified and harmonious movement of the entire manufacturing process, it is necessary to learn from Prigogine's theory of self-organization in dissipative structures. With the combination of the technology of the continuous/quasicontinuous operation of entire manufacturing process with the theory of nonequilibrium thermodynamics, the discipline of *Metallurgical Process Engineering* appeared.

2.3 PHYSICAL ESSENCE OF DYNAMIC OPERATION OF THE MANUFACTURING PROCESS SYSTEM

In the production operation of process manufacturing industries (including metallurgical industry), there are mechanical interactions, thermal action, and mass action in its complicated mass flow, energy flow, and related information flow. Such interactions appear as the processes of different types, performances, spatiotemporal scales like fluid flow, chemical reactions, energy conversion, heat exchange, phase transformations, deformations, etc.

It will not be certain that all points of macro characteristics of the actual engineering system can be deduced from constitutive units and their combination. The characteristics of the engineering system can hardly be regarded as a static structure, yet be regarded as a nonlinear interaction among constitutive units as well as the dynamic coupling between the engineering system and its external environment. The impacts of those dynamic factors on the system's structure, function, and efficiency should be taken seriously. Through investigations on the nonlinear interactions and the dynamic coupling of an open system, the complicated engineering system becomes observable and confirmable. The essences of an engineering system (including the manufacturing process engineering system) are *process* and *process structure*. In other words, the manufacturing process system is integrated by the different types of processes with their own properties and spatiotemporal scales via dynamic coupling. Among processes, there are structures and these structures are dynamic, orderly, and hierarchical. Because of their openness, fluctuations, and nonlinear interactions, these structures are complicated. The basic parameters to which a process structure can refer include mass, energy, time, space (even life), and many derived information parameters.

2.3.1 Features of Manufacturing Process

The goal of process engineering research is to discover the organization principles in different levels of the manufacturing process as well as to improve the structure of the process, to promote the system innovation and its evolution, to raise the efficiency of the process system, to reduce the energy dissipation, and to expand the functions of the manufacturing process through integration-synergy.

The following are the points that need be taken seriously in the research of process engineering:

1. The systematic integrating of the manufacturing process. A manufacturing process is an organic system with dynamic structures, and the organic system has different properties and rules from its composition units.
2. The dynamic running of the manufacturing process. A manufacturing process is a "living" dynamic operation system which opens to the external environment. The features should be organization-integration and the interdependence with the environment.
3. The hierarchy of the manufacturing process. A manufacturing process is an organic hierarchical system. The constitutive units with different levels and spatiotemporal scales get different structures and boundary conditions. Thus the order of their running parameters are different but relevant to each other.
4. The evolvability of the manufacturing process. A manufacturing process is constantly evolving, and the fundamental mechanism of the process system's evolution is the unity of inner contradictions of the whole to parts or a part to another. It is embodied in three optimization sets: the analysis-optimizing set of procedures' functions, the synergy-optimizing set of procedures' relations, and the restructuring-optimizing set of procedures in manufacturing process (Yin, 2009).

So, the structure and running characteristics of manufacturing process are reflected as follows:

1. The systematic integrity and hierarchy;
2. The macroscopic ordering via feedback and amplification of random fluctuations and nonlinear interactions;
3. The interaction of selectivity and adaptability;
4. The system is open, and its stability is founded on mass and energy exchange with the external environment.

From the point of view of the structures and running characteristics, the manufacturing process has the following characteristics: the self-organization of the system; the dissipation in the operation; and the adaptation to the environment. Such characteristics have a great influence on the evolution and development of the path of technology integration (including the theory and method of engineering design), of engineering operation (including the production plan projecting, and time scheduling), and of enterprise management (including the departments' establishment organizing, and management mode).

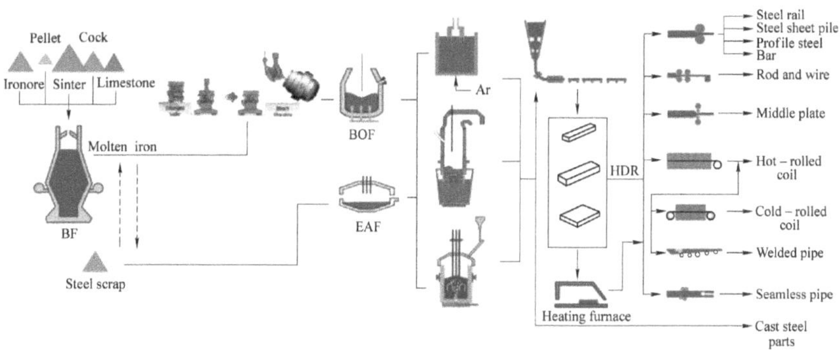

FIGURE 2.4 The schematic of two types of steel manufacturing processes.

2.3.2 Essence and Functions of Steel Manufacturing Process

The manufacturing processes of modern steel plants have been developed into two types, as shown in Fig. 2.4.

1. One is the process of blast furnace–BOF–CC–hot rolling–deep processing or the process of smelting reduction–BOF–CC–hot rolling–deep processing. It is based on the use of natural resources (such as iron ore and coal). It includes the storage, transport and blending of raw materials, sintering–coking–ironmaking (include smelting reduction) procedures, steelmaking–secondary refining–solidification procedures, and the reheating–hot rolling and cold rolling–surface finishing procedures.
2. Another one is the manufacturing process of electric arc furnace–secondary refining–CC–hot rolling process with steel scraps and electric power as raw materials and energy. The resources are recycled metals, such as returned scrap, shredded scraps, social recovery scraps, machinery residual scraps. This kind of manufacturing steel plant is known as a "mini-mill."

From a superficial understanding, the manufacturing process of a steel plant consists of a stock yard, coking, sintering (pelletizing), ironmaking, steelmaking, steel rolling, and so on. Attention is usually paid only to the equipment and the operation of each unit procedure and its own capacity in a static design. The production flow chart is simply constructed by series or parallel connections. For example, if a steel mill has an ironmaking capacity of 3×10^6 t/a, a steelmaking capacity of 3×10^6 t/a, and a rolling capacity of 3×10^6 t/a, it should be a steel plant of 3×10^6 t/a by superposition. The aforementioned opinion seems correct in the view of the partial working procedure/device abilities and the superficial technologic presentation. However, in the actual production operation process of a steel mill, when the upstream procedure can hardly obtain synergy with downstream or is in batch type operations, there will be a phenomenon of capacity lacking or "lost pass." Based on this recognition, the "capacity spare coefficient" for different production units is applied during steel plant design and performance. As different production units deal with problems separately, it often brings a new imbalance, or even leads to more problems like overcapacity, discordance, and too much investment in production units. It will result in enormous waste, as steel production requires heavy equipment with intensive investment.

Therefore the manufacturing process should hardly be regarded as the superposition of the different procedures together. The superposition of different procedures is at most the static appearance of the manufacturing process of a steel plant. Actually, the manufacturing process is a process of dynamic operation, which is the running/evolution of the mass flow, energy flow, and the relevant information flow in the rational spatiotemporal scales. Its physical essence is dynamic-orderly operation of mass flow (mainly ferruginous flow) driven by energy flow (primarily carbonaceous flow for a long time) under a given program (such as the instruction of operation) and specific process network (eg, general layout). The dynamic-orderly operation provides the multiobjective optimization: improving product quality, reducing cost, raising efficiency, reducing energy consumption, lowering emissions, being environmental friendly, and so on. Hence, at the core of the operation of the manufacturing process is evolution and flow—the multifactor mass flow and energy flow following a given program in the process network.

In thermodynamics, the steel manufacturing process is an open, nonequilibrium, irreversible process, and is a dissipative structure composed of some procedures with different structures–functions by nonlinear coupling. The essence of the dynamic-orderly operation is self-organization in the dissipative structure. In order to decrease the dissipation during operation, the manufacturing process should tend to run dynamic-orderly, synergetic-continuously, and steady-compactly.

From the physical essence of the dynamic operation of the steel manufacturing process, the functions of steel manufacturing process include (Yin, 2008):

1. The operational function of ferruginous mass flow—the manufacturing function of steel products;
2. The operational function of energy flow—the function of energy conversion and the function of waste heat recovery and related waste material recycling;
3. The interactional function of ferruginous mass flow and carbonaceous energy flow—the function to achieve the objectives of the manufacturing process and the function of treatment and recycling of wastes.

2.4 OPERATION PROCESS AND PHYSICAL LEVELS OF DYNAMIC PROCESS SYSTEM

In the summer of 1947, Chian (1948) pointed out in his speech of "*engineering* and *engineering science*" to students in Zhejiang University, Shanghai Jiao Tong University and Tsinghua University, China: "the real progress about metallurgical engineering science hardly exceeds the application of Gibbs phase rule... In other words, there is a large gap between practical engineering and scientific research. The connection between them should be set up. In the field of metallurgy, the physical theory can not only provide the systematic explanation for the abundant accumulated empirical data, but also reveal new possibilities in the field of developing materials." He also indicated that "There is an essential difference between the view of physicists and engineering scientists. The physicists aimed at the pure science, and their major interest is to simplify a problem to a degree by which the exact answer can be found, while engineering scientists are more interested in solving the complicated problem and their solution may be approximate but accurate enough for the purpose of engineering."

The discussion about engineering science by H.S. Chian is still very useful for modern engineering. Actually, as the fundamental of the national economy, process manufacturing industries (eg, metallurgical industry, chemical engineering industry, building material industry, papermaking industry) have common problems or similar issues regarding engineering and engineering science—engineering issues of *process engineering*. The operational essence of those industries is just the dynamic integration of the process structure and the process running.

2.4.1 Physical Features of Dynamic Running of Manufacturing Process

The production and operation of the process manufacturing industry (including the metallurgical industry) is always accompanied with closely related and complex mass flow, energy flow, and relevant information flow. When the process system is running, the operation state at different levels might be disordered, chaotic, or ordered. An ordered state means a regular interrelation among objects whereas a disordered state means an irregular interrelation among objects. Pure order or disorder is only a theoretical abstract, and in the practical system, the ordered state and disordered state are relative. For a complicated system, they are always showing together. But how do we understand *chaos*?

Harken (1977) gave a definition of chaos as "irregular motion stemming from deterministic equations." Chinese scholar Hao (1988) thinks that chaos is an internal randomness of a decisive system. Chaos also can be referred to "intrinsic stochasticity," "spontaneous stochasticity," and "dynamic probability."

"Chaos" does not equals "disorder." The chaotic state has essential distinctions from the disordered confusional state. Confusion and disorder is defined on the scale of molecular kinematics, for example, molecules in an ink-solution are in the extra-disordered motion, while the random state concerning chaos is defined on a macroscopic scale, such as turbulent fluid flow. Hence, a state that is disorderly in macro but orderly in micro can be called chaos. Bailin Hao pointed out: "Chaos is not a simplified disorder, but more like an orderly state without some periodicities or obvious symmetry. Under ideal condition, there is infinite internal structure in the state of chaos. As long as precise means to observe are available, the periodic motion or quasi periodic motion in the chaos state and the repeated chaotic motions in smaller scale can be found."

Chaos is a common phenomenon in nature, and also exists during the operational running of a steel manufacturing process.

2.4.2 Three Kinds of Physical Systems

Systematic methodology and evolutionary philosophy may help us to overcome the limitations and to break away from the one-sidedness of reductionism. Jantsch (1980a) had introduced Prigogine's opinion that in physics at least three classes must be distinguished, explored, and described. That is:

1. Classical or Newtonian dynamics: Newton described the velocity or the relative position of a particle using mechanics. In Newtonian mechanics, anything in the universe can be simplified and reduced to the trajectory of particle motion or the space–time curve, indicating that the movement of a particle from point A to point B is completely reversible. For example, the process described by Newton's motion equation

$$F = m\frac{d^2x}{dt^2}$$

This equation is symmetric under time reversal. The motivation for particle motion is provided externally but is hardly self-organizational. Newtonian mechanics is a precise science; it can describe and forecast the motion of an object accurately. Time can be accurately measured by applying the cycle motion. Time becomes absolute yet its arrow direction can hardly be distinguished. Classical mechanics provides an idealized description for the "pure" motion of a particle. However, the "dirty" reality contains a variety of conflicts, collisions, exchanges, mutual excitement, challenge, and coerciveness, and must be accompanied with complexity and collectivity.

2. Classical thermodynamics deals with equilibrium systems or systems near equilibrium: Clausius established the second law of thermodynamics in 1852 on the base of Carnot's study. The second law indicates that all spontaneous processes are irreversible, and always proceed to the direction of entropy increase. The macro orderly generation of process was used in the description of the process. In other words, entropy is not a direct physical property, but it is a measure to characterize the energy distribution of the system. In an isolated system, the entropy of the future always appears higher than that of past, which includes the internal logic of time orientation and irreversibility. The entropy change describes the evolution direction of process, and this description expresses the arrow of time as an inherent characteristic of a process. This is the obvious distinction between classical thermodynamics and Newtonian mechanics.

   In classical thermodynamics, Boltzmann attempted to discuss the second law through numerous collisions of system components (eg, atoms and molecules) and exchanges with each other, and the entropy of the macro state always tends to increase. The entropy increase in an isolated system possesses irreversibility. An increase in entropy means to enhance the degree of disorder leading to the limitless increase of symmetry of the system structure and then the breakage of it.

   Because of the irreversibility, the evolution of a system can only proceed toward the direction of the thermodynamic arrow which is deduced from the cascade entropy increase, namely, a "nonenvironment" isolated system (without mass and energy exchange between the system and surroundings) will become a specific self-organization. The attraction factor of this type of self-organization is the equilibrium state, that is, the system, develops toward equilibrium spontaneously and terminates in equilibrium.

   Sometimes, classical thermodynamics discusses a closed system or an open system. For example, the free energy in isovolumetric (or isobaric) processes:

$$F = U - TS$$

   where F is free energy, U is inner energy, T is temperature.

   The essence of this formula is the competition between inner energy and entropy. Free energy mostly depends on the inner energy of the system at low temperature, but entropy runs the leading role at high temperature. Although free energy could describe the process direction in a closed system, even an open system, the exchange of mass/energy between surroundings and the system is mainly to maintain the system with constant parameters (such as temperature, pressure). The equilibrium structure after exchange hardly needs any more mass and energy from the surroundings to keep its steady state.

3. The nonlinear irreversible process thermodynamics or the thermodynamics far from equilibrium: In the open system which has continuous energy and mass exchange with surroundings, another essential physical system is found—a specific nonequilibrium system. With further study, there comes into being the third level that physics explores and describes and which is the level of the coherent evolutionary system—the level of dissipative structure (Prigogine and Nicolis, 1967; Prigogine and Lefever, 1968).

In the open system, mass and energy (negentropy) can be continuously input from the environment outside of the system and there is an output of mass and energy at the same time. Different from the isolated system, the entropy production in the open system can be compensated from the negative entropy flux (negentropy) input from environment.

Entropy is an extensive quantity, it can be partitioned into two parts. Namely, the total entropy change dS in a time interval can be split into the entropy production d_iS and the entropy flux d_eS. d_iS is the entropy production during the irreversible process in a system and d_iS can only be " + " or "0" and can never be " − ," that is, $d_iS \geq 0$; while d_eS is entropy flux from the environment accompanied by mass/energy transfer, d_eS can be " + " or " − ", and $d_eS = 0$ only for an isolated system.

Thus the appearance of an orderly and steady state in a system far away from equilibrium can hold only if the system is open. The system must incessantly exchange mass and energy with the external environment, and thus it is called dissipative structure. At this level, it is synchronous with the existence of evolution.

In conclusion, it is important that the description of the motion and the evolution of a substance needs at least three levels of physical investigation.

As Prigogine (Coveney and Highfield, 1990) once said, "the world is richer than it is possible to express in any single language. Music is not exhausted by its successive stylisations from Bach to Schoenberg. Similarly, we cannot condense into a single description the various aspects of our experience." The reversibility in mechanics and the irreversibility in thermodynamics are the two sides of a coin. The structure of the world is very abundant and it can hardly be summarized by just one analysis. Combining the explorations on different levels helps us know the rules which decide human behavior and nature evolution.

2.5 EVOLUTION OF THERMODYNAMICS

Mechanics and thermotics are two of the essential and widest fields in classical physics. After the appearance of the steam engine, there came into being a technical proposition about how to improve the efficiency of the heat engine. This proposition prompted scientific research on the thermal properties and the thermal phenomena of related substances. The efficiency of the

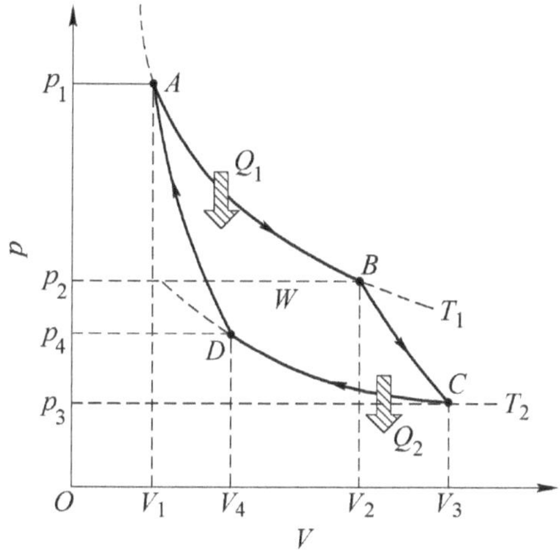

FIGURE 2.5 Carnot cycle.

steam engine (heat engine) was studied and correspondingly the horizons of science were broadened, and thermodynamics was established.

2.5.1 From Thermomechanics to Thermodynamics

Carnot (1796–1832) performed an in-depth on the conversion of heat into work to improve the efficiency of the heat engine. In 1824 he accurately recognized that in the heat engine, work is not only in need of consuming heat, but also relates to the heat transfer from high-temperature objects to low-temperature objects. Physical abstract methods (such as particle, rigid body, perfect fluid) were applied during the study of the heat engine by Carnot, which abstracted the research targets to an ideal sample and summarized the substantive characteristics of objective things.

Through the abstract and ideal Carnot cycle (Fig. 2.5), the extremely important Carnot theorem was deduced: in all kinds of heat engines that work on an isothermic hot source and cold source, the reversible engine has the best efficiency, and the efficiency of the reversible heat engine is proportional to the temperature difference between the hot source and cold source. Though Carnot's theorem in 1824 hardly overcame the constraints from the theory of phlogiston, its perception on heat–work conversion had a great prospective meaning. When the theory that thermokinematics decides heat was considered as the only correct theory, Joule (1818–89), who was well known for researching the problem about mechanical equivalent of heat, raised doubts about Carnot's theorem: since work can transform into heat with an equivalent value, is the opposite correct in turn? Why does the part of heat transform into work, and where does the heat go? This query raised by Joule was solved by Kelvin (1824–1907) and Clausius (1822–1888). They recognized that the heat from the high temperature was transformed as work partially, and the other was delivered to the low temperature. Kelvin said that even though this kind of heat hardly disappears, it is not used and is of no use for human beings.

In 1850 Clausius had demonstrated that it was not able to achieve a heat transfer from a low-temperature object to a high-temperature object without other effects. In 1851 Kelvin raised the theory that it is impossible to absorb heat from a single heat source and completely to turn it into useful work without other effects. Their studies both concerned the problem of process direction, and can be summarized to the irreversibility of a spontaneous process. To formulate the irreversibility by the way of a thermodynamic function of state, Clausius created a new notion in 1854—state function S to represent quantity of conversions. Considering that it is necessary to embody the immutability on language, he named it with a Greek alphabet εντρωπη (development), its corresponding German homophone is Entropie (entropy in English). The function entropy is quantified as

$$dS \geq \frac{\delta Q}{T}$$

where the equality sign expresses reversibility and means that the equilibrium (reversible process) is the measurement for the tendency of process; the inequality sign expresses nonequilibrium and means that all the spontaneous processes are irreversible. The mathematical expression with an equal sign and an unequal sign together is the effective way to show the irreversibility rule.

2.5.2 Classification of Thermodynamic System

Thermomechanics only discusses the transformation between heat and work, while thermodynamics extends to cover energetics and energy conversion in systems.

In thermodynamics, to define suitable thermodynamic functions for different objects and the conditions for the variation of component concentration and temperature in a system, the relationship between the object and environment in three systems is always considered: isolated system, closed system, and open system. The isolated system exchanges neither energy nor mass with surroundings; the closed system only exchanges heat with surroundings but no mass exchange; the open system is open to the external environment and there are energy and mass exchanges between them incessantly.

In the isolated system, after any actual process, the total energy value in the system remains unchanged according to the first law of thermodynamics, while the total entropy value in the system constantly increases according to the second law. In other words, in all real processes, although the total energy remains unchanged, its availability always reduces due to the increase of the entropy.

Furthermore, someone introduced the concept of ordered energy and unordered energy and pointed out the rule of the direction energy conversion: ordered energy can entirely and unconditionally translate into unordered energy, and it is impossible or conditional for unordered energy to entirely translate into ordered energy (Feng and Feng, 2005).

Actually, the isolated system hardly exists in nature since it is difficult to find fully adiabatic and insulating materials and there are cosmic rays and high-energy particles radiated to the earth constantly, by which heat enters into the system through the boundary. So, an isolated system is only a theoretical and abstract concept. However, it is of importance and the entropy increment can be defined by the concept of isolated system. Since there is little material or energy exchange in the isolated system, the entropy change in the isolated system is

$$dS \geq 0$$

This formula shows that in an isolated system, the irreversible process occurs spontaneously only if the entropy increases. The $dS = 0$ means that the entropy increases to maximum while the process comes to its equilibrium. Equilibrium seems to keep still while the equilibrium is hardly a static state. On the contrary, motion just finds its measurement from its opposite—stillness. The direction and limitation by which the reaction proceeds are exactly judged by the equilibrium calculation.

Since it is hard to build a true isolated system in laboratory; the closed system can reach an isothermal state depending on proper energy exchange (Fig. 2.6). Investigations on physical chemistry in metallurgy measure the free energy change of each metallurgical process of a closed system. Free energy is a function of temperature and pressure (or volume). Now, sufficient data of metallurgical thermodynamics are available. Concerning the solutions

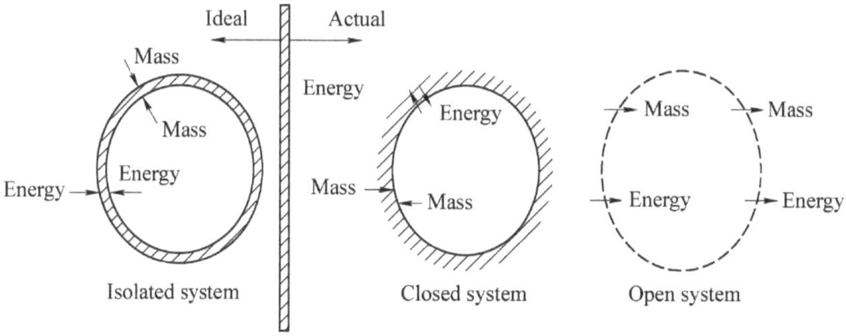

FIGURE 2.6 Isolated system, closed system, and open system.

of liquid metal and molten slag, the concept of chemical potential (partial molar free energy) is needed. The influence of mole increment could be verified in the open system. No matter what kind of system it corresponds to, only an equilibrium state can be accurately determined in classical thermodynamics. Equilibrium is important since it is the direction of process and the leading mark for evolution. Practical processes must proceed to equilibrium spontaneously.

If the entire steel manufacturing process is considered as one system, in order to know the rules of different procedures, the procedures have be to studied separately, such as the process of hot metal desulfurization pretreatment. The separated subsystems also might be open or closed systems. However, if they are only separately studied, the proper knowledge can hardly be obtained. It is necessary to insert the subprocedures into the entire process for thorough investigation. The local optimization of one procedure is not actually the optimization of the entire process, and, therefore, separate studies on subprocedures may lead to wrong results in process engineering. However, separate investigations and isolated systems are two different concepts. An isolated system is one of the basic concepts in thermodynamics. The principle of entropy increase was deduced with an isolated system. As a system, the entire manufacturing process has an exchange of mass, energy, and information with external environment, which is an open system. Therefore the process should be studied as an whole rather than each procedure/device separately.

2.5.3 Irreversibility

Based on the second law of thermodynamics, the entropy increment makes it impossible for energy to be completely converted into useful work. Heat engine maximum efficiency is a result of entropy increase so that the entropy production can be regarded as the energy dissipation. Energy dissipation and entropy consumption are the source of all natural processes (including human activities). The law of energy conservation does not limit itself to heat energy and mechanical energy. The interconversion among various forms of energy, such as electromagnetic energy, chemical energy, and nuclear energy, also obey the law of energy conservation. Entropy is produced not only by the temperature difference in heat–work conversion, for example, in a fuel cell, chemical energy transforms into electric energy without the intermediate stage of heat energy; a nuclear reaction is not a result of the temperature difference between atoms, yet entropy is still produced. Inevitable entropy production is a universal principle that is associated with all natural processes. Using solar energy is hardly a free lunch either. When the energy transforms into another form, there must be entropy production, which causes the specific influence on environment.

The universality of the law of entropy production is related to irreversibility of the natural process. The irreversibility shows the time asymmetry (the irreversibility of the arrow of time). The inequation of $\Delta S \geq 0$ means that the process of entropy production goes in the positive time direction. Although in the formula variable t does not appear, the formula has incorporated the time asymmetry by using the unequal sign. The theorem shows that the entropy function has a particular characteristic—only increasing with time.

Process asymmetry can be discussed regarding a macroscopic system composed from a lot of molecules. For instance, in a vessel, the possibility that gas molecules under free motion distribute throughout an entire vessel is greater than that they just distribute in a corner. L. Boltzmann (1844–1906) realized the generality between the macroscopic continuance and with the discontinuous microstructure. He contributed a famous physical equation as follows.

$$S = k \cdot \ln \omega$$

where the ω is the thermodynamic probability, representing the numbers of microstate of numerous molecules. The $\omega = 1$ means evenly distributed at equilibrium, and the $\omega < 1$ for other states.

This equation was carved on Boltzmann's gravestone. The Boltzmann equation can conclude that "The production of entropy means an increase in the degree of disorder."

The Boltzmann equation is an important formula, connecting the macroscopic entropy with the microstates. But it also led to the paradox of "This formula was derived from the reversibility of kinematic laws at the micro level but ended with the process irreversibility at the macro level." The state of a molecule depends on its motion, and the law of motion for a single molecule obeys Newtonian mechanics. Newtonian mechanics is the determinism theory and has strong predictive ability. With the differential equation of mechanical motion, the past and also the future can be worked out as long as the initial conditions (namely the present) are decided. The variable t is indistinguishable for the past and the future. When t is turned into $-t$, the Newtonian formula remains unchanged. Newtonian mechanics breaks the absoluteness of space, considering all positions are relative. But it still maintains the absoluteness of time. Newtonian mechanics considers the time measured in different sites is isochronism; wherever, as long as time measurement technology is accurate enough, the watches of two people hardly show the same time forever. This is suitable for all kinds of motions on earth. However, when the motion is close to the speed of light in space, there is no longer the absoluteness of time. But relativity theory still retains the symmetry under time reversal.

2.5.4 Processing Within Steady State—Near Equilibrium Region

Equilibrium is the basic concept of thermodynamics. However, equilibrium has limitations. Equilibrium can describe only the endpoints of the process. But, it can hardly describe the process evolution quantitatively. Therefore, from this point of view, classical thermodynamics should be considered as thermostatics. In order to overcome this limitation, it is necessary to surpass equilibrium into nonequilibrium. The investigation of the nonequilibrium irreversible process starts from the study of the near equilibrium system. To study the relationship between the driving force and the rate of processes, the linear nonequilibrium thermodynamics is established.

Nonequilibrium thermodynamics (also known as the thermodynamics of irreversible processes) discusses states and processes that are not in equilibrium and the isolated system is ignored. In a closed system and an open system, there is energy exchange or heat flow between the system and the external environment. Since entropy is an extensive variable, it can be divided into two parts in using the balance method as introduced from hydrodynamics. The entropy increment of a system can be divided into inner entropy change and outer entropy change.

$$dS = d_i S + d_e S$$

where $d_i S$ is the entropy production and a measure of the process irreversibility in the system; $d_e S$ is the entropy flux and the entropy change caused by mass/energy exchange between system and surroundings. Entropy production and entropy flux work differently. Entropy production determines the direction of the process in the system, hence, $d_i S$ is always positive. Entropy flux $d_e S$ can be positive or negative, but it can never make the entropy change of the process in system become negative.

Since it will not concern equilibrium, the entropy conservation ($dS = 0$) can be excluded. The time t as a variable may be inserted into the formula. The time derivative of entropy $d_i S/dt$ may be called entropy production rate σ. The multiple product of σ and T is called the dissipation function Φ.

$$\Phi = T\sigma = T\frac{d_i S}{dt}$$

The dissipation function shows that the greater the entropy production within the system, the greater the energy loss.

Entropy flux can be positive and also negative. Namely, entropy can be entered into the system or the system can give out entropy. The physicist E. Schrödinger (1887–1961) proposed a term "negative entropy" (negentropy). The measurement of order may be indicated with negentropy.

When the value of $d_e S$ is below zero, it means that there is negative entropy input to the system or the existence of negentropy flux. Input of negative entropy is a figurative statement, indicating that it can have mutual compensation with the entropy production of the system.

When the amounts of negentropy flux and entropy production are equal, the system would get

$$dS = 0$$

$$d_iS > 0$$

$$\frac{dS}{dt} = \frac{d_eS}{dt} + \frac{d_iS}{dt}$$

where the $d_iS > 0$ indicates that the process continues, however $dS/dt = 0$ means that the system state does not change with time. This condition is called *nonequilibrium steady state*, indicating the process proceeds smoothly neither exhibiting the faster and faster outbreak type, nor decaying gradually to the equilibrium. The entropy production rate of a nonequilibrium steady state is at the minimum. In a near equilibrium region, the existence of a local equilibrium is assumed.

$$\sigma = \frac{d_iS}{dt} \geq 0$$

That means that under outside influence the boundary condition has prevented the system from reaching equilibrium for a near equilibrium region. The system becomes stable under minimum dissipation. That is $\Phi > 0$ and $d\Phi/dt = 0$. When the dissipation increases, $d\Phi/dt > 0$, the stability has broken, so that the process deviates from steady state. Steady state is the state where entropy production reaches the minimum.

The existence of negentropy flux is the necessary condition to maintain steady-state evolution. When the negentropy flux input breaks, steady state tends to collapse. There is no entropy flux for an isolated system so that steady state never appears in an isolated system. Equilibrium always terminates the process in an isolated system.

For engineering, the concept of steady state is more practical than equilibrium. For example, when the slabs pass on the roller table, they must gradually be cooled until room temperature—equilibrium if there was no outside influence. Trends of going to equilibrium state are spontaneous. However, if there is an appropriate external influence, it will deviate from equilibrium and turn to a steady state. Strand heating especially edge heating by electromagnetic induction will establish a condition that the slabs can be held at a high temperature at a steady state. This is favorable to hot/direct charging-rolling even semiendless rolling. Therefore steady state is important to the continuous dynamic-orderly operation of processes.

2.5.5 Linear Irreversible Process

The linear process can irreversibly proceed into a steady state near equilibrium. The so-called linear feature is that the flux/flow rate of the process is proportional to the driving force. In other words, as the impetus increases many times, so does the caused effect. For many phenomena in nature, especially in transport phenomena, linear relations are common, such as relationships between heat conduction and temperature gradient; between diffusion flux and concentration gradient; between convection mass flux and difference of interface concentration and bulk concentration; between the electric current and electromotive force; between the reaction rate and chemical affinity; and between the shear stress in fluid and velocity gradient. Such linear relationships are also called the phenomenological relation. A general formula expresses it as

$$J_i = \sum_{k=1}^{n} L_{ik}X_k \; (i = 1, 2, \ldots, n)$$

where, J represents rate of processes, named generalized flow, in brief as flow or flux; X represents the driving force of each process, named generalized force, in brief as force; L is the proportional factor, named phenomenological coefficient. i and k denote items of process: when $i = k$, L_{ii} (or L_{kk}) is called the self-phenomenological coefficient; when $i \neq k$, L_{ik} is called the mutual phenomenological coefficient.

Different kinds of forces and flows interfere with each other. For example, matter flow can be induced by temperature gradient, that is, Soret effect; and electro current can be caused by temperature difference, that is, Thomson effect. These mutually interfered phenomena were summarized to a famous reciprocal relationship by Lars Onsager, that is, $L_{ik} = L_{ki}$. The Onsager reciprocity law shows that phenomenological coefficients form a symmetric matrix, self-phenomenological coefficients L_{ii} are ordered in the main diagonal, and mutual phenomenological coefficients L_{ik} are arranged in the remaining positions. Because of symmetry, the choice of force and flow is arbitrary, and a flow can be affected by another force. Onsager relations have been examined by many trials to prove that the Onsager reciprocity law is a universal law which is suitable for a variety of conditions near to equilibrium.

Pyrometallurgical reactions are chemical reactions at a high temperature and the chemical reactions proceed at a high rate. In a sense, these reactions proceed mostly at the rate dependent on mass transfer. Therefore the steady process matter flow has great significance to the efficiency of metallurgy. So the issues of metallurgical engineering and technology that extend into the linear irreversible process of the near equilibrium region should become necessary. The mass and heat transfer and fluid flow in metallurgical reactors must be studied and their spatiotemporal scales will expand from the molecule/atom level of the reaction itself to the processing reactor level.

However, for the entire manufacturing process level, the structures of procedures/devices are more complex, and their operation may not be steady, and the relations among them are also not linear. Therefore the development

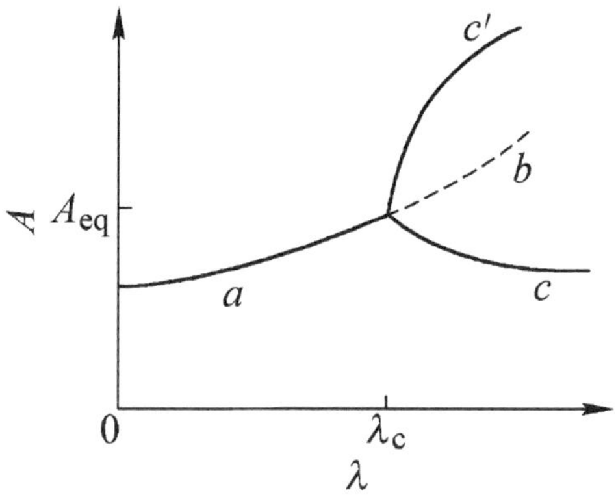

FIGURE 2.7 Schematic diagram of bifurcation and singular point.

of nonlinear nonequilibrium thermodynamics—self-organization in dissipative structure—has great significance for understanding and clarifying metallurgical process engineering on a larger scale.

2.6 OPEN SYSTEM AND DISSIPATIVE STRUCTURE

The dissipative structure theory was established by Ilya Prigogine (1917–2003) of the Brussels school in the 1960s and 1970s. The theory of process orderly evolution—self-organization far from the equilibrium of an open system was proposed. It is the rule of how the open system evolved from the chaotic state to an orderly structure. The conditions and the behavior near the critical point of phase change in the system were also described.

2.6.1 What Is the Dissipative Structure

When the state of a system is far from equilibrium, the steady process starts to lose stability. Fig. 2.7 shows a scheme of conditions far from equilibrium.

In Fig. 2.7, the ordinate A represents a state function of the system, such as concentration; the abscissa λ represents the distance from equilibrium, which is the degree of deviation from equilibrium controlled by outside parameters. When $\lambda = 0$, the system is in the state equilibrium. The state of $\lambda = \lambda_c$ is the critical point and it is a singular point of the trajectory of evolution. The minimum entropy production principle ensures the stability of steady process within the λ value between 0 and λ_c. The system may return to the state compatible with external conditions. Both the equilibrium state and steady state have the characteristics of spatial uniformity, time invariance, and stability to various disturbances. Therefore it is impossible to form spatiotemporal order structures spontaneously. At that time where $\lambda > \lambda_c$, the steady state is no longer stable, and the process is no longer extending along line b. In the critical state, small perturbations or fluctuations will induce the system change to a new state, that is, trigger the nonequilibrium phase transformation. The new stable state may be disordered (c'), but also may be spatiotemporal ordered (c) under the guidance of the order parameters. The coexistence of different stability is called the bifurcation phenomenon. Bifurcation can occur in secondary or in higher order, namely multilevel bifurcations. The disorder before the bifurcation point is equilibrium, characterized as space–time high symmetry. However, the emergence of a new ordered state is the result of the breaking of the space–time symmetry structure. Maintaining this space–time structure must rely on negentropy flux from the environment, so it is called a *dissipative structure*. Formation of a dissipative structure is related to three aspects. Functions determined by physical and chemical changes in the system breed instability that is known as internal cause. Space–time structure results from the instabilities. Under certain "threshold" critical conditions, the interrelation of fluctuations of slow relaxation enlarge themselves to order parameter to lead the old phase of instability transforms into a space–time asymmetric new phase. A time and space ordered state must rely on negative entropy dissipation to be maintained. Therefore, when the system is out of equilibrium, there are a variety of possibilities to choose and there may be different dissipative structures, there may be chaos or there may be an explosive reaction. This thorough complexity makes it difficult to use decisive causality to reason.

The basic understanding about dissipative structure theory proposed by Prigogine is that: for any system near equilibrium, if it has no mass and energy exchange with surroundings that induces instability, the development trend will obey the second law of thermodynamics, that is, $dS \geq 0$, that means the entropy increases gradually in the system and the process tends to equilibrium and disorder. However, the system is growing far from equilibrium and in an open system, that is, exchange mass flow/energy flow with the external environment (input and output), the development of the open system will increase the degree of order and make a new structure via local fluctuations and nonlinear interrelation. In other words, only through inputting negentropy flux to compensate for the dissipation of entropy production in the system, can the system operate in a dynamic-orderly manner.

For an open system (either mechanical, physical, chemical, or biological) after it enters the nonlinearity region far from the equilibrium state, once a certain parameter of the system reaches a certain threshold, through "fluctuations" and nonlinear interrelation, the mutation (namely nonequilibrium phase transformation) will occur. The chaotic state will be changed into a new state of time–space or function in an orderly manner.

As a state of the system, *chaos* can be understood as a nonlinearity and a bifurcation state and can switch with others. *Chaos* is not disordered and messy. Of course, *chaos* is different from orderly, but it likes a position without a periodic sequence (Sun, 2003). Feigenbaum pointed out that: *Chaos* is an irregular, aperiodic, detail intricate and unexpected nonlinearity effect appearing in the decisive system.

The steady state will lose stability of itself as the system is far from equilibrium. The process evolution no longer follows deterministic mechanical equations. The external disturbances and the internal fluctuations can cause the system to deviate from uniformity, and break the spatiotemporal symmetry. When the critical threshold is approaching, the ability to recover symmetry by internal fluctuations is smaller, and the relaxation time of fluctuations is more and more. One or several slow-relaxing fluctuations will become (alternately) the order parameter.

Fluctuation is a random factor, and it not only promotes the formation of the dynamic-orderly structure, but also leads to chaos. These are the two possibilities that may occur when the system evolution reaches the critical point.

The three elements of *Fluctuations*, functions of system, and the space–time structure caused by instabilities interact with each other, and result in the emergence of a dynamic-orderly dissipative structure. The dissipative structure is a product from the original instability state, however, once it is generated, its relative stability exists.

The new dynamic-orderly state needs to constantly exchange energy and matter with the environment to sustain itself. It will withstand certain stability and can hardly disappear under small outside disturbances. Prigogine (1978) pointed out that nonequilibrium may become a source of order and that irreversible processes may lead to a new type of dynamic states of matter called "dissipative structures." It will generate organization and coherence spontaneously in the open system and be called *self-organization*. For example, the life phenomenon and celestial body evolution are considered as self-organization. Therefore the theory of the dissipative structure is also known as self-organization in a nonequilibrium system. Dissipative structure theory is the theory of studying the change of a class of open system in some special conditions. From this theory the evolution of different systems can be discussed.

2.6.2 Features of the Dissipative Structure

The concept of the dissipative structure is related to the equilibrium structure. The equilibrium structure is a static orderly structure, for example, crystals. This kind of orderly structure has many essential differences with the orderly dissipative structure. Characteristics of dissipative structure are reflected in the essential differences of these two kinds of "orderliness" (Yan, 1987a).

1. The category of orderliness in the space scale of these two structures is different. The orderliness of the equilibrium structure is mostly referring to the microscopic and the orderly scale is characterized in the microstructure unit as atoms and molecules. However, the orderliness of the

dissipative structure is characterized in the macroscopic scale. The characterization of magnitude is exhibited in orderliness of the long-length space correlation and a large time cycle.

2. The stable and orderly equilibrium structure is a "dead" structure but the orderly dissipative structure is a "living" structure.

   The word "dead" means that once the stable and orderly equilibrium structure is formed, it will hardly change with time or space. The thermal motion inside the crystal can only induce the vibration in the vicinity of equilibrium place of the molecule/atom. The variation of conditions can only damage the equilibrium structure and go to the disordered state.

   The word "living" means that the orderly dissipative structure is a kind of dynamic and changeable orderliness. It will change with time or space and shows itself regularly and periodically. When achieving new mutant conditions, the system can go to another new orderly structure.

3. The conditions of continuous existence and sustenance of the two kinds of structures are different. In the equilibrium structure, once the stable orderliness is formed it can be maintained in an isolated environment, and without the need for the mass/energy exchanges with the surroundings. The equilibrium structure is an orderly structure which hardly dissipates energy. The dissipative structure must be formed in an open system and also must be sustained in an open system. It must have a continuous exchange of energy, matter, information with the external environment, and it must dissipate the negentropy flux input from outside environment. Only in this way can the dissipative structure maintain an orderly state, so named the *dissipative structure*.

2.6.3 Formation Conditions of Dissipative Structure

The dissipative structure in a dynamic-orderly operation would be formed in the system far from the equilibrium state under the following condition: the system must be open and far from the equilibrium state, with *fluctuations* and a nonlinear interaction mechanism (Yan, 1987b).

1. The system exists only in an open system.

   Open system has the following characteristics:

   a. The system exchanges matter, energy, and information with the external environment constantly. Without such exchanges, the occurrence and development of an open system are impossible.
   b. The system has the ability of self-organization, it can regulate automatically through feedback in order to adapt to environment changes.
   c. The system has sticking capabilities to ensure the structural stability and functional stability of the system, having certain antiinterference.
   d. The system takes evolution tendency of transformation and perfection when it is interacting with the external environment.

e. The system is constrained by various parameters of its own structure and function or the external environment.

The first condition for the formation and maintenance of the dissipative structure is that the system must be in the open state. Namely, the system must exchange matter, energy and information with the external environment constantly.

2. Nonequilibrium is the source of dynamic orderliness.

The system may be in three different states: (1) thermodynamic equilibrium state; (2) near equilibrium state; and (3) far from equilibrium state, that is, nonlinear nonequilibrium state. The equilibrium state hardly results in a dynamic orderliness; on the contrary, it reflects disorderly movement. The state near equilibrium tends to form a nonequilibrium steady state under modest energy exchange with the external environment, but the dissipative structure can hardly occur. Only when the system is far from equilibrium and in open conditions can the new stable, dynamic-orderly dissipative structure be possible to form. "Nonequilibrium is the origin of orderliness."

3. Fluctuations introduce orderliness.

Openness and nonequilibrium are referred to as the external conditions of the orderly structure mainly. Nevertheless, a qualitative change of a system evolution needs internal conditions. The internal incentives of making the system produce an orderly structure are *fluctuations*. Fluctuations are referred to as the deviations from the statistical average value of the system variables. Most of the fluctuations decay gradually (relaxation) and the system restores the original state. But when the state is far from equilibrium, some fluctuations decay very slowly, but they may be fed back and amplified as "giant fluctuations," and when the critical value is reached the system jumps into a new orderly state from an unstable state and a dissipative structure appears. The slow decaying fluctuations can be called order parameter.

Any kind of stable and orderly state can be seen as a result of a loss of stability of a disorderly state leading to some "fluctuations" being amplified.

4. Nonlinearity mechanism.

Openness, nonequilibrium, and fluctuations are important external and internal causes of the formation of the dissipative structure. But they are not sufficient conditions for the system to be able to form the dissipative structure spontaneously and maintain it. Only through the nonlinear interaction of each component within the system and dynamic coupling, can the elements within the system generate synergies and coherence effects, resulting in feedback, enabling the dissipative structure to occur in the system.

It is noted that the dissipative structure shows two different behaviors (Jantsch, 1980b):

1. Near their equilibrium, order is destroyed (as it is in isolated systems).
2. Far from equilibrium, order is maintained or emerges beyond instability thresholds.

The latter is called coherent behavior. Processes in a dissipative structure will operate with entropy produced inevitably. However, this entropy production will hardly accumulate in the open system, energy/matter is exchanged with the environment continuously. It is different in that the free energy and reactants belong to the input, whereas the entropy production and the reaction products belong to the output. This is the simplest kind of metabolism. With the help of the exchange of matter and energy with the surroundings, the open system sustains its inherent nonequilibrium. Conversely, this nonequilibrium, in turn, maintains the exchange.

Hence, the characterization of a dissipative structure is not the statistical measure of entropy distributed from total energy, but is the dynamic measure of the entropy production rate and exchange with the surroundings. It is the intensity of the conversion and compensation of input energy (from matter) for the open and dynamic system.

Two things to be understood about the process character of the dissipative structure are as follows:

1. The dissipative structure is understood to be the matter structure due to the organization of energy flow (such as coal-fired power plants). This is the theory for organizing energy flow in operation and conversion under material dissipation (eg, power generation processes and equipment, specific coal consumption).
2. The dissipative structure is understood to be the energy structure due to the organization of matter (such as in a steel plant). This is the problem in the view of organizing mass flow in operation and conversion under energy dissipation (eg, metallurgical manufacturing processes and devices as well as the specific energy consumption).

Concerning self-organization, each of the two processes have their particular emphasis, and both of them have special meaning. These two kinds of understanding constitute two complementary aspects of the characteristics of dissipative structure. The following relationships in engineering systems should be involved:

Mass flow-mass flow network (space structure)—conversion and running program of mass flow (dynamic space−time structure);
Energy flow-energy flow network (space structure)—conversion and running program of energy flow (dynamic space−time structure).

2.6.4 Fluctuations, Nonlinear Interaction, and Self-Organization in Engineering System

2.6.4.1 Fluctuations

Fluctuation phenomenon is the deviation from the mean value of the process parameters in a nonequilibrium steady state. A bifurcation point will appear during certain characteristic parameter increases. Before the bifurcation point λ_c, the function maintains a monodromic solution. When the bifurcation point has passed, the function solutions will be increased. The number of solutions will increase corresponding to the higher level bifurcations. The probability method must play an important role. The fluctuations may be amplified, and decide which bifurcation the system goes to.

Fluctuations are ubiquitous. Fluctuations quickly decay and disappear near to the equilibrium point. Only in the case of nonequilibrium, do fluctuations emerge. For example, metal solidification occurs at the temperature below melting point, that is, a sufficient supercooling, because the nucleus of crystals should be formed under the additional energy and this extra energy comes from energy fluctuations in the liquid metal. If there are no energy fluctuations, it can only become a supercooling liquid state, and the solidification will hardly occur.

Fluctuations are discrete characteristic events. In the equilibrium area or the local equilibrium hypothesis, the distribution of fluctuations shows as Poisson's distribution. When the mean value is ξ, the probability of the variable x is as follows:

$$P(x) = e^{-\xi}\frac{\xi^x}{x!}(x = 0, 1, 2, 3, \ldots)$$

Poisson's distribution characterizes both mean and variance equal to ξ. When ξ is determined, the Poisson's distribution is determined. The probability decreases rapidly with the increase of variable x. Fluctuations near to equilibrium show Poisson's distribution and they rapidly decay to zero. When the system is far from equilibrium, the distribution of fluctuations deviates from Poisson's distribution. At the bifurcation point the fluctuations of non-Poisson can magnify themselves. The long-range correlations as well as the breakdown of space symmetry will lead the unstable old phase to convert into a new phase with another structure. Then, the nonequilibrium phase transformation occurs. Fluctuations are a kind of confusion and reveal themselves as the factor for triggering instability. But in a critical region, fluctuations with long-range correlations promote the new space–time structure to show macro orderliness. Fluctuations will be effected by the relaxation time—the time from peak to valley. The fluctuations of slow relaxation can be amplified and become the order parameter of the new phase. Therefore the dissipative structure is a contradictory unity.

However, it should be noted that fluctuations are accidental, with dual properties, namely, the coexistence of construction and destruction. Reasonable and modest fluctuations induce the system's evolution to a new dynamic-orderly structure. On the contrary, unreasonable fluctuations can also weaken the stability of the original dynamic-orderly structure, or even lose efficacy. The evaluation of reasonable and modest fluctuations lies in their constructive to structural orderliness. Therefore the determinative theory expressing the system should be simultaneously studied. The world is colorful, determinative and fluctuations are complementary. Attention should be paid to both the certainty and probability. The achievement of nonlinearity thermodynamics lies in the possibility of orderly structures appearing in the dissipative system. Fluctuations are the origin of orderliness—"Order comes from chaos." Considering how the orderly mechanism of complex and open engineering systems come from fluctuation phenomena, fluctuations in a complex and open system have multifactors (eg, temperature/energy, concentration/mass, time, space), and multiunits (eg, procedure/device of different function and structure).

2.6.4.2 Nonlinear Interaction

The nonlinear interaction and the linear interaction exist correspondingly. The linear interaction means that the sum of the effects of each part is equal to the superposition of the effect of all parts (algebraic sum). The effects of each part are independent and satisfy with superposition in mathematics. The linear effect item of a dynamic system should be expressed as the first-order differential equation mathematically. However, the nonlinear interaction hardly meets the superposition rule. Namely, the overall effect of the system is not equal to the superposition of the effects of each part.

Compared with contradictions in philosophy, the interaction means exclusion versus attraction, competition versus cooperation of the contradictory opposites. Under the condition of the linear interaction, the association of various elements of the thing or everything is relatively simple. Namely, the synergetic relationship of the system or of correlation among units (subsystems) in a system is uncomplicated. But under the nonlinear interaction conditions, the exclusion versus attraction, competition versus cooperation of the subsystems within the system may produce a new emergence in its entirety. The nonlinear interaction will lead to an amplification of local fluctuations of the system, causing its mutation, and promoting the evolution of the system.

During the running of a complex and open system, the nonlinear interactions (convergence, matching, coordination, etc.) among different units on the same level are often characterized by nonlinearity coupling, as shown in Fig. 2.8. The nonlinearity coupling "force" comes from the synergism among the multifactor fluctuations (eg, temperature, time) and among the multiunit

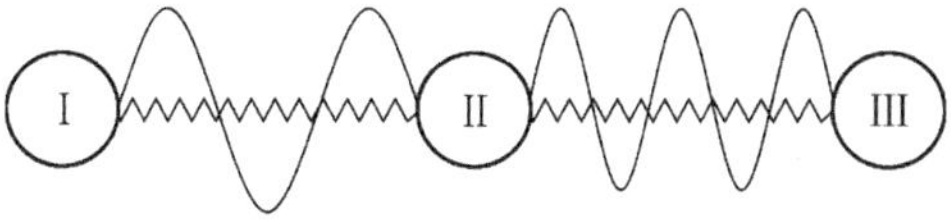

FIGURE 2.8 Nonlinearity coupling among the same level's units I, II, III—unit.

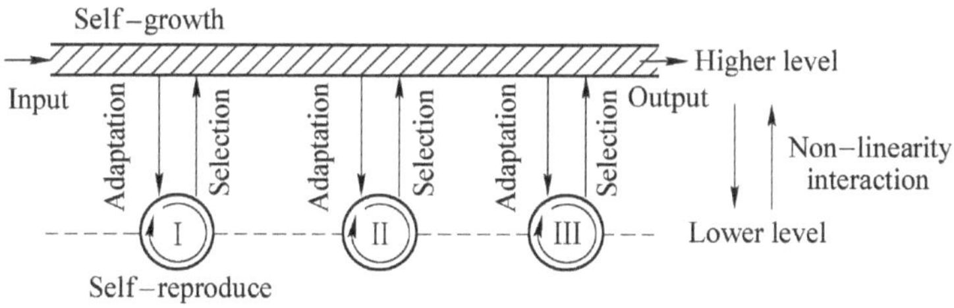

FIGURE 2.9 Nonlinear interaction between high level and lower levels (Yin, 2012).

fluctuations (eg, the heterogeneous units like procedures, reactors). This is due to the existence of multifactor competition or cooperation relations in the open system. That is, the fluctuations induce the generation of the order parameter. The order parameter promotes the nonlinear interaction or nonlinearity coupling.

In a complex and open system, there are nonlinear interactions among different levels, as shown in Fig. 2.9. The nonlinear interactions are mainly reflected in the adaptability and obedience of lower level units to the high-level system, as well as the selection and regulation by the high-level system with regard to the lower level units. The nonlinear interactions among different levels are often manifested in the adaptability and obedience of the different heterogeneous units to the entire system, as well as the selection and regulation by the entire open system with regard to different heterogeneous units.

The openness is the premise of dynamic-orderly operation of the system. Fluctuations lead to orderliness, and the nonlinearity effects are synergetic mechanisms of orderliness.

In engineering systems, the nonlinear interactions among units on the same level and the nonlinear interactions (selectivity−adaptability) between system and units are the orderly synergism mechanisms of the open system. The origins of these mechanisms are the unit fluctuations and multifactor fluctuations.

2.6.4.3 Self-Organization

A complex, open, irreversible, and far from equilibrium system exists in inherent self-organization, which originates from the multifactor fluctuations (temperature, time, etc.) and the multiunit fluctuations (different procedures/ devices, etc.) in the system. As the different units not only possess heterogeneity, but also have the same factor, the self-organization should be induced from the nonlinearity coupling among units and/or the nonlinear interaction between the system and unit.

The self-organization of an open system is a performance. Open systems with self-organization take different self-organization degrees in different orderly states. The improvement of the self-organization degree depends on two major respects: (1) to improve the orderly state and self-organizing mechanisms within the open system, for example, the moderate and reasonable fluctuations of different units and higher nonlinearity coupling degree among units as well as the selectivity–adaptability between units and system; (2) to use information inputted from the environment for the regulation of an orderly state and on improvement of the degree of self-organizing in the system. In short, the self-organization degree of open systems would be further enhanced through the combination of the additional information control "force" and internal "flow" of self-organization in the system. The self-organization phenomena of manufacturing processes have a variety of forms: self-generation, self-reproduction, self-growth, self-adaptation, etc. (Xu, 2000).

Engineering design is the beforehand plan and implementation plan to build an artificial existence. Through the integration process of selection, assembly, reciprocation, coordination and evolution, engineering design reflects self-organization features characterized by entirety, openness, hierarchy, processing, dynamic-orderliness, etc. of the engineering system. Through the combination with basic economic factors (resources, capital, labor, land, markets, environment, etc.), social civilization elements and ecological elements, engineering design should design and construct a reasonable structure. Then, the functionality and efficiency of the engineering system will be presented in operation. In this sense, engineering design must be in accordance with the objective laws of running evolution. Not only does each unit accord its own rules, but also these units can be inserted into the entire engineering system effectively, enabling the entire process to be a system with a high degree of self-organization. Therefore, the process network, mass flow network, energy flow network, and information flow network should be designed, in addition to the function structure space–time structure and information structure.

Information is the characterization property obtained in matter, energy, time, space, life, etc. Information cannot exist outside of matter and energy. The self-organization of mass flow and the self-organization of energy flow are the basis of the self-organization of information flow. Conversely, information flow can improve the self-organization degree of the mass flow and energy flow, namely:

1. On the basis of self-organization occurring in the mass flow, the self-organization degree of the mass flow can be improved through the regulation function of anthropogenic input information flow. The operation

efficiency of mass flow will increase and the losses of matter in the dynamic operation will be minimized.

2. On the basis of the self-organization for the mass flow and energy flow (concretely in the mass flow network and corresponding energy flow), the conversion efficiency of energy flow and the degree of secondary energy utilization can be improved through the synergetic optimization of mass flow with energy flow and then the regulation function of the anthropogenic input information flow. The energy dissipation of the mass flow/energy flow at each running will be minimized.

Information does not obey the law of conservation. The information can be copied and enlarged during its transmission/transfer, and may also be lost due to random noise. It is the difference between information and matter/energy. The information flow network and the regulation program of information flow are the self-organization ability established to achieve a minimum dissipation of the open system (manufacturing process system) in its dynamic-orderly running of the mass flow, mass flow/energy flow, or energy flow.

2.6.5 Critical Point and Critical Phenomenon

The restructuring of the manufacturing process somehow can be considered as a phase transformation. With the system being out of equilibrium, the steady state loses its own stability and the temporal and spatial symmetry falls suddenly into symmetry breaking instabilities. The motion and the structure of matter will be changed with the symmetry breaking and its aggregate state becomes another so that phase transformation occurs. There are two kinds of phase transformation. For the first type, thermodynamic potential itself remains continuous but the first derivative is discontinuous. The phase transformation is accompanied by latent heat and volume change. Nuclei of the new phase can be formed in the old phase and grow up. Most phase transformation in physics is this type. For the second type, the thermodynamic potential and its first derivative are continuous but its second derivative is discontinuous. There are no latent heat and volume changes in this type of phase transformation. There are no nucleation and growth phenomena, that is, no phenomena of two-phase coexistence. Phase-change occurs suddenly. The phase-change has no definite position and boundaries. It presents partly hidden and partly visible, arising one after another, inserting other dynamic images within each. This higher level phase transformation can also be called a continuous phase transformation. The phase change point is called the critical point that is a singular point on the trajectory of evolution. The singular point is a specific point which does not present a single slope on the function curve. At this point, the function is not differentiable, and it cannot be described by the state equation of analysis. The state corresponding to the manufacture process restructuring is a nonequilibrium phase transformation, which is similar to the higher level phase transformation.

The space–time symmetry breaking of the process can be expressed by the order parameter. The variable with slow declining fluctuations is called the slow relaxation variable. This is the order parameter which decides the degree of ordering of the system. The order parameter governs the behavior of the subsystem. Near the critical point, the recovering force lowers the fluctuation becoming smaller and smaller, and thus the relaxation time becomes longer and longer. Furthermore, near the critical point, fluctuations in different locations are correlated with each other. The length of the correlation increases gradually. The length tends to infinite through the whole system at the critical point. Thus the new phase and the new structure emerge instantly.

It has emerged as a series of historic critical phenomena at the macroscopic scale of the metallurgical production process. The criticality affected the structure, feature, and efficiency of the manufacturing process. The different macrostructures, features, and efficiencies are connected with the microscopic (mesoscopic) orderliness in metallurgical processes. Replacement of the open hearth by BOF, replacement of ingot teeming–blooming by CC, improvement on slab casting into thin slab casting and rolling, as well as emergence of the thin strip direct casting, caused fluctuations of the order parameter in the steel manufacturing process and induced a new integration and reconstruction of the process structure. Thus a number of engineering effects were presented.

In the metallurgical manufacturing process, the engineering effect means that the effect on the change of macroscopic structure, feature, and efficiency of entire manufacturing process is due to the appearance of the critical value of order parameters of the process/procedure after the application of some new technology (Yin, 2000).

In the steel manufacturing process, the engineering effect on "critical-optimization" or "critical-compact-continuous" is inspired by some or a certain order of parameters reaching their critical values. It results in some functions of a procedure being substituted by others and eliminating certain procedures/devices, and thus the entire manufacturing process can be simplified (see Fig. 2.10).

For example, the casting speed has great influence on the annual production per strand and the annual output of a caster. Only if the casting speed reaches a certain critical value, can the 100% CC process be realized and the proper economic benefits are achieved. The blooming mill with an annual capacity of 3×10^6 t may be replaced by continuous caster only if the slab thickness reaches 220–300 mm, which is the critical value of the slab thickness. The "critical flow rate" consisting of the "critical casting speed" and "critical thickness" means the critical order parameter to replace ingot

Dimension of Ingot/Slab (mm)	Process Pattern	Manufacturing Process
(About 500mm) even more (Ingot)	Ingot teeming Period: 10 days	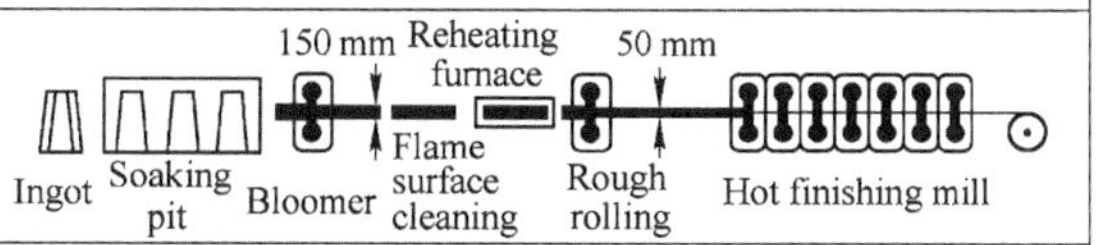
250 mm (Castslab)	Slab casting Period: 12 h Energy saving: 0.84 MJ/t	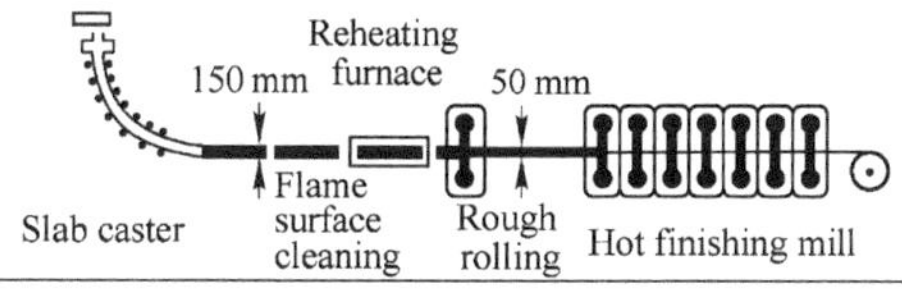
70–50 mm (Thin slab)	Thin slab casting - rolling Period: 1 hour Energy saving: 2.72 MJ/t	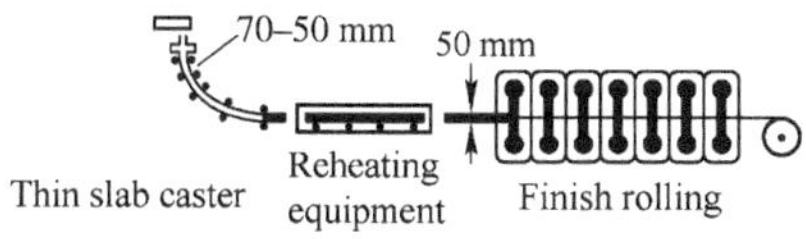
3 mm (Strip)	Strip casting Period: 5min Energy saving: 2.93 MJ/t	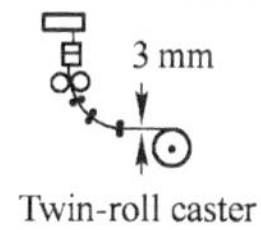

FIGURE 2.10 Evolution of the steel manufacturing process and critical-compactness-continuation effect (Yin, R., 2000. Structural analysis on steel manufacturing process and some problems of engineering effectiveness. Iron Steel 35 (10), 1–7 (in Chinese)).

teeming–blooming by CC. Once the 100% CC process replaces the ingot teeming–blooming, the structure, feature, and efficiency of the steel plant change significantly. Especially the "department store" type steelworks, symbolized by huge ingot-thick slabs and the blooming-break down as the keynote, can be replaced by a specialized steel plant with a CC-rolling process. This is the development trend for the steel industry. As can be seen over a long historical period, the solidification processing is the procedure with the greatest relevancy in the steel manufacturing process.

If the slab thickness for strip rolling is reduced from 250 mm (220 mm) to 180 mm (150 mm), the roughing mill can still not be completely canceled. However, once the thickness of the slab is reduced to 50–70 mm, the roughing mill not only can be eliminated, but also the long-scale slab heating-rolling and even the semiendless rolling with tunnel type roller hearth furnace may be practiced. Moreover, the near-net-shape casting makes the process structure more compact and continuous (Fig. 2.11).

The effect of "critical-compact-continuous" is also reflected in the "bottleneck effect" of a certain device that needs to be solved in the manufacturing process. The "bottleneck" phenomenon may occur in the procedure capacity and may also appear in its function strength. For example, the furnace campaign of BOF hardly meets the requirements of the sequence

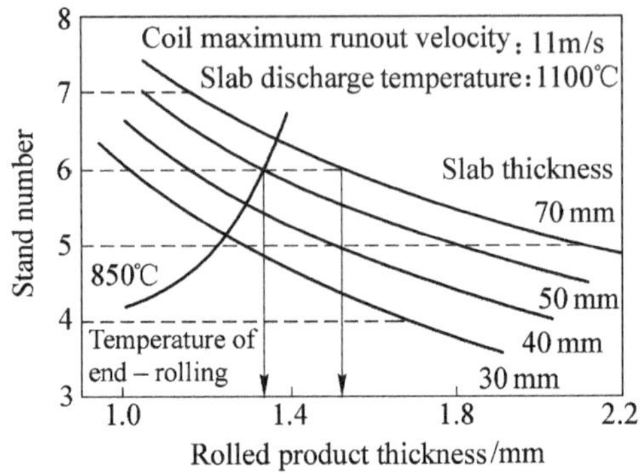

FIGURE 2.11 Relationship between critical slab thicknesses and number of rolling stands (Flick, A., Schwaha, K.L., 1993. Das Conroll-Verfahren zur flexiblen und qualitätsorientierten warmbanderzeugung. Stahl Und Eisen 113 (9), 66 (in German) (Flick and Schwaha, 1993)).

casting. Particularly, the hot gunning or the flame gunning of lining refractory takes quite a long time. It is difficult to achieve the coordinated operation of BOF and CC, further affecting the coordination and rhythm of the steelmaking workshop and rolling mill. The BOF slag splashing technology successfully eliminates the time "bottleneck" of BOF—CC synergism in the continuous running. The BOF slag splashing technology enables the BOF campaign to be controlled with an appropriate target. Firstly, the furnace life is kept within the time of the caster periodical maintenance. Hence, the two repairs are coordinated. Secondly, the furnace life should coordinate the maintenance cycle of the steelmaking shop with the rolling mill and the reheating furnace. Thirdly, the coordination of the overhaul of the production and operation of the steelmaking shop, the rolling mill, and the air separation plant should be considered. These are the critical values of the life of a BOF lining refractory.

Above all, the critical point, the critical effect, and the critical threshold have great importance to the formation of a dissipative structure and the dynamic-orderly operation.

REFERENCES

Chian, H.S., 1948. Engineering and engineering sciences. J. Chin. Inst. Eng. 6, 1–14.

Coulson, J.M., Richardson, J.F., 1997. Chemical Engineering, third ed. Pergamon Press, New York, NY, pp. vii–xii. (Coulson J.M., et al., 1997).

Coveney, P., Highfield, R., 1990. The Arrow of Time: A Voyage Through Science to Solve Time's Greatest Mystery. W.H. Allen, London, p. 281.

Feng Duan, Feng Shaotong, 2005. The World of Entropy. Science Press, Beijing, pp. 36–39. (in Chinese).

Flick, A., Schwaha, K.L., 1993. Das Conroll-Verfahren zur flexiblen und qualitätsorientierten warmbanderzeugung. Stahl Und Eisen 113 (9), 66 (in German).

Grossmann, I.E., 2004. Challenges in the new millennium: product discovery and design, enterprise and supply chain optimization, global life cycle assessment. Comput. Chem. Eng. 29 (1), 29–39.

Hao Bailin, 1988. Study of Chaotic Phenomena, vol. 1. China Academic Journal Electronic Publishing House, pp. 5–14. (in Chinese).

Harken, H., 1977. Synergetics. An Introduction. Nonequilibrium Phase Transitions and Self-Organization in Physics, Chemistry, and Biology. Springer-Verlag, Berlin-New York, p. 319.

Jantsch, E., 1980a. The Self-Organizing Universe: Scientific and Human Implications. Pergamon Press, New York, NY, p. 24.

Jantsch, E., 1980b. The Self-Organizing Universe: Scientific and Human Implications. Pergamon Press, New York, NY, p. 31.

Prigogine, I., 1978. Time, structure, and fluctuations. Science 201 (4358), 777–785.

Prigogine, I., Lefever, R., 1968. Symmetry breaking instabilities in dissipative systems. II. J. Chem. Phys. 48 (4), 1695–1700.

Prigogine, I., Nicolis, G., 1967. On symmetry-breaking instabilities in dissipative systems. J. Chem. Phys. 46 (9), 3542–3550.

Richardson, J.F., Harker, J.H., 2002. Chemical Engineering Volume 2, fifth ed. Butterworth-Heinemann, New York, NY, pp. xxv–xxvii.

Sun Xiaokong, 2003. The Philosophical Debate of Modern Science. Peking University Press, Beijing, pp. 92–107. (in Chinese).

Xu Guozhi, 2000. Systems Science. Shanghai Scientific and Technological Publishing House, Shanghai, pp. 196–202. (in Chinese).

Yan Zexian, 1987a. Dissipative Structure Theory and System Evolution. Fujian People's Press, Fuzhou, pp. 79–80. (in Chinese).

Yan Zexian, 1987b. Dissipative Structure Theory and System Evolution. Fujian People's Press, Fuzhou, pp. 72–78. (in Chinese).

Yin Ruiyu, 1997. The multi-dimensional mass-flow control system of steel plant process. Acta Metall. Sin. 33 (1), 29–38 (in Chinese).

Yin Ruiyu, 2000. Structural analysis on steel manufacturing process and some problems of engineering effectiveness. Iron Steel 35 (10), 1–7 (in Chinese).

Yin Ruiyu, 2008. The essence, function of steel manufacturing process and its future development model (translated). Sci. China Ser. E Technol. Sci. 38 (9), 1365–1377 (in Chinese).

Yin Ruiyu, 2009. Metallurgical Process Engineering, second ed. Metallurgical Industry Press, Beijing, pp. 139–146. (in Chinese).

Yin Ruiyu, 2012. Integration technology of high efficiency and low cost clean steel "production platform" and its dynamic operation. Iron Steel 47 (1), 1–8 (in Chinese).

Yin Ruiyu, Wang Yingluo, Li Bocong, et al., 2007a. Philosophy of Engineering. Higher Education Press, Beijing, pp. 9–11. (in Chinese).

Yin Ruiyu, Wang Yingluo, Li Bocong, et al., 2007b. Philosophy of Engineering. Higher Education Press, Beijing, p. 7. (in Chinese).

Yin Ruiyu, Li Bocong, Wang Yingluo, et al., 2011. Theory of Engineering Evolution. Higher Education Press, Beijing, pp. 28–30. (in Chinese).

Chapter 3

Basic Elements of Dynamic Operation of the Steel Manufacturing Process

Chapter Outline

From the physical essence of the dynamic operation of the steel manufacturing process, the process operation contains three elements: "flow," "process network," and "running program," as described below.

3.1 "FLOW" IN THE MANUFACTURING PROCESSES—MASS FLOW, ENERGY FLOW, INFORMATION FLOW

In the manufacturing process, the "flow" can be reflected with three carriers: the mass flow with materials and semiproducts as the carrier, the energy flow with energy resources as the carrier, and the information flow with the information message as the carrier.

Theory and Methods of Metallurgical Process Integration.
DOI: http://dx.doi.org/10.1016/B978-0-12-809568-3.00013-9

The mass flow in the manufacturing process is different from the general sense of Logistics. According to the definition of the national standards of China, "Logistics term" (GB/T18354-2006), logistics refers to the entity flow process of goods from deliverers to receivers. According to actual requirements, the basic functions of the transport, storage, handling, conveying, packaging, circulation processing, distribution, and information processing are combined organically. According to the definition above, logistics involves little physical and chemical changes of materials, or the manufacturing processes of goods. The mass flow is the dynamic movement with physicochemical conversion in the manufacturing process, such as the steel manufacturing process. Different physical conversions, chemical changes, and various processing occur in the manufacturing process. In the steel manufacturing process, there are both mass flow and logistic flow. For example, when a ladle is lifted by a crane, the logistic flow considers only the transportation and movement of the ladle; while mass flow in the manufacturing process considers the chemical reactions in the molten steel, the temperature drop of molten steel, and the removal of nonmetallic inclusions from the molten steel.

In the production process of a steel plant, the ferruginous mass flow (eg, iron ore, steel scrap, pig iron, slab billet, finished steel product) is the processed subject. The carbonaceous energy flow (eg, coal, coke, gas, power) works as the driving force, chemical agent, or thermal medium. Materials in mass flow are processed and handled according to process requirements, leading to the changes, such as displacement, chemical and physical conversion, deformation and phase transformation, achieving the multiobjective optimum of high production efficiency, low cost, high product quality, low energy consumption, low process emissions, and environment friendliness.

The mass flow of the steel manufacturing process is the "flow" of multifactors (eg, chemical composition factor, physical phase state factor, geometry factor, surface feature factor, energy-temperature factor, time-sequence order factor, space-location factor). In such a complex "flow" system, the connecting-matching, coordinating-buffering, even heredity variation of these factors among units (procedure, reactor, etc.) are all nonlinear interactions. The key role is played in forming a quasicontinuous/continuous process system, and characterizing the optimal operation dynamics of manufacturing processes. In a sense, process optimization is the coupling (synergetic) of various factors within the "flow" on the respective network nodes of the process system. Preferably, coupling favors the expected target state of the "flow," and promotes the operation efficiency.

The "flow" of the manufacturing process is a different concept from the generalized flow discussed in the condition of linear nonequilibrium. "Flow" and fluid flow must be distinguished too. The generalized flow refers to the flow rate of mass, energy, momentum, or electricity per unit area and is named as flow or flux. Fluid flow refers to the motion of continuum (fluid).

"Flow" broadly refers to the dynamic evolution of various types of resources or events that are orderly operated in the open system. In some cases, "flow" can be analyzed and simulated with the concept of fluid flow (Yin, 2011).

"Flow" generally refers to the dynamic operation/conversion of various forms of "resources" and/or "events" moving in an open system. "Flow" possesses the following characteristics:

1. Dynamic-openness of input/output;
2. Timeliness of dynamic operation/conversion process;
3. Spatiality of input/output direction;
4. Disorder, orderliness, or chaos during running;
5. Running process coupling with transformation of matter and/or energy.
6. "Flow rate"/"throughput," etc., characterizing the running process.

Generally, the basic feature of the "flow" is dynamic openness. In the steel manufacturing process, the energy flow and information flow usually coexist with the mass flow. The evolution of steel manufacturing processes and their developing direction are often reflected in the pursuit of quasicontinuous/continuous "flow." The quasicontinuity/continuity, synergism, and compactness are attractive goals for the steel manufacturing process and of the process operation since the goals reflect the optimum of matter, energy, time, space, and information in the "flow." Namely, they reflect the coordinative arrangement and optimal operation of mass flow and energy flow of the steel manufacturing process in a reasonable spatiotemporal scope.

As the dynamic operation behavior of the "flow" is attractive for the goals of quasicontinuity/continuity, synergism, and compactness, the optimization on the function set of units (procedures/devices), the optimization on the relation set of units about mutual interaction, mutual support, and mutual restraint in the process, the reconstruction optimization on the function set of units and the relation set of units in the manufacturing process are very realistic and very significant issues.

In view of technology, the production process of steel plant on the one hand is essentially technical processes of state transformation and the property control of matter, such as, converting and control of the material composition and structure, control of variety and quality of products, control of the shape, size, and surface quality, and control of the product performance. Another aspect is the regulation and control of mass flow/logistic flow. The mass flow control requires not only the control of throughput, but also the convergence of major parameters of the material flow and optimization, such as the reasonable coordination and matching of material quantity index, temperature, and time-cycle in mass flow; the coordination of material throughput of input/output among related procedures; the harmony and buffer of the time-rhythm; the optimization of mass flow path, direction, distance, and routes; the compression of the time-cycle and the compactness of the route of mass flow/logistics. These are very important parameters having great influence on the steel plant mode, the investment and investment return, the environmental burden, etc. From the view of engineering science, the characteristics of the steel manufacturing process are the coordination-optimization of matter state transformation, matter property control, and mass flow control in the process systems. Only through synergetic optimization of improving technology, optimizing of equipment and regulation with information technology in a dynamic operation process, can they play a timely and effective guiding role on the mode of steel plant, investment direction, investment steps, investment intensity, etc. For the steel manufacturing process, the transformation of matter state, the control of matter properties, and the control of mass flow are carried out simultaneously with or separately from the operation of mass flow. Generally, the mass flow process of the steel plant includes material state change, matter property control, and movement of mass flow with vectorial characteristics. The mass flow of the steel plant is obviously different from that of the rail transportation system, the post-mail distribution system, the circulation-distribution system into department stores, and the logistics system of the automobile assembling line or machine workshop. The process mass flow may be summarized as the multifactor mass flow control mode (Fig. 3.1).

As a multifactor mass flow control engineering in steel production, the connection, matching, continuity, and steadiness of such parameters should be considered as follows:

1. Conversion, transfer, connection, matching in states (oxide state or metallic state), and quantity of mass flow;
2. Changes of liquid metal to solid state with a certain geometric size, followed by deformation, transport, connection, matching in geometric section and dimensions of solid metal blanks;
3. Conversion, transfer, connection, and saving in energy or temperature of mass flow;
4. The transformation, heredity, and regulation in surface quality, macrostructure, microstructure, and property of mass flow;

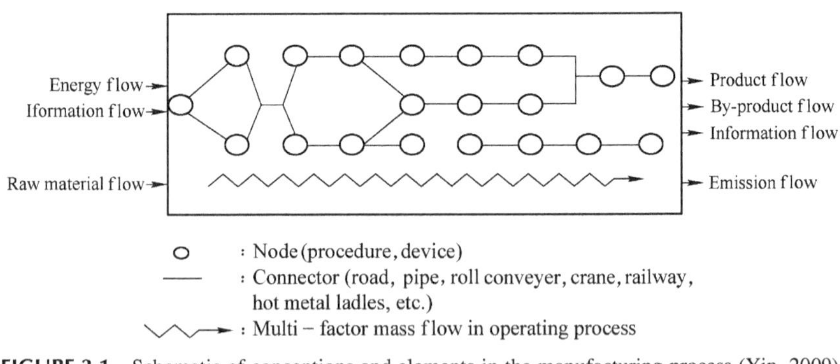

FIGURE 3.1 Schematic of conceptions and elements in the manufacturing process (Yin, 2009).

5. Selection, regulation, and connection of transportation ways and pattern of the materials in mass flow (and logistics);
6. Harmony, buffer, and acceleration in time-rhythm of mass flow.

3.2 RELATIONSHIP BETWEEN MASS FLOW AND ENERGY FLOW

The modern integrated steel enterprise is a kind of iron-coal chemical process engineering and deep-processing system. The steel manufacturing process can be abstracted into the input-output processes of ferruginous mass flow and carbonaceous energy flow as well as the interaction process between mass flow and energy flow, which favors analyzing the dynamic behavior and the effect of mass flow (mainly ferruginous flow), energy flow (mainly carbonaceous flow), and their interaction mechanisms in steel production and promotes the new breakthrough in function extension, that is, energy conservation, emission reduction, and waste recycling.

It is necessary to have a further study on the behavior and relationship of energy flow and mass flow in the steel manufacturing process. Mass flow plays the main role in the steel manufacturing process, and is always associated with energy flow. In view of energy flow, energy flow could not operate with the mass flow all the time in steel production process. A part of the energy flow will run relatively away from mass flow. Therefore, mass flow and energy flow have a kind of partner relationship of interacting with each other when together and respectively respecting their characteristics during separated. In general, mass flow and energy flow are sometimes combined or sometimes separated. The operation of the input-output of the entire plant is considered in the form of a mass flow network and an energy flow network, respectively, which are correlated but do not fully overlap (Figs. 3.2–3.4).

For the local process or device, mass flow and energy flow are inputted at their respective inputs, and interact with each other inside the device. At the output, mass flow tends to be outputted together with some energy flow. At the same time, there is the secondary energy in different forms outputting separately. The output of surplus heat and residual energy is unavoidable in the operation process. For example, in the procedure of the blast furnace (BF) ironmaking process, sinter (pellet), coke, pulverized coal, and blast air are imported, respectively, at input; in other words, the input of mass flow and energy flow is separated. Inside the BF, sinter (pellet), coke, pulverized coal, and blast air interact with each other through various processes like combustion, heating-up, ore reduction, softening-smelting, slag formation, desulfurization by slagging, and carburization of liquid iron, resulting in the production of hot metal. At the output points, the mass flow as hot metal and liquid slag carrying the major energy is exported. Meanwhile, the BF gases carrying kinetic, heat, and chemical energy are exported, which appears to be mass flow in its form but actually is secondary energy flow. A similar

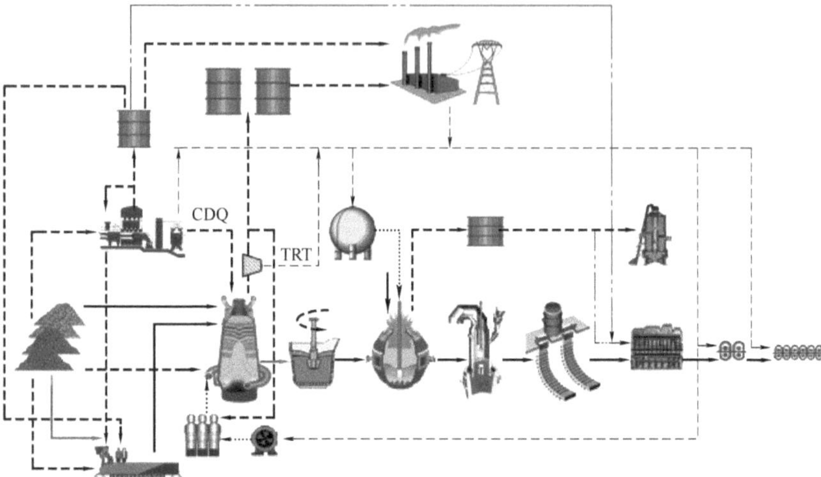

FIGURE 3.2 Schematic of operation network and path of mass flow and energy flow in steel plant.

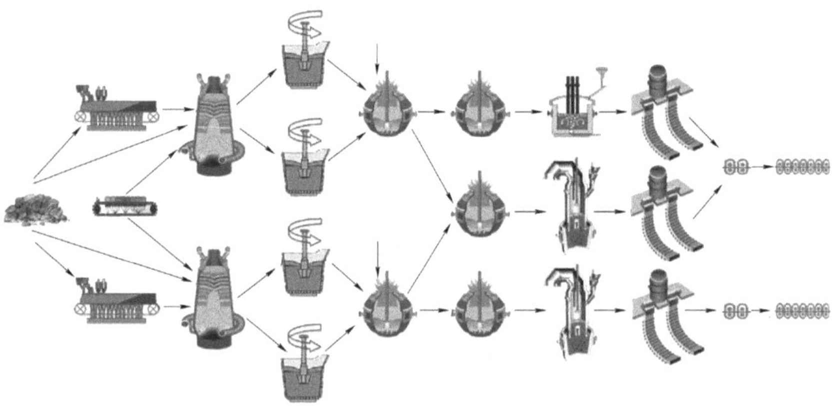

FIGURE 3.3 Operation network and path of mass flow (ferruginous flow) in steel plant.

phenomenon also exists in the sintering process and coking process, the steelmaking process, and the reheating furnace for steel rolling.

As a result, for the behavior of the mass flow and energy flow, attention should be paid not only at the input, but also at the output.

In the study of the behavior of energy flow during the steel production process, the balance calculation of mass and energy cannot be limited to a closed condition because the calculation is just at certain condition, which is a lack of concept of the dynamic operation and the correlation of upstream and downstream. To analyze the behavior of the manufacturing process, it is necessary to further establish the concept of the dynamic operation of the

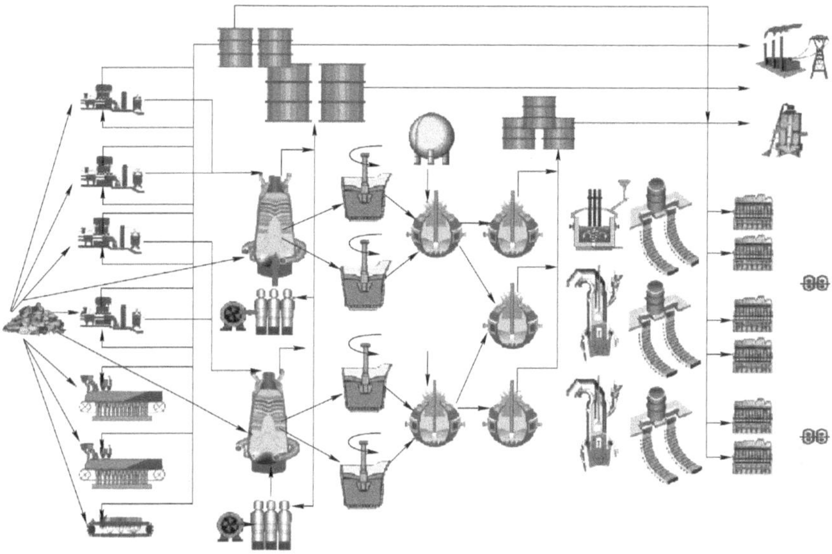

FIGURE 3.4 Operation network and path of energy flow in steel plant.

inlet "flow" and outlet "flow" under an open system condition. Concerning the concepts of energy input and energy output, these are not only related to energy itself, but also related to such factors as energy level, energy grade, time-space distribution, running program of energy flow, etc. It favors the building of an energy flow network (or energy conversion network) to further improve the efficiency of energy utilization and to establish the corresponding information control system.

In the steel manufacturing process energy flow includes chemical energy, heat, power, pressure, etc., in which chemical energy of carbon (carbonaceous flow) is the main type of energy resources. The energy flow is composed of several kinds of energy medium and operates independently from the mass flow (ferruginous flow) sometimes, but also can operate together and interact with the mass flow. In the raw materials yard, iron ores and coals are stored separately, becoming the starting points of the ferruginous mass flow and the carbonaceous energy flow, respectively. In the following working procedures, the ferruginous mass flow operates dynamic-orderly under the action and drive of energy flow along the given "process network" (eg, general layout). It presents the various conversions of mass and energy accompanied with displacement processes. The ferruginous mass flow plays the main role of processing, while the energy flow provides moving power, chemical reaction agents, and heat sources. Most of the energy medium runs along with the mass flow and enters the procedures, and output with the discharged mass flow. However, there is some energy running independently from the ferruginous mass flow, and much

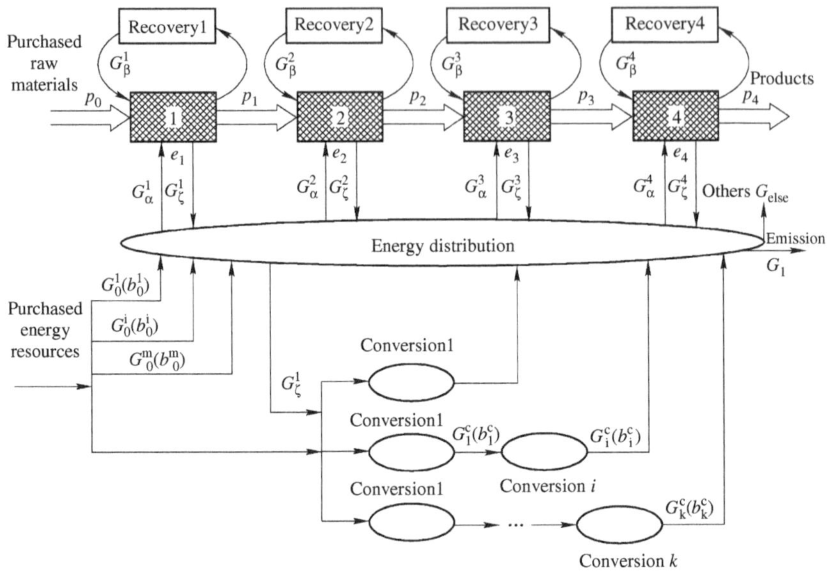

FIGURE 3.5 Interaction diagram of mass flow and energy flow in a steel plant (Wang, 2007).

independent residual energy flows out of each procedure. If the energy emission (secondary energy) from each procedure can be organized by a "program" with a certain necessary additional primary energy and is fully used, the energy conversion network-energy flow network should be constituted in the steel plant (Fig. 3.5).

Fig. 3.5 shows the relationship between mass flow and energy flow in a steel plant, where the hollow thick arrow lines represent the mass flow, single solid arrow lines represent the energy flow; boxes with the No. 1, 2, 3, 4 represent the production procedures (simplified to list only sintering, ironmaking, steelmaking, and rolling); p_1, p_2, p_3, p_4 and e_1, e_2, e_3, e_4 respectively represent coefficients for steel ratio and energy consumption of each procedure; G_α^1, ..., G_α^4 represent input energy for each procedure; boxes marked "recovery 1" and "recovery 2" represent the energy recovery reactor for each procedure, such as waste heat boiler, top gas pressure recovery turbine (TRT) of blast furnace (BF), gas recovery equipment; G_β^1, ..., G_β^4 represent the self-consuming energy of each procedure; the ellipse marked "conversion *i*" represents the energy conversion procedure, such as coking, power generation, air separation, blower, water supplying.

The purchased raw materials pass through sintering/briquetting, ironmaking, steelmaking, and rolling until the output of steel products, forming the mass flow operation. To practice the conversion (state transformation, structure advancing, chemical refining, etc.) and the transportation (displacement, conveying, etc.) of the mass flow, the energy flow should be organized.

Most of the purchased energy resources G_0 enters into the network of energy conversion to produce energy flows ($G_1^c, \ldots, G_1^k$ taking energy grades $b_1^c, \ldots, b_1^k$, respectively). Some purchased energy resources ($G_0^1, \ldots, G_0^m$) are supplied to each procedure and in coupling with mass flow and produced energy flows they interact with each other in these procedures. The energy flow promotes the changes in mass flow effectively to produce steel products. On the other hand, the entered energy flow can be converted into secondary energy flow ($G_\varsigma^1, \ldots, G_\varsigma^4$ and $G_\beta^1, \ldots, G_\beta^4$). This part of the energy flow runs separately from the mass flow and enters the network of energy (energy flow network) as the beginnings (input) of the secondary energy flow.

Meanwhile, part of the energy flow runs together with mass flow. The sensible heat and chemical heat of hot metal, the sensible heat of molten steel and latent heat of liquid steel solidification, and the sensible heat of cast semis are taken into the next procedure with ferruginous mass flow in all, but the sensible heat of the sinters or pellets are separated from mass flow partially. Limited by the charging system and top feeder, only the cold burden material can be charged into BF, so it is necessary to convert the sensible heat of hot sinter/pellet into steam or electricity by heat recycling. At this situation, the synergy of mass flow and energy flow is shown as the full use of the sensible heat. When the energy flow enters the next procedure together with mass flow, the synergy displays full use of the effective energy contained in the mass flow, such as using the sensible heat and the chemical heat of the hot metal in the converter, charging hot continuously cast semis into reheating the furnace. Moreover, the effect of energy flow on mass flow is shown as driving the mass flow ($G_{tran}^1, \ldots, G_{tran}^3$), such as the sinter belt driven by power, the hot metal ladle car driven by power or diesel engine, the roller table for cast semis driven by power, etc. The driving force of the energy flow on the mass flow, of course, should not be too much, otherwise, the excessive energy may be beyond the ability of the waste heat recycling and lead to energy emission (G_l).

3.3 MASS FLOW/ENERGY FLOW AND INFORMATION FLOW

Information is a characterization property contained and reflected in matter, energy, time, space, and life. The composition, structure, surroundings, condition, behavior, function, nature, becoming tendency, etc. of objective things can be expressed with information. Information must belong to a certain reflected thing called the information source. In the nature world and human brains there are no information that are absolutely isolated from matter, energy, time, space and life. A certain material, energy, or other forms of information are named as the information carriers. Therefore, the information flow, mass flow, and energy flow are closely related.

Upon confirming the dependency of information on matter or another source, the separability of the information and information source should not be ignored. It is very important that the information can be obtained without directly contacting with things. The collecting, exchanging, processing, accessing, and using of information will not bring any change to things itself. Information transfer can also pass over time and space. Information does not possess the characteristic of conservation, implying that information may be increased by copying during transmission. It also may be decreased due to the noise interference in identifying, processing, sending, receiving, and translating. This is how information differs from matter or energy.

To identify the information contained in mass flow/energy flow, first, various sensors are needed. Then the information is transmitted to the secondary apparatus or computers, after translation and programming to form the information flow model of the behavior of mass flow/energy flow. Of course, the application of artificial intelligence and expert systems can also strengthen and improve the mass flow/energy flow operation regulation, optimize the self-organization degree of mass flow, and improve the operation efficiency. The input of hetero-organization information is equivalent to the input of negentropy flux. It should provide a benefit by promoting the degree of order in the dissipative system.

For the mass/energy, information is used to characterize their attributes, such as structure, state, behavior, function, etc. For instance, in the manufacturing operation process, if the dynamic operation of mass flow/energy flow does not run smoothly, or is under a chaotic state, the information is undoubtedly confused and it is difficult to form a dynamic orderly information flow model, so the application of information technology in the manufacturing process system for effective controlling is unimaginable. Valuable information flow and intelligent control systems should be based on a reasonable network of mass flow/energy flow and on the dynamic orderly operation of the corresponding "flows."

Therefore, the self-organization and hetero-organization in the process manufacturing industry would be carried out through the information identifying, processing, feedback, etc. to constitute the dynamic operation instructions (measures) of mass flow/energy flow in different levels. The process parameters of matter, energy, time, and space of "flow" in the running may be controllable with fluctuations in a reasonable and moderate range and at nonlinear interaction in the process. The self-organization degree of an open system, that is, the orderly degree of function, space, time, and space-time will be improved. The entropy production rate of the dynamic operating in an open system will decrease. The dissipative structure will be improved to achieve optimal process dissipation.

In the steel production process, information flow is the combination of the behavior information of mass flow, the behavior information of energy flow, the reflectional environment information, and the artificial controlling information.

3.4 "NETWORK" OF MANUFACTURING PROCESS

3.4.1 What Is the "Network"

From the viewpoint of graph theory, the "network" is a graph composed of "nodes" and "lines" (path arc) to present a particular structure. Thus, a "network" is the sum of "nodes," "lines" (arcs), and their relations. A process network is the moving path and trajectory of circulation, and the space-time boundary of the operation carrier.

The study of "networks" is very important. It may be applied in many aspects, such as transportation, information and communications, process manufacturing industries, as well as culture, education, finance, etc. "Network" is the moving path and space-time boundary of the related operation carrier in each case. The carrier mentioned here includes all kinds of goods and various groups of people in the transport industry. Besides, it includes all kinds of messages, such as text, picture, audio, and video information contained in the information and communications. Moreover, it includes different levels of electric current and voltage in the power substation and distribution, and even the circulating currency and capital in the financial industry. For the process manufacturing industry, such as heavy chemical plants and steel plants, it includes all kinds of mass flow, energy flow, and information flow. The effective operation of different types of an operating carrier needs a necessary and reasonable matched "network" to optimize its "functions" and "efficiency."

In short, the "network" is a general concept and tool in the modern world. It has been used as an entity in engineering, and networks have gradually been formed with reasonable structure, fitting functions, and high efficiency in engineering systems.

3.4.2 How to Study "Network"

In the topics of process engineering, "network" is not only a concept, but also the engineering entity.

3.4.2.1 "Network," "Flow," and "Program"

A study of a "network" can not only investigate the structure, function, and efficiency of the "network" itself, but also study the resources of the "network." That is, the study of all kinds of "flows" with different natures and different types, such as the logistic flow, mass flow, energy flow, information flow, cash flow, human flow, is important. The "flows" are dynamic-orderly run with different characteristics, different operation modes in the corresponding "networks." "Flows" with different characteristics, types, and operation modes put forward different requirements on the structure and function of the "network." So, the study of a "network" must be considered with the carried "flow" included. It is meaningless to perform isolated studies on a "network" without considering the nature and the requirement of "flows."

Patterns of "flows" in the "network" are varied, such as reversible, irreversible, orderly steady, random, seasonal, laminar type or stochastic type running, series connection, parallel connection, or series–parallel connection. In order to satisfy the requirements of efficiency, security, stability, and comfort of different "flows," the design, building, and running of network must match the requirements in the structure and function of the network. Meanwhile, the "program" of "flow" in the network should be noted. The "program" involves varieties of rules, tactics, as well as the function order, space order, time-order, and time-space order.

As a consequence, the dynamic operation system under a specific surrounding condition is constructed with "flow," "network," and "program." This operation system has its own structure, function, and pursuit of efficiency. It plays the basic elements in the dynamic operation of the manufacturing process engineering system.

3.4.2.2 Structure and Function of "Network"

To understand a "network" well, the structure, the function, and its efficiency must be studied first.

In order to know the "flow," "network," and "program" as the three basic elements of dynamic operation of process system, it is of importance to achieve a clear understanding of the entirety, dynamic, effectiveness, and related hierarchical structure of the "network."

The methods of graph theory and operations research have been used to study the structure of "networks." Due to the requirement of "flows" with different natures, types, and operations, each "network" is different, so the corresponding structure is varied. For example, in the specialized steel plant with high efficiency, it tends to require the "ferruginous mass flow network" to be a minimum directed tree structure (Fig. 3.6). This type of structure is adaptable to the simple efficient and dynamic orderly irreversible operation of ferruginous mass flow. Whereas, an energy flow network requires the

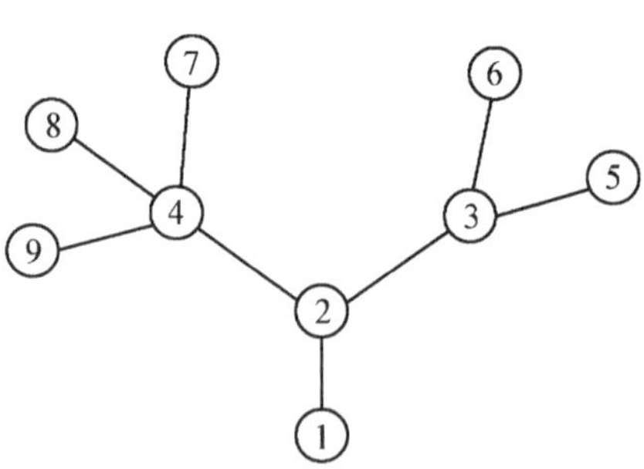

FIGURE 3.6 Diagram of the minimum directed tree network (it contains set of nodes with lines between node-assemblies).

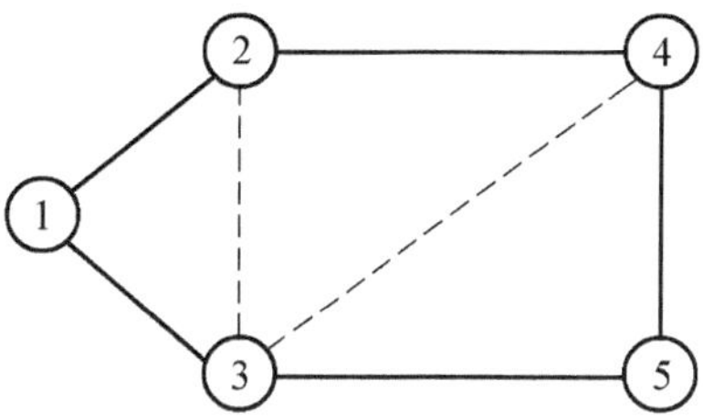

FIGURE 3.7 Diagram of primary circuit network: primary circuit (1,2,4,5,3,1); simple circuit (1,2,3,4,5,3,1, point 3 is repeated but none of the lines).

structure of a primary circuit (Fig. 3.7) in order to facilitate primary energy conversion efficiently, to recycle secondary energy, and to fully utilize and integrate optimization in time. Thus, requirements of the mass flow network and energy flow network of a steel plant are originated from the mass flow running efficiency and lowering loss, as well as optimizing the energy flow dissipation and maximizing efficiency.

The "network" structure in the transportation industry requires achievability, high speed, stable operation, safety, comfort, etc. The structure of the transport network is different from that of the process manufacturing industry. The network of transportation is a complex circuit. However, as a tool, a mathematical method, such as graph theory, is used to study these network structures in operational research. This can be used to determine reasonable "node" number, "node" layout, "connector" feature, "connector" length, series connection, parallel connection and series-parallel relationship, the characteristics of network graphics, space-time boundary condition, etc.

Any study of a "network" function must start from the nature, type, and operation mode of a "flow." The nature of the "flow"—either the mass flow, or energy flow, or information flow, or cash flow, or personnel flow—should be determined first. Then, according to the operation character of the "flow," the type of "flow" can be analyzed, for example, continuous flow, quasicontinuous flow, batch flow, and/or high frequency periodic flow. Then, based on the time factor of operation, it may be divided into steady flow, stochastic flow, and seasonal flow. Of course, the nature, type, and operation mode of "flows" require different structures and functions of the "network" (Diao et al., 2009).

3.4.2.3 Efficiency of "Network"

After fully understanding the requirements of a specific "flow" for the structure and function of the "network," it is easier to put forward the clear goal for the efficiency of the "network."

As "flow," "network," and "program" form a dynamic operation system, requirements for the "network" efficiency should be satisfied with multiobjective optimization. It is actually a multiobjective optimal system under different surrounding conditions.

While the efficiency of the "network" is studied, attention must be paid to maximize efficiency (concise and high efficient), minimize dissipation (energy dissipation, consumed materials, and information loss), environment friendliness (near zero emissions, zero pollution, environment protection, etc.), and safety protection (wealth and health safety, stable and comfortable operation). This can involve the minimum directed tree problem, the shortest directed path problem, the maximum flow rate problem, and the minimum cost flow problem in applied research and technological development (Diao et al., 2009).

3.5 "PROGRAM" OF MANUFACTURING PROCESS RUNNING

The operation program is the integrated instructions of a series of inner self-generated and outside entered information in the process of a dynamic operation. The operating program is closely related to the dynamic characteristic of "flow." This correlation with the dynamic operation is reflected in the logical relationship between the various kinds of "order" and "quantity" of "flow" and the change of contrasted "order" and "quantity."

Furthermore, the operation program embodies the logical integration and coordination relationship of various kinds of relevant "order" and "quantity." It is the integrated methods and results of dynamic ordering measures of various "order" and "quantity," especially the relationship between order parameters.

"Order" is an arrangement and presentation of *early* or *late*, and *big* or *small* according to the sequence of arrangement. That is, "order" can be regarded as the arrangement of a queue of units according to certain rules or regulations. In accordance with the above features, there is a strict definition of partial order (ie, order in general), in that "order" is a kind of binary relation with transitivity, antisymmetry, and reflexivity.

Study of the operation program and programmatic synergy inevitably involves the *order parameter*, a concept derived from synergetics, in which the order parameter is regarded as the result of the synergistic effect among subsystems. Any subsystem has two tendencies of movement: one is random motion spontaneously (such as thermal kinematic motion), leading to chaos or disorder; the other is coherent and coordinated motion, derived from the fluctuations in the subsystem and nonlinear interaction between subsystems, encouraging the generation of a specific structure with macro-ordering. Obviously, if coherent and coordinated movement is dominant in the entire system, then it results in the order parameter. The order parameter dominates most of the behavior of the subsystem in the system. The basic features of the order parameter are listed as follows: It is a macro-variable derived from the competition and coordination of subsystem, dominating a lot of the subsystem's behavior; it exists in the subsystem's evolution process; it is able to measure the orderly degree of the system; finally, it relates to other state variables as commanding and subordination, that is, it meets the "slaving principle" (Haken and Wunderlin 1988). It should be further noted that in an open and dynamic running system, one or few order parameters appear alternately through competition.

Therefore, the operation program possesses the logic correlation, integrated synergism, and dynamic orderliness with respect to "order," "quantity," "order parameter" of the system and the subsystems. The operation program is an important method and measure to adjust and control the self-organization system using information.

As the information instructions, the operation program involves matter, energy, time, space, life, and other information sources. The program can be of different types, for example, matter-dominated-program instruction, energy-dominated-program instruction, or time-space-dominated, even cost-dominated-program instruction. Due to the different levels and scales involved in the system, the operation program has different levels of hierarchy or scopes of scale. Different levels, scales, and types of operation program also can be integrated together to form the network of the operation program. The combination of "software" and corresponding "hardware" constitutes the network of information flow.

3.6 DISSIPATION IN DYNAMIC-ORDERLY OPERATION SYSTEM

The dynamic operation of a steel plant is an evolution process in an open system far from equilibrium. According to the theory of dissipative structure, the process cannot sustainably exist without the input of negentropy flux. The entropy production in the system drives the process toward equilibrium, or the process becomes static. However, even too much negentropy flux input will not lead to the entropy production in system shifting into negative. On the contrary, too much negentropy flux input increases the entropy production rate in system and increases the tendency from low entropy to high entropy. Life, for example, as a typical dissipative structure, has a daily metabolism through the input of nutrition and bodily exercise, but aging cannot be avoided. If there is too much nutrition inputted into the body, it will prompt more frequent multiple disease attacks at a younger age, such as hyperglycemia and high lipid diseases. In terms of industry production, too much mass and/or energy input can mean the process departs from running in a dynamic-orderly and coordinate-continuously manner, resulting in much more dissipation.

3.6.1 "Flow" Patterns and Dissipation

The manufacturing process is a kind of open and dynamic-orderly operation process integrated with a number of units (procedures/devices). To ensure the operation process runs dynamically and orderly with a lower internal

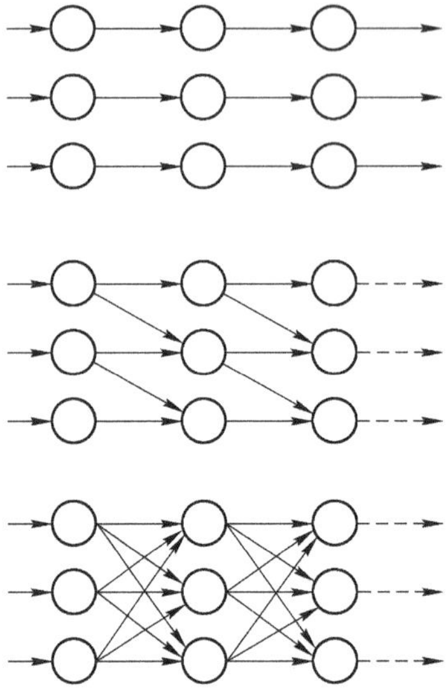

FIGURE 3.8 Laminar, chaotic, and stochastic types of operations of the manufacturing process.

dissipation, it is crucial to reduce the rate of entropy production in the system. As referenced by fluid flow, dynamic-orderly operated "flow" can be divided into laminar type running and stochastic type running. With the former, it is easy to achieve the orderly and steady state with little dissipation, while the latter characterizes high entropy production rate and much dissipation. It can be sorted into two substates: chaotic state or extra stochastic state (Fig. 3.8).

The "flow" and the fluid flow have different topical objects. The types of "flow" of the manufacturing process only borrow the appearance of the pattern of fluid flow. The characteristics of movement are not fully the same. "Laminar type running" refers to the "flow" without interference and crossing in the operation, but laminar flow means fluid at low speed with large internal friction and visible velocity gradient. Turbulent flow refers to the fluid flow at a Reynolds number higher than a critical one, which shows instantaneous fluctuation velocities and turbulence eddies in the fluid. However, the time-averaged velocity remains constant at a certain location. Hence, turbulent flow can hardly be considered as stochastic. In fact, the interference and crossing of "stochastic type running" result in higher dissipation in the system. The dissipation has a source of "flow" character in itself, instead of artificial interference.

Fluid flow is driven by force. The generalized flow (momentum, mass, or heat transfer) is driven by the generalized force (ie, gradient). To determine the direction and extent of a gradient, the knowledge of the "field" must be understood.

The "field" refers to the continuous distribution of a physical parameter along space and time. If at a certain time every point in space corresponds to a given physical quantity value (eg, mass, heat, velocity, momentum, concentration), there is a physical field in the space. If the values in the field hardly change with time, it can be called a steady-state field. Otherwise, for an unsteady-state field the distribution of a physical quantity is an instantaneous value. The distributed physical quantity can be scalar and vector. Components of a vector in every axial direction can be used as scalar. In the scalar field, the change rate of physical quantity along a certain direction (directional derivative) is called the gradient. In an unsteady-state field, the gradient of a time-point can be determined in time (instantaneous gradient). Generalized flow should be in proportion to the gradient (generalized force).

In the manufacturing process, "process network" (plan and vertical arrangement) and "operating program" are identical to "field," however, the "field" is not continuous. It could be understood as the space itself also belongs to the field and the distribution of the parameter value in the field is uneven and also has the feature of mutation. The discontinuous function is not differentiable so that the gradient could never be used as a "driving force." The nonlinear interaction, nonlinear coupling effect among subsystems can be considered as the driving force in the manufacturing process. For example, the BF tapping becomes the pushing force to the downstream procedures; sequence casting presents the pulling force to the upstream procedures and the pushing force to the downstream procedures, etc. Under the condition of "force" and "field," "flow" goes in different types of running of dynamic operation. The process dissipation is determined by the dynamic behaviors of "force," "field," and "flow."

There are interference and coupling phenomena between the heterotypical "force" and "flows." In the zone of an almost equilibrium (closed system), we have:

$$J_k = \sum_{i=1}^{n} L_{ki} X_i \quad (k = 1, 2, \ldots, n)$$

according to the Onsager reciprocity relation:

$$L_{ki} = L_{ik}$$

so

$$\sigma = \frac{d_i S}{dt} = \sum_{i=1}^{n} L_{ki} X_i X_k$$

where σ is the entropy production rate per unit volume of medium, W/m^3/K; X_k, the generalized force of number k in the irreversible process; J_k, the generalized flow or rate of number k in the irreversible process; L_{ki} and L_{ik}, the linear phenomenological coefficients; and n, the number of independent generalized flows or forces.

Interfering and coupling between flows increase the entropy production rate, which is also the situation if applying this rule to an open system far from equilibrium, that is, the nonlinear area. For example, the unreasonable layout will make the mass flow difficult and lead to the increase of the heat dissipation. Another example is that the far distance or unmatched capabilities between the reheating furnace and continuous caster will cause an off-line cold slab or even slow the casting speed, leading to more energy consumption and slab scaling. The types of "flow" types have influences on the dissipation of the dynamic operation of the manufacturing process.

3.6.2 Operation Rhythm and Dissipation

Events of each procedure/device are different in the production process. Their running speed and evolution rhythm are also much different. To realize the smooth operation, not only the operation route of the "flow" should be undisturbed, but also the time-cycle and frequency of procedures and devices ought to be coordinated. Therefore, a time-scheduling plan is needed in production to regulate and control the development and processing of the dynamic running.

For the metallurgical production, the time-scheduling plan is based on "minute flow rate" (processing amount per minute)—matter throughput of ferruginous mass flow every minute is the basis for controlling the production. Namely, the operation should follow the principle of equal "minute flow rate," neither is the equal second flow rate, nor is the equal hour flow rate. Actually, most of the procedures and devices (except rolling mill) do not need to control the accuracy of the running throughput at the level of the equal second flow rate, which is of little real significance for production. The hour flow rate is excessive slack to realize dynamic-orderly and synergetic-continuous operation. The equal hour flow rate does not mean that there is continuous steady operation in the metallurgical process. The discussed throughput here is a kind of macroscopic flow. Ferruginous mass flow will undergo a series of physical and chemical changes during the production process. The influence of these changes on the flow rate is not included in macroscopic flow—flow throughput discussed above.

The metallurgical process consists of different units (procedures, devices) through tandeming, paralleling, and integrating in order to make them match and coordinate with each other (some procedures or devices with larger elastic time values are usually used as the flexible buffer components), which is necessary to maintain the continuous running of the entire process. However, the excessive time elasticity of some procedures may increase the dissipation of the entire dynamic process. Therefore, each working procedure and device in the process should have reasonable fluctuations of an optimized running rhythm to realize a dynamic-orderly operation for a long time.

In addition to the running cycle and frequency in the production period of each procedure and device, the maintenance cycle should also be coordinated, otherwise poor quality maintenance or an unforeseen accident will result in unplanned downtimes which will have a detrimental influence on the running rhythm of the operation and inevitably increase the dissipation.

3.6.3 Distribution of Procedure's Functions and Dissipation

Development of technology in the steel manufacturing process often leads to analysis-redistribution of the function set of unit procedures and promotes the rationalization of functions of different unit procedures. The optimum coupling and coordination among procedures will promote the manufacturing process to develop in the direction of dynamic-orderly continuous/quasicontinuous.

The functions of procedures in the manufacturing process are usually multivariate, while a certain function can be achieved in several unit procedures. For example, desulfurization can be performed in sintering, ironmaking, hot metal pretreatment, steelmaking, and ladle refining. Generally speaking, hot metal pretreatment is the procedure prior to desulfurization. However, with changes of the production requirement, for example, steel grades, raw material conditions, goods delivering time, the optimized distribution of the desulfurization function among the procedures is required to be adjustable according to the actual situation. Since partial optimization is different from complete optimization, the entire process optimization of the function set of procedures and distribution of functions of unit procedures may lower the dissipation in the manufacturing dynamic operation.

3.7 FORMS AND CONNOTATION OF TIME FACTORS IN STEEL MANUFACTURING PROCESS

In the production process of steel plants, especially in compact steel plants, the coordination of time factors among procedures/devices is very important. So the expression forms of time factors in the production process are becoming more and more plentiful. In order to coordinate and control the process easily, the time is expressed not only as the time length of some processes, but also as the time-point, time-domain, time-position, time-order, time-rhythm, and time-cycle. It is important to analyze the expression forms of the time factors to establish an effective information controlling system in steel plants. In order to achieve the continuous/quasicontinuous operations of the manufacturing process, time factors must be studied as the objective function. The meaning of time in the coordination or integration of unit procedures in the steel manufacturing process should be analyzed. In fact, the steel manufacturing process is incessantly developing toward continuation. The continuation of the manufacturing process is more complicated than that of a simple continuous reactor. For example, when intelligent scheduling and control of steel production is studied, all kinds of influence factors and mechanisms must be investigated first. The shortest time of the entire process and the highest continuation degree, including high temperature linkage, compatibility, and production efficiency, should be reached in the case of all requirements and boundary conditions. Here, time is regarded as an objective function. Obviously, with a view to the regularity of temperature change, processing time and transportation time among procedures have great influence on the temperature change in the ferruginous mass flow (hot metal, molten steel, slabs, and rolled pieces). In these moments, the time of processing, transporting, and waiting are treated as independent variables. Thus, it is clear that the time factor in steel production processes comprises the role of independent variable and dependent variable (objective function).

For a procedure/device, the time process value is composed of the main operation time, supplement operation time, transporting time, loading/unloading time, and waiting/buffering time. The concrete expression forms are the scheduling and controlling of the time-point of the procedure (starting time or leaving time), time-domain of the process (time length and start-to-end period), and time-rhythm (frequency of procedure operation time and process cycle).

For the relationship between upstream and downstream procedure, time factors behave as the arrangement of the time-characteristic order, the "embedability" control of time-position, the linkage, buffering, and coordination of time-domains in long and/or short range. The short-range control exists between two procedures and the long-range control exists among three or more procedures. It also behaves as the arrangement of transportation, length of waiting time, and scheduling between transporting and waiting time.

The concept and mathematical presentation of time forms has been described in *Metallurgical Process Engineering* (Yin, 2009). Here is a précis.

Time-point: The start−end time of operation k in procedure o corresponding to ferruginous mass flow, such as ore, scrap, hot metal, molten steel, slab, rolled piece, is defined as the time-point, which is expressed as $[t^o_{ks}, t^o_{ke}]$ (Fig. 3.9a). The meaning of time-point is reflected not only in the procedure, but also in the entire manufacturing process (Fig. 3.9b). However, there is difference between the scheduled time-point and the real time-point in the production process.

In Fig. 3.9: $t^o_{1s}, t^o_{2s}, t^o_{3s}, \ldots, t^o_{ks}, \ldots, t^o_{ns}$ are the start time-points of the operations 1, 2, 3, ..., k, ..., n in procedure o;

$t^o_{1e}, t^o_{2e}, t^o_{3e}, \ldots, t^o_{ke}, \ldots, t^o_{ne}$ are the end time-points of the operations 1, 2, 3,..., k, ..., n in procedure o;

$t^{I}_{S}, t^{II}_{S}, t^{III}_{S}, \ldots, t^{N}_{S}$ are the start time-points of procedures I, II, III, ..., N in manufacturing process;

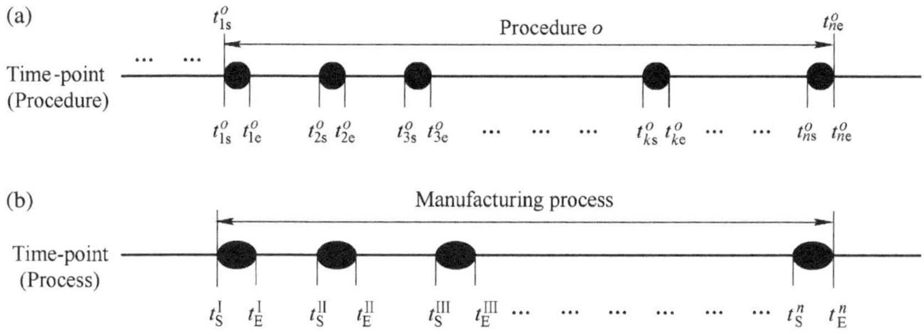

FIGURE 3.9 The schematic diagram of time-point.

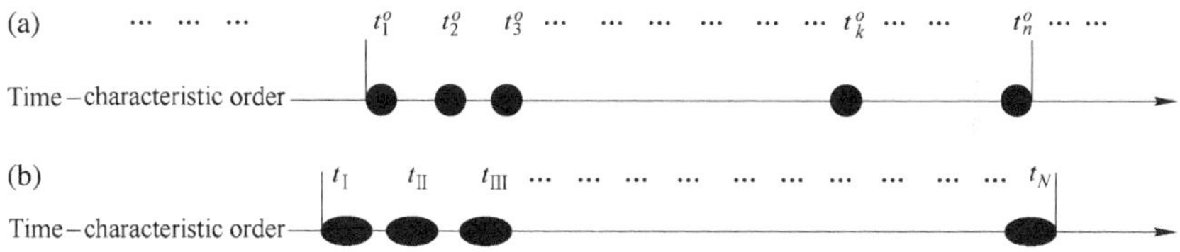

FIGURE 3.10 The schematic diagram of time-characteristic order. (a) Time-characteristic order of different operations in a procedure $\{t_1^o, t_2^o, t_3^o, \ldots, t_k^o, \ldots, t_n^o\}$ and (b) time-characteristic order of different procedures in the manufacturing process $\{t_{\mathrm{I}}, t_{\mathrm{II}}, t_{\mathrm{III}}, \ldots, t_N\}$.

$t_{\mathrm{E}}^{\mathrm{I}}, t_{\mathrm{E}}^{\mathrm{II}}, t_{\mathrm{E}}^{\mathrm{III}}, \ldots, t_{\mathrm{E}}^{N}$ are the end time-points of procedures I, II, III, …, N in manufacturing process;

Time-characteristic order: To improve the technical and economic indexes, such as cost, productivity, the time of every procedure in the mass flow should be arranged in series and scheduled coordinately according to the sequence of the technological process. Hence, the time-characteristic order of the manufacturing process integrated running would be formed. The time-characteristic order contains two concepts, that is, time-scheduling order of some operations in procedure o and the array order of some procedures in the manufacturing process (Fig. 3.10). It can be described as follows:

$$\{t_1^o, t_2^o, t_3^o, \ldots, t_k^o, \ldots, t_n^o\}$$
$$\{t_{\mathrm{I}}, t_{\mathrm{II}}, t_{\mathrm{III}}, \ldots . t_N\}$$

$\{t_1^o, t_2^o, t_3^o, \ldots, t_k^o, \ldots, t_n^o\}$ are the time-scheduling order of the operations 1, 2, 3, …, k,…, n in procedure o;

$t_{\mathrm{I}}, t_{\mathrm{II}}, t_{\mathrm{III}}, \ldots, t_N$ are the time-array order of the procedures I, II, III, …, N in manufacturing process;

Time-domain: Generally, the time length of mass flow located in the procedure is composed of times for main operating, supplementary operating,

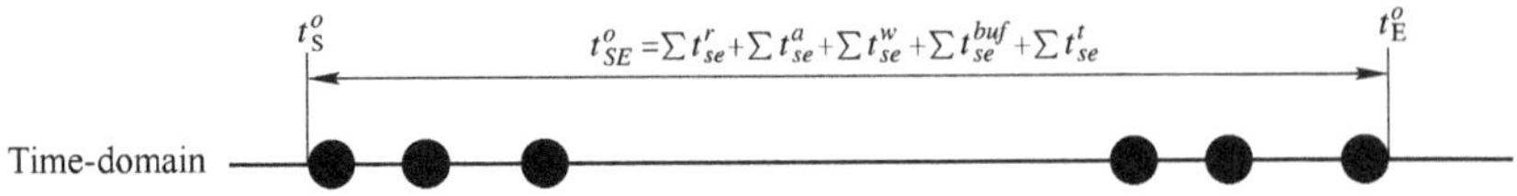

FIGURE 3.11 The schematic diagram of time-domain for a procedure.

transport, waiting, buffering, etc. The time-domain contains the total time above and the start and end time-points. It can be expressed as follows:

$$t_{SE}^o = \Sigma t_{se}^r + \Sigma t_{se}^a + \Sigma t_{se}^w + \Sigma t_{se}^{buf} + \Sigma t_{se}^t$$
$$\left[t_S^o, t_E^o\right]$$

where, t_{SE}^o is the total time-domain of procedure o, min; Σt_{se}^r, the processing time of procedure o, min; Σt_{se}^a, the auxiliary operating time of procedure o, min; Σt_{se}^w, the waiting time of procedure o, min; Σt_{se}^{buf}, the buffering time of procedure o, min; t_{se}^t, the transport time of procedure o, min; t_S^o, the start time-point of procedure o, min; and t_E^o, the end time-point of procedure o, min.

The time-domain of the procedure characterizes not only the start time-point and end time-point but also the time length of different procedures (Fig. 3.11). In addition, the concept of the time-domain could be extended to a series of procedures or even more.

Time-position: Due to the supplementary operation time, waiting time, and buffering time, it is observed from the definition of the time-domain that the main operating time should be embedded at a rational position in the corresponding time-domain, which would directly influence the scheduling, control, and optimization of the entire manufacturing process. Therefore, the time-position means the position of a certain time-domain. The time-position is described by following expressions mathematically:

$$t_{SE}^r = t_E^r - t_S^r$$
$$t_{SE}^r = \Sigma t_{se}^r$$
$$t_{SE}^o = \left\{\Delta t_{BE}, t_{SE}^r, \Delta t_{AF}\right\}$$
$$\left[t_S^r, t_E^r\right]$$

where, t_{SE}^r is the processing time-domain of a procedure, min; t_{SE}^o, the total time-domain of a procedure, min; t_S^r, the start time-position of processing, min; t_E^r, the end time-position of processing, min; t_{SE}^r, time-domain of a certain operation in procedure, min; Δt_{BE}, auxiliary time before processing operation, min; Δt_{AF}, auxiliary time after processing operation, min.

The relative values of Δt_{BE}, Δt_{AF}, and t_{SE}^r have direct influence on the position of t_{SE}^r in certain t_{SE}^o (Fig. 3.12). The time-position characterizes the time length of the ferruginous mass flow through the procedure, the rational time-position of the operation, and its start–end time-point.

Time-cycle and time-rhythm of the production: The time-cycle is the total time length of the mass flow passing through all the procedures in the

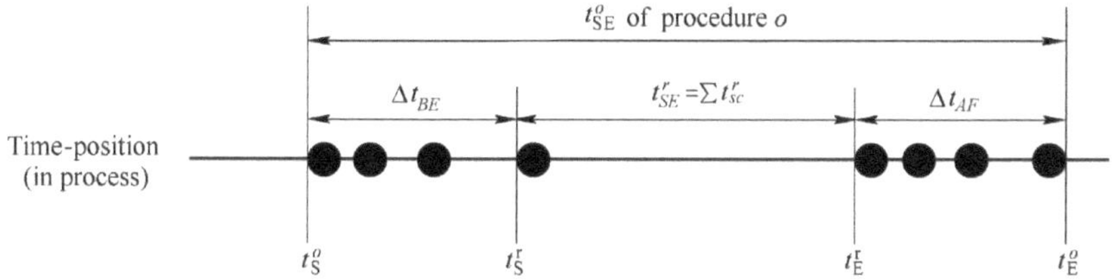

FIGURE 3.12 The schematic diagram of time-position.

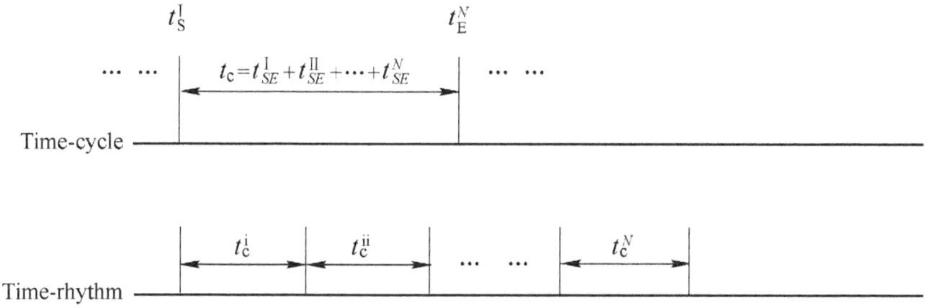

FIGURE 3.13 Scheme of time-cycle and time-rhythm.

manufacturing process. It contains the main operation time, supplementary operation time, transport time, waiting time, and buffering time (Fig. 3.13). It can be expressed as follows:

$$t_c = t^{I}_{SE} + t^{II}_{SE} + t^{III}_{SE} + \ldots + t^{N}_{SE}$$

where, t_c is time-cycle, min; and $t^{I}_{SE}, t^{II}_{SE}, \ldots, t^{N}_{SE}$, time-domain of the procedure I, II, III, …, N respectively, min.

When a number of production time-cycles proceed continuously and the time-cycles are approximately equal to each other, the time-rhythm is established as following and is shown in Fig. 3.13:

$$t^{i}_{c} \approx t^{ii}_{c} \approx t^{iii}_{c} \approx \ldots \approx t^{N}_{c}$$

A regular and ordering time-rhythm would be formed in this way.

Of course, this type of manifestation of time-cycle and time-rhythm of the production expresses not only in operation running of the entire manufacturing process but also in the running of procedures and workshops.

3.8 CONTENTS AND OBJECTIVES FOR DYNAMIC OPERATION OF STEEL MANUFACTURING PROCESS

3.8.1 Recognition Thinking Way

In process manufacturing industries, such as metallurgical industry and chemical industry, Newtonian mechanics and classical thermodynamics have been the theoretical methodologies to determine various phenomena for a long time. Topics based on determinism lead to the analysis of problems separately and simplistically. However, by these methodologies, the theoretical analysis lacks consistency with the practical operation of the production and is unable to adapt to the development requirements of the industry. Thus a study on manufacturing process engineering is needed to deal with the dynamic complex problems of the open system. The development of three kinds of physical systems has favored the recognition of these issues in the metallurgical process engineering. The theory of dissipative structure in the open system coordinated with the theories of systems analysis, synergetics, and engineering evolution is applied to study the complex problems of the process manufacturing industries (process engineering). Hence, the substance displacement, mass transfer, conversion of energy, chemical reactions among atoms or molecules, processes in procedures, the process in a workshop, as well as the running of entire plant can be inserted into each other and integrated as a whole. Finally, a continuous/quasicontinuous, dynamic-orderly and coordinated operation system can be achieved by selecting-integrating and interacting-coupling. The formed dissipative structure promotes the process optimization of the manufacturing process.

In other words, process engineering is developing in the field of comprehensive-integrated optimization. It has little focus on the correlation in local problems and static structures, but emphasizes the thermodynamics correlation of self-organization among multiple levels. Process engineering should be a dynamic-related structure constituted by varying kinds of processes with different properties and different time-space scales, and also a dissipative structure gathering processes and dynamic structure together.

The evolution and development of process engineering can be understood as the efficient and orderly organization of substances and energy. The optimized structure of the processes will gain from minimum energy dissipation, near-zero emissions, and rational material consumption, and then correspond to the optimization of the organization of information.

3.8.2 Research Contents of the Discipline

As a discipline branch, the following aspects of process engineering need further in-depth study:

1. Physical essence of a dynamic metallurgical manufacturing process;
2. States, properties, and flow dynamics of ferruginous mass flow as well as network and operation program of mass flow;
3. States, functions, and conversion mechanism/efficiency of energy flow in the metallurgical manufacturing process; and the formation of network

and operation program of energy flow, the energy flow network to be coordinated with mass flow;

4. Space factor study of the dynamic operation of the metallurgical manufacturing process, including: space scales of processes, space position, operating direction, and structure of space-levels;
5. Time factor study of the dynamic operation of the metallurgical manufacturing process, including: time-point, time-domain, time-position, time-order, time-cycle, and time-rhythm of processes, and the dynamic-structural relationship among these time factors;
6. Theory and method of the engineering design of steel plants, the theory, method, and tools of the dynamic-tailored design;
7. Rules of the dynamic operation of steel plants, and the methods and tools of simulation and information regulation of steel plants designing, etc.

3.8.3 Strategic Objectives of Research

In order to establish the optimized dissipative structure in the open system of dynamic metallurgical process engineering and to develop the theory of engineering science, consequently to promote the innovation of engineering design, and to promote the process restructuring and multiobjective optimizing of metallurgical plants based on the theories above, the strategic objectives of metallurgical engineering research should generally investigate the following:

1. Low cost and high efficiency clean steel production platform (including for sheet products, plate, common long products, alloy steel long product, stainless flat products, tubular products, etc.);
2. Establishment of energy flow networks and system optimization of energy flow networks;
3. Theory, method, and case studies of dynamic tailored design in the steel manufacturing process;
4. Optimization study of modes of new generation specialization steel plants with high efficiency;
5. Information characteristics and computer simulation of dynamic-orderly and compact-continuous process operation.

REFERENCES

Diao Zaijun, Liu Guizhen, Su Jie, et al., 2009. Operations Research, third ed. Higher Education Press, Beijing, pp. 198–216 (in Chinese).

Haken, H., Wunderlin, A., 1988. The Slaving Principle of Synergetics—An Outline. Order and Chaos in Nonlinear Physical Systems. Physics of Solids and Liquids. Springer US, pp. 457–463, http://dx.doi.org/10.1007/978-1-4899-2058-4_17, Print ISBN: 978-1-4899-2060-7, Online ISBN: 978-1-4899-2058-4.

Wang Jianjun, 2007. Study on Material Flows, Energy Flows and Material Flow-Energy Flow Relationship in Iron and Steel Factory. Thermal Engineering Department of Northeastern University, Shengyang, p. 91 (in Chinese).

Yin Ruiyu, 2009. Metallurgical Process Engineering, second ed. Metallurgical Industry Press, Beijing, p 151. 205–209 (in Chinese).

Yin Ruiyu, 2011. Metallurgical Process Engineering. Springer, pp. 126–127, 128–129.

Chapter 4

Characteristics and Analysis of the Dynamic Operation of Steel Manufacturing Process

Chapter Outline

Theory and Methods of Metallurgical Process Integration.
DOI: http://dx.doi.org/10.1016/B978-0-12-809568-3.00014-0

From a superficial point of view, the manufacturing process is composed of the upstream and downstream series–parallel connections of multiple procedures. However, the manufacturing process is essentially a multifactor, multidimension, and multiscale coordination-integration of many procedures in a dynamic operation which includes not only the mass and energy's flow and conversion/transformation, but also these phenomena's orderly and efficient insertion, in a spatiotemporal scale, into different dimensions and procedures. Based on the above, the static structure and dynamic operation structure of the manufacturing process will be established. The characteristics of the dynamic operation structure plays so important a role in the functions and efficacy of the manufacturing process in dynamic operation that these characteristics should be closely studied.

4.1 RESEARCH METHOD OF DYNAMIC OPERATION PROCESS

4.1.1 Evolution of Vision and Conception

From the viewpoint of engineering science, the new generation steel manufacturing process should be studied by exploring the entire manufacturing process and its dynamic operation behaviors instead of the unit processes, such as heat transfer, chemical reaction, phase change, deformation, or elementary procedures, such as ironmaking, steelmaking, and rolling. The study method must be upgraded from fundamental science and technological science up to the level of engineering science. The achievements of fundamental science and technological science in metallurgy should be integrated with the ideas of engineering science. The integration results of the manufacturing process will require new issues of fundamental science and technological science studies. The integration process can be further optimized.

The research of metallurgical process engineering has evolved from the traditional exploration of the microstructure and physicochemical behavior of atoms, molecules, or continuum media in metallurgy to the macroscopic field of mass flow, energy flow, and information flow, as well as environmental friendliness. It places an emphasis on multiobjective optimization and coordinated integration. Using the idea of top level design, research subjects and directions, including problems referring to certain behaviors of mass, energy, or information, as well as the structure, function, efficiency, and spatiotemporal arrangement of procedures/devices, have been proposed.

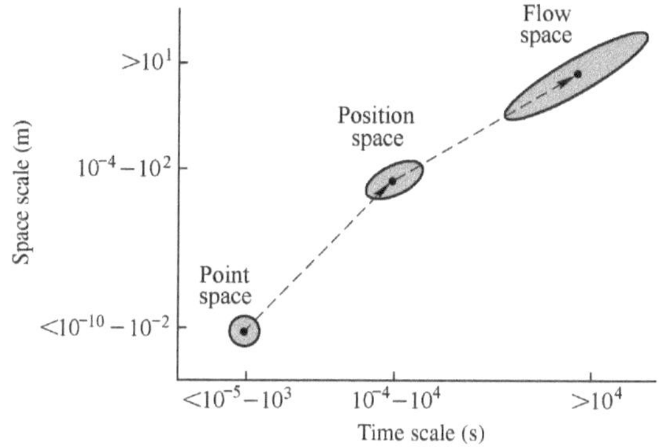

FIGURE 4.1 The scheme of time-space field of different scales in manufacturing process.

In other words, in order to realize the dynamic operation of the manufacturing process, the vision and conception of the spatiotemporal domain have been expanded from a *point* scale of an atom/molecule, to a *position* scale of a procedure/reactor and then to a *flow* scale of a workshop/plant (see Fig. 4.1).

Process engineering places an emphasis on the integration, innovation, and optimization of the entire process, rather than only on the invention of a unit process or single procedure. The entire process optimization is based on the scientific and reasonable arrangement of static structures, including process networks and function structures, etc. It even places an emphasis on the realization of efficient regulation and multiobjective optimization. The integration and innovation of the entire process are mainly based on the function-analysis of procedures/devices, the mutual replacement and mutual supplement among them, the synergism and unification of their operational ideal, as well as technological normalization. The reform and innovation of any unit process and single procedure plays an important role in process operation. Meanwhile, the related process, upstream and downstream procedures/devices, should be reformed correspondingly. Generally, the operation control of the metallurgical manufacturing process should be established in the realization and stabilization of dynamic-orderly-coordination-continuous running of the process system.

4.1.2 Research Method of Process Engineering

With regard to methodology, theory, and program or step, the methods to study process engineering include:

1. From the concept, the main target is to realize the entire process optimization, rather than to strive for advanced technologies or the leading position of a single procedure partially and separately. The optimization

should be adapted to the contemporary background of the entire process system. Of course, a "bottleneck-like" in a certain procedure or technology should hardly be ignored, the process of which can promote the multiobjective optimization for the entire process for a factor, structure, function and efficiency, etc.

2. Attention should be paid to the complexity of the process system. The prediction by mathematical analysis or mathematical model and by nondeterminism methods should be studied. The popular method for solving equations with initial and boundary conditions is closely related to determinism. However, in the complex process system, opportunity (probability) and certainty (decided) coexist and the evolution is controlled by randomness and inevitability.
3. Based on qualitative and quantitative analysis, the scientific system integration methods should be applied to the process optimization. Firstly, the top level design should be carried out. Applying its concept, idea, principle, and regulation, the analysis optimization of local procedures or devices and the coordination-optimization among them will be performed. Through the mutual feedback about analysis-integration and integration-analysis between locality and entirety and reiteration of each other, the structure of the manufacturing process is improved.
4. The dynamic running coordination, including coordination-optimization of a set of procedures' relations and study on interface techniques, should be emphasized. Innovation and reinnovation of a process dynamic operation structure can be promoted.
5. The application of software engineering and model engineering should be emphasized.
6. Regarding engineering design, the design methods should be transferred to realize the dynamic tailored design for dynamic-orderly and coordination-continuous operation of the manufacturing process instead of traditional design of procedure/reactor structure and the static approximate calculation.
7. Theoretically, the relationship of fundamental, technological, and engineering science and their correlation, integration, and mutual insertion on microscopic-mesoscopic-macroscopic levels should be focused upon.

4.2 DYNAMIC OPERATION AND STRUCTURE OPTIMIZATION OF PROCESS SYSTEM

In recognition of engineering activities, especially the dynamic operation of the process system and the corresponding engineering design, the establishment, optimization, and upgrade of the process structure are very important, which is also an important manifestation of the engineering design level. Since the design innovation has a close connection with the structure optimization and structure promotion, the "structure" should be discussed and recognized.

4.2.1 Process System and Structure

The manufacturing process is an emerged entire system, rather than a simple superposed system. The emergence of the process is embodied by the function and structure of itself. The structure is the basis of the functional emergence of the entire process. The structure is also the summary of procedures/devices with relatively steady, regular, and various connections, including not only chemical, physical, spatial, and temporal connections but also permanent or transitory connections and decided or uncertain correlations. Generally, the summary of connection modes among procedures or devices in the entire process is called the process structure.

For the design and operation of the manufacturing process, attention needs to be paid to the reasonable spatial arrangement (eg, general layout, plant layout), and the physical–chemical correlation in production and their function structures (eg, mass transformation and energy conversion). However, the time structure of the dynamic operation of the entire process (eg, time-rhythm description, time-scheduling, time-controlling) has been usually superficially studied or ignored. Actually, in the production of the manufacturing process, matter state transformation, matter property control, and mass flow control are dynamically carried out along the same time-axis. The arrow of time is irreversible. The time structure is very important for a nonequilibrium and open dynamic process system.

Therefore, it has been deduced that structures of process can be divided into hard structure and soft structure. The selection of function of units, their space arrangement, and frame establishment are the hard structure. Once selected, they are difficult to change. The coordination and restriction of the time factor, time-rhythm, and operating order are the soft structure. Operation modes, especially information connections, are the soft structure, too.

For a long time, people engaged in the manufacturing process have traditionally paid attention to the hard structure design and have ignored in-depth study of the soft structure. Hence, they possessed a lack of the understanding of the soft structure. In practice, the design, rebuilding, and reformation of the hard structure seldom occurs. However, the change, adjustment, control, and optimization of the soft structure often occurs. All kinds of information, especially the dynamic adjustment of time information, have played more and more important roles in the optimization of the soft structure. In the manufacturing process, the soft structure of mass flow and energy flow has been controlled and optimized by means of information flow, information network, and information program. They have also affected the correlation and functions of procedures in the manufacturing process.

In order to understand the process structure, the relationship of formation and running of a manufacturing process should be well grasped. The relationship has reflected a large amount of information. The correlation of the static framework (network) with the dynamic operation (program) occurs among procedures.

As an entire system, the manufacturing process has also been related to the external environment, which is very important and necessary for the formulation of the entire process. The external environment has an influence on the manufacturing process more or less. It also leads to the change of relevant modes between process and environment. Meanwhile, the functions and efficiency of the manufacturing process have been influenced by the environment (Miao, 2006a).

Wang (2010) pointed out that "*Structure can be regarded as the space-time correlations and interaction orders of space and time among elements of system.*" Therefore, the so-called structure consists of the mode of connections or interactions and the organization forms under certain circumstances. Its intention includes not only the static quantitative relationship among procedures in the system but also the qualitative advancement, reasonable organization, dynamic coordination, and internal activity among elements in the system.

4.2.2 Connotations of Steel Plant Structure and the Trend of Steel Plant Restructuring

The structure of steel plants refers to the methods of intrinsic connection and interactions among various elements in steel plants. These methods are relatively steady and have conformed with such rules as socioeconomic, technological progressing, enterprise organizational, and market competition rules. Fundamental elements of the structure of steel plants include the market, capital, energy and resources, transporting conditions, quality of human resources, and environment status.

The structure of steel plants and its various elements are interconnected and interact. It plays a decisive role in the overall function and performance of the steel plant. The elements can be combined in an effective way and embody comprehensive advantages to the entire plant. The structure of steel plants, once formed, will have comparative stability and independence. Therefore, what determines the competitiveness of a steel plant, as the main body of market competition, is its own product mix, process flow sheet structure, equipment levels and capabilities, rational economic scale, and the group quality and the organizational structure of its staff. Meanwhile, the market competition among steel plants will considerably facilitate the structural change in steel industries. Therefore, the structure optimization of steel industries is based on the structure optimization of steel plants.

The structure of the industry and the structure of the enterprise are interrelated, but they are problems at different levels of structure. Indexes of industry, such as plate-and-pipe to rolled product ratio, cannot be applied to evaluate an enterprise. The target of structure optimization of the new generation steel plant is to balance the structure based on advanced quality and to exert the three new functions of manufacturing process (advanced steel product manufacturing, high-efficiency energy conversion, and waste recycling). The multiobjective group optimization of the entire process should be realized in this way.

Looking at the development trends of the international steel industry since the 1980s, we can see that, under different circumstances, sometimes the structure optimization in steel plants is accompanied by sharp increases in scale and quantity, for instance, steel plant shifted from mainly producing long products to producing flat products. But some enterprises reduced their numerous, jumbled, and ultralarge scale to an appropriate competitive one in the course of structure optimization. For example, large integrated steel plants started specializing in producing flat products, abandoning the production of long products or moving the long production plant out to another place as a separate entity with independent management. Steel plants attach great importance not only to low-cost, high-quality, and high-efficiency, but also to energy conversion function as power generation and to recycling and processing bulk wastes as scrap and worthless plastics.

4.2.3 Dynamic Mechanics and Rules of the Macroscopic Operation in Manufacturing Process

Considering the operation features of different procedures/devices in the steel plant, it is obvious that different procedures/devices have different running characters and operation modes. For example, the DL sintering machine operates continuously but a certain amount of minus sieve products have to return; BF ironmaking operates continuously but hot metal tapping and transport are batch-type; BOF or Electric Arc Furnace (EAF) steelmaking is batch-type operation and the liquid steel tapping and shifting are batch-type too; the CC process is quasicontinuous. With a generalized observation on the coordinative running of the steel production process, different procedures and devices play different roles in the operation dynamics of the entire coordinative running of the steel manufacturing process. From the point of view of the running time of mass flow, different procedures and devices play different roles, such as of pushing source, buffer, or pull source, in the mass flow of production in order to advance the time-scheduling fluently, coordinately, and continuously. The characteristics of upstream and downstream sections in the running strategy for steel production are shown in Fig. 4.2.

To meet the requirements of dynamic-orderly and synergetic-continuous/quasicontinuous operation of procedures/devices in the manufacturing process, the running rules and regulations should be obeyed as follows:

1. Batch-type procedures/devices should be adapted to and obey the quasicontinuous/continuous operation procedures/devices. For example, steelmaking and ladle refining should be adapted to the requirements of

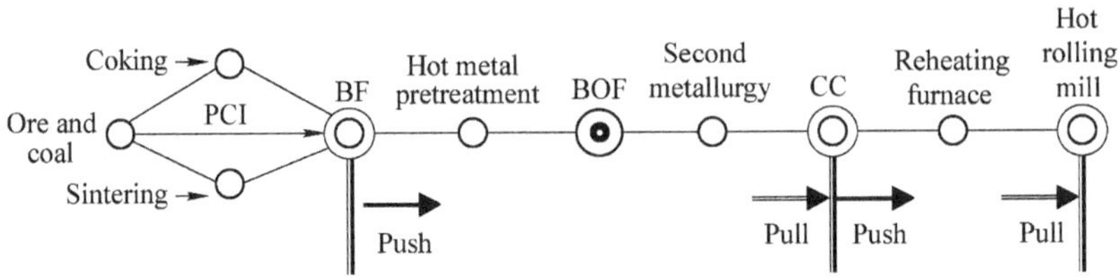

FIGURE 4.2 The scheme of the running dynamics of steel production process (Yin, 2009).

sequence continuous casting on temperature and composition of liquid steel and especially time-rhythm factor.

2. Quasicontinuous/continuous procedures/devices should guide and regulate the behavior of batch procedures/devices. For example, the high efficient-constant speed CC process requires mass flow rate, liquid steel temperature, cleanliness of the molten steel, and the time-rhythm of hot metal pretreatment, steelmaking, and ladle refining.
3. Low temperature operation procedures/devices should be obedient to high temperature operation. For example, sintering or pellet production should be adapted to the requirements of BF dynamic operation.
4. In a series–parallel process structure, the process should be operated according to a laminar type operation rather than a stochastic type operation. For example, a kind of relatively stable steel product can be achieved in a steelmaking-second refining-CC line.
5. Matched capacities and compact layout of upstream and downstream procedures/devices are the basis of the laminar type operation. For example, slab hot charging and rolling (HCR) rely on the corresponding capacities and coordinated operation among CC, reheating furnace, and hot rolling mill.
6. The dynamic-orderly, synergetic-continuous/quasicontinuous operation mechanism of pushing, buffer, and pull force in the mass flow of the entire manufacturing process should be organized.

4.2.4 The Relationship Between Dynamic Operation and Structure Optimization of Process

According to the understanding of the steel manufacturing process as an entire process, every unit (eg, reactor, procedure, and layout plan) should be regarded as a subsystem of the whole. The regulations and specifications of dynamic operation for the entire process system should be followed in order to achieve the engineering effect of the manufacturing process. The engineering effect includes integrity, emergence, hierarchy, openness, progressive, dynamism, nonlinearity, etc.

The structure optimization of the entire manufacturing process is the basis of process dynamic operation. Attention has been paid to the optimization of the static structure and single procedure/reactor for a long time. The requirements of the dynamic operation to the whole process optimization were previously ignored. Actually, the dynamic operation and structure optimization are correlated and dependent on each other. So far as the engineering problem is concerned, an item of engineering must integrate nontechnological factors, such as land, funds, resources, market, and labor, in its design and operation. All integration goals of the entire process should be realized in a process dynamic operation. The connotation of "integrate" is colorful, meaning "judge, decide, select, adjust, reciprocate, coordinate, evolution." Integration is different from the superposition of all factors and aims to generate a rational system of factor-structure-function-efficiency in order to achieve the dynamic-orderly and synergetic-continuous/quasicontinuous operation and the minimum dissipation in the process.

As discussed above, the concepts of "*flow*," "*process network*," and "*program*" are related to process dynamic operation. In engineering problems, the basic fact is that *all things are "flow," "motion,"* and "*changing*." "*Flow*" is a dynamic process, that is, an irreversible evolution under time asymmetry. The dynamic operation of "*flow*" possesses multilevel, multiprocedure, multiscale, and multifactor dynamic integration characteristics. Multilevel means the relations of structure. Multiscale concerns mass, energy, and their spatiotemporal scales. Multiprocedure and multifactor depend on types and characteristics of industry. The practical problems of "*flow*" can hardly be effectively solved with the methodology reductionism, isolation, and decomposition.

The process dynamic operation should include ferruginous mass flow operation, carbonaceous energy flow operation, and the corresponding information flow operation. In the early years, the target of steel plants was to produce metal products. High-efficiency conversion and the best application of energy based on the carbon carrier were ignored. However, in modern steel manufacturing processes, ferruginous mass flow and carbonaceous energy flow must be interacted and promoted mutually. The dynamic and organic combination between ferruginous mass flow and carbonaceous energy flow is the feature of the steel manufacturing process.

Meanwhile, the relationships among procedures, including the link interfaces and corresponding control techniques, are not nominal or separable. The relationships are the parts of the entire manufacturing process. To a certain extent, the relationships and the linkage structure among procedures are more important than a single procedure/reactor.

In order to further optimize the steel production process, to solve the "weakness problem," and to promote the steady, synergetic, and high efficient continuous manufacturing process, the interface techniques should be studied.

Before World War Two, the main procedures, such as ironmaking, steelmaking, blooming, and hot rolling, ran separately according to their own rhythm and schedule. The dynamic relationship and matching between different procedures were barely of concern, leading to a situation where each procedure ran individually while different procedures usually waited for each other or linked disorderly. The connection between different procedures was loose, and mainly included a simple connection of inputting, waiting, storing, transforming, and outputting. For example, in the process: "BF tapping-mixer-OH steelmaking-ingot teeming-blooming-hot rolling," the procedures of heating, cooling, reheating, invalid cooling, repeated removing, and repeated hoisting led to problems of long process time, high energy consumption, low yield, and low efficiency as well as unstable production quality, large floor space, poor economic performance, and heavy environmental load.

After World War Two, there appeared a series of enclosures and relevant generic technologies, such as BOF, CC, wide strip tandem mill, large bulk BF, and UHP-EAF, and these generic-key technologies had a huge effect on the manufacturing structure and the process of macroscopic dynamics. For example, the technologies of BOF, CC, and wide strip tandem mill accelerated the production rhythm and boosted continuity; the technology of large bulk BF increased the flux of mass flow and energy flow as well as efficiency. Based on this, the development and integrated-combined application of generic-key technologies caused the evolution of the function integration of each procedure, and then demanded functional reallocation of each procedure (devices) in the entire manufacturing process, as well as changed the relationship integration between each procedure. The analysis-optimization of each procedure's (reactor's) function integration and coordination-optimization of the relationship integration between procedures (devices) afforded technical supports to the reordering and high efficiency of each procedure (devices) in the steel manufacturing process, and also promoted the reconstitution-optimization of the procedure group. The evolution and optimization process of procedure function integration and procedure relationship integration caused a series of evolution and optimization of interface technologies in the steel manufacturing process, and even brought out several new interface technologies, which could be efficiently combined in different production sections. The appearance and efficient combination of the series of interface technologies had a direct influence on the revolution of the steel manufacturing process structure, including process technology structure, reactor structure, plane layout (space structure), running time structure, and production structure.

The interface technology is established on the basis of process design innovation such as unit procedure function optimization, running program optimization and process network optimization, etc. and it is the integration optimization technique of the relationship among different procedures. It is also established on the basis of rational procedure relationships including the relationship of integration-optimization of adjacent procedures or multiprocedures.

In modern steel plants, the plane layout presents many new characteristics due to the continuous revolution and improvement of ironmaking–steelmaking interface, steelmaking-secondary refining-CC interface, CC-reheating furnace-hot rolling mill interface, and hot rolling-cold rolling interface, especially the development of the thin slab casting-rolling procedure. The BF-BOF layout requirements tend to lead to shorter distance between BF and BOF, shorter delivery time of hot metal, lower number of hot-metal bottle, and faster hot-metal bottle return-cycle speed. In addition, the hot metal mixer should be eliminated and the torpedo ladle can be deleted. Due to the domestic and international establishment of the CC system, serialization between CC and hot rolling has been boosted. Therefore, the high temperature connection that processes span from the hot metal pretreatment to hot rolling have been developed in various ways, leading to the change of plane layout (space layout):

1. The distance between the steel plant and hot rolling mill becomes shorter and shorter. The link between each other should mainly depend on the roller bed;
2. The matching of CC and hot rolling production capacities should be established and even be developed in a tendency of one-to-one or integral correspondence between CC and hot rolling;
3. The distance between CC and reheating furnace becomes shorter and shorter. The link between each other should mainly depend on the roller bed, while the railway transportation should be no longer adapted.

The intention of establishing interface technology is to improve the dynamic-orderly and continuous-compactly running of process system. The interface techniques can be divided into the following styles: spatiotemporal interface technology of mass flow, interface technology of material character transfer, and interface technology of energy/temperature transfer, etc.

Interface technology should dynamic-orderly and continuous-compactly combine and integrate all factors in the manufacturing process, including physical phase state, chemical component, temperature-energy, geometry-size factor, surface feature factor, space-position factor, and time-time sequence factor. The optimization of multiple objects could be realized (including working efficiency, "minimum" of material/energy dissipation, stable product quality, optimization of product character, and environmental friendliness). During dynamic operation of ferruginous mass flow and carbonaceous energy flow, the relationships among procedures may be coordinated by use of interface techniques in the steel manufacturing process. Thus, the dependence of the dynamic operation on the process structure appears.

4.3 SELF-ORGANIZATION OF MANUFACTURING PROCESS AND HETERO-ORGANIZATION WITH INFORMATION

4.3.1 Self-Organization and Hetero-Organization of Process

Concerning logical conception, the word *organization* belongs to the upper concept—*genus*. Self-organization and hetero-organization belong to the lower concept—*species*. The process system is an engineering group organized with procedures/devices. It is named self-organization when the organizing ability comes from within itself, while it is hetero-organization when the organizing ability comes from its surroundings. Haken (2006) said: "A system is self-organizing if it acquires a spatial, temporal or functional structure without specific interference from the outside." Similarly, if the system has got the space, time, or function structures under special interference from external surroundings, it is hetero-organizing. The action of external special interference is the ability of hetero-organization. The hetero-organization with information is performed mandatorily and consciously for the manufacturing process.

The complex process system with open, irreversible, far-from-equilibrium characteristics embodies internal self-organization. The mechanism comes from its own multifactor fluctuations (eg, temperature, time) or fluctuations of units (eg, operations of procedures/devices). Since different units have heterogeneity, the emerging effect and then the self-organization are obtained by nonlinear interactions and dynamic coupling among procedures, or by the nonlinear interactions between system and unit. The openness, fluctuations, nonlinear interaction, feedback, and emergence are the mechanism of the self-organization for the system. The bottom-up style, spontaneity, and emergence are the important characteristics of self-organization.

For the self-organization generated from the system, the open and far-from-equilibrium characteristics are the prerequisites of dynamic-orderly structure. The negentropy flux input (ie, input of energy/matter or information) keeps the system maintaining an orderly state. Because of the heterogeneity of every unit in the process system, the heterogeneous units in an open system fluctuate temperately and reasonably. The different types of nonlinear interaction and dynamic coupling among procedures have been produced by the competition-cooperation of heterogeneous units of the same levels. There are also nonlinear interactions between a system and its units. These interactions embody the selection of units by the open system and the adaptation of units to the open system.

In process systems, the nonlinear interaction and dynamic coupling of heterogeneous units of the same levels as well as the nonlinear interaction (selection and adaptation) between the system and units are the synergetic mechanism of the open system for ordering. The source of synergism is fluctuations of units or the multifactor fluctuations. However, these fluctuations have duality, including destructive and constructive. Reasonable fluctuations can induce a new emergence of a dynamic-orderly structure. Adversely, non-reasonable fluctuations may reduce or eliminate the stability of the original dynamic-orderliness. To study the reasonable fluctuations of procedures/devices, the optimization of dissipative structure of process operation should be promoted. Without fluctuations, the correlation effect of nonlinear interaction would hardly be enlarged and the order parameter will hardly appear. It is also impossible for a new order structure to evolve.

4.3.2 Self-Organization Phenomenon in Steel Manufacturing Process

The steel manufacturing process is an open, far-from-equilibrium, and irreversible complex process system. Its self-organization (ie, dynamic-orderly, coordination-continuous operation) is from the integration of different processes. This type of complex process system holds a number of units with different functions, complicated structures, and complex operation behaviors. The dynamic operation proceeds at multilevels (atoms, molecules, fields, reactors, procedures, entire processes) and multiscales, taking in connotations of orderliness or chaos (function, time, and space), connection-matching static, buffer-coordination dynamic, etc. The complex system pursues the goal of being dynamic-orderly synergetic-continuous/quasicontinuous to decrease the dissipation in operation of the manufacturing process.

In the open and complex system of the steel manufacturing process, the resonance effect of elastic chain/semielastic chain has been acquired by controlling the reasonable fluctuations of each procedure/reactor (ie, nonlinear action/dynamic coupling) to realize the self-organization and order operation (Yin, 2009) as in Fig. 4.3. Actually, this type of resonance effect of elastic chain/semielastic chain is the reflection of rational self-organization under control of hetero-organization for an open system far from equilibrium.

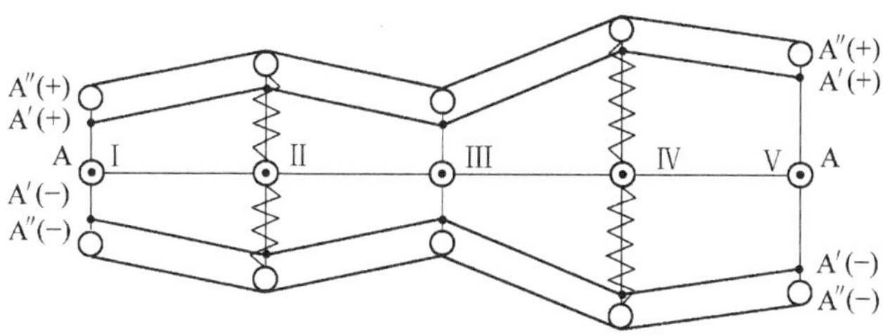

FIGURE 4.3 EAF process route, steady resonance state of elastic chain/semielastic chain and its different types.
A–A, designed common state; A′(+)–A′(−), normal flexible control range; A″(+)–A″(−), maximum flexible control range; I, EAF; II, LF; III, CC; IV, reheating furnace with store; V, hot rolling mill

4.3.3 Self-Organization and Hetero-Organization in Process Integration

Engineering design of a manufacturing process is building a system of an artificial reality. The process engineering design should embody the self-organization characteristics of integrity, openness, hierarchy, progressive and dynamic-orderliness by means of selection, adjustment, reciprocation, coordination, evolution, etc. Furthermore, by combining economical essential factors (resource, funds, labor, land, market, environment, etc.), social civilization factors, and ecological factors, a project of a reasonable structure, its functions, and efficient engineering systems can be designed. The dynamic operation of the process is the implementation of the plan. In this sense, process engineering design and dynamic operation have the meaning of hetero-organization. In other words, the design having hetero-organization ability (ie, process network, function structure, spatiotemporal structure) is projected for a system based on the self-organization of relevant elements. Correspondingly, the program and management methods can be established to promote the hetero-organization of process dynamic operation. The degree of organization is controlled under the management of hetero-organization, and is the external performance of the engineering system of self-organization. It is not the same as self-organization. It is the embodiment of dynamic orderliness and coordinating continuousness under an external aid of control.

4.3.4 Impact of Informatization on Self-Organization and Hetero-Organization

Information is the characterization of substance, energy, time, space, and life, and can be described as follows (Miao, 2006b):

1. Information possesses characterization. The action of information is to express the composition, structure, environment, state, behavior, function, property, and developing trend of objects. Where the information comes from is called the information source.
2. Information and the information source can be separated. The information can be obtained without contacting the information source directly and can process the information, including collecting, transferring, arranging, storing, utilizing, and delivering, without changing the information resource.
3. Information is immateriality. In the world, there is opposition and unity between material and nonmaterial. Nonmaterial information exists in material as its property.
4. Information depends on material. Nonmaterial information can hardly exist apart from material. All information operation, including collecting,

fixing, delivering, processing, storing, extracting, controlling, utilizing, eliminating, etc., can hardly be performed without material and energy.

5. Information has the feature of nonconservation. Conservation laws are suitable for matter and energy, but not for information. Information can be copied and shared. Furthermore, information can be lost or annihilated during collection, processing, delivery, receiving, and translation.

There is no doubt that the cognition of the dynamic operation is an activity of information, including communication. The cognizance of behaviors in dynamic operation by means of communication allows new concepts about process mechanism to be put forward. However, cognition of the dynamic operation cannot be done by communication activities alone. Cognition of the process involves the entirety of the information activities, including sensing, collecting, formulating, delivering, processing, storing, extracting, controlling, and eliminating. Among them, the most important activities are sensing, formulating, and processing rather than communication. From the most important activities, the operation rules of the process would be known from the formulating and processing of enough information. All activities above must be carried out on the basis of the understanding of the physical essence of the manufacturing process and the establishment of the rational operation network, operation program, and operation model for process factors. In other words, these activities should be carried out with the formulation-evolution of production technology for the manufacturing process with high orderliness.

The self-organization/hetero-organization of information flow is related closely to the behavior of mass flow and energy flow. The self-organization nature of mass flow and energy flow is the source of the self-organization/hetero-organization with information flow. On the contrary, inputting information of mass flow and energy flow to promote their order means inputting negentropy flux into the system. The interaction is as follows:

1. To the mass flow with self-organization capacity, the information enters as means of hetero-organization, the degree of self-organization and orderly operation of the mass flow has been enhanced. The material consumption and energy dissipation in the mass flow operation process will decrease.
2. On the basis of mass flow and energy flow with self-organization (eg, the mass flow network and corresponding energy flow network), the synergetic optimization of mass flow and energy flow will be promoted by inputting information as a means of hetero-organization. The rational limit of fluctuations of units, the nonlinear interaction, and dynamic coupling will increase the conversion efficiency of energy flow and the utilization of waste heat energy. The consumption of material and energy in the operation will decrease.

Information has characteristics of nonmaterial and nonconservation. It can be added by copying or lost by hindering. This is the diversity in quality of information from material or energy. The control program for the information network and information flow is the hetero-organization system established to achieve the minimum dissipation of the open system (process system) in its dynamic-orderly operation of mass flow, mass/energy flow, or energy flow.

To hold the information-characteristic parameters by means of sensing, processing, and feedback within the dynamic operation of mass/energy flow and then to construct the hetero-organizing steps (instruction) to enhance the orderliness of the system by rational fluctuations of nonlinear interaction and dynamic coupling of parameters (mass, energy, time, space), the self-organization and hetero-organization functions with informatization are achieved. By increasing the degree of order in the functions, space, time of the process system as well as procedures, the stable coupling of the open system is achieved and the dissipation in the process is minimized.

4.4 DYNAMIC OPERATION OF MASS FLOW AND TIME-SPACE MANAGEMENT

4.4.1 Dynamic Regulation of the Time and the Dynamic Operation Gantt Chart

The Gantt chart originates in the time management of items, that is, schedule management, which refers to the management of the project schedule and the final completion date within the process of the project implementation. In the first half of the 20th century, the Gantt chart was the main tool of a project schedule arrangement. The production of a Gantt chart involves listing all types of project activities on the ordinate and list the time on the abscissa. The horizontal line expresses the time-interval of project activities and also the time-point of start and end. In a Gantt chart, relationships among activities are also shown so that the different activities can be arranged and coordinated in time. The Gantt chart can solve problems such as time conflicts, time-characteristic order, time-cycle, and time-rhythm in the construction (or production) program. A simplified Gantt chart is shown in Fig. 4.4 (Kerzner, 2001).

The essence of the dynamic operation of the process manufacturing industry (chemical engineering, metallurgical industry, building materials industry, etc.) is the continuous system rather than the discrete system. Continuous system refers to a system with state variables changing with time continuously. The process industry pursues continuous running as much as possible. So the system dynamics and its regulations of process operation should be designed perfectly. The operation process of different procedures/devices in the manufacturing process should be arranged rationally. The

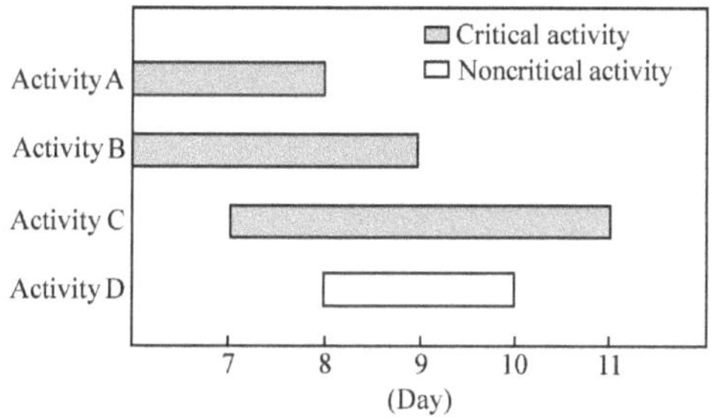

FIGURE 4.4 The schematic Gantt chart.

time-characteristic order should be rationalized. The production process can be performed to be continuous for a long time by match-coordination measures.

The dynamic-orderly and synergetic-continuous operation of the process system and then the multiobjective optimization can be achieved by use of a dynamic Gantt chart.

The dynamic Gantt chart is an operation diagram of dynamic integration of multilevels, multiprocedures, and multifactors. Furthermore, the operation process of all procedures/devices is related to the comprehensive optimization of matter-energy-time-space-information. Especially, the degree of orderly operation through information hetero-organization should be promoted. The dynamic Gantt chart shows the operation dynamics of the process in a macro-scale, including a road map, mechanism, limited steps, and the rate of different objects.

How can the dynamic Gantt chart of the steel manufacturing process be understood?

The dynamic Gantt chart embodies the over-level selection-adaption relationships among unit operations, procedures/devices, and manufacturing process, and then the emergence of the integration optimization phenomenon. For the events in the same level, the dynamic Gantt chart indicates dynamic-orderly, synergetic-continuous operation integration of multiprocedures and multifactors on the time-axis. In a steel plant or a workshop of steel manufacturing, the relevant nonsynchronous procedures with different functions should be coordinated by means of a dynamic Gantt chart to make a synergism, continuum, and stabilization of the entire manufacturing process (Yin, 2010a).

In practical terms:

1. Dynamic Gantt chart is applied to control the technology and unit operation of procedures/devices. The concrete embodiments are as follows:
   a. Reasonable selection and control of chemical composition, metal temperature, and input/output of mass flow;

   b. Reasonable arrangement of time-characteristic factors, projecting and readjusting of time-scheduling plan;
   c. Vector rules of matter-energy input/output between upstream and downstream procedures/devices.
2. Dynamic Gantt chart can be applied to guide and coordinate the scheduling and regulation of manufacturing process. The concrete embodiments are as follows:
   a. Batch operation reactors should be adapted to and obey the requirements of continuous operation procedures. For example, steelmaking and refining should be adapted to the requirement of CC;
   b. In series–parallel process structure, operation should be laminar type running rather than stochastic type. For example, relatively stable steel products can be achieved in efficiently steelmaking-ladle refining-CC line;
   c. Continuous procedures/devices should guide and regulate the behavior of batch operation reactors. For example, high efficient-high speed CC requires mass flow rate, liquid steel temperature, liquid steel cleanness, and time factors of steelmaking and refining reactors.
3. Dynamic Gantt chart provides the direction of the design of the layout plan and the restructuring of mass flow/energy flow network. The concrete embodiments are as follows:
   a. The principles of simplicity, compactness, and stability for logistic routes of materials;
   b. The principles of orderliness, coordination, and stability for mass flow operation;
   c. The mutual correlation and optimization for mass flow network and energy flow network;
   d. The mutual correlation and rationalization for mass flow network–energy flow network–information flow network.

4.4.2 Conception of Clean Steel and the High-Efficiency and Low-Cost Clean Steel Production Platform

The effective production of clean steel at a lower cost is an important function of the steel manufacturing process.

The concept of "clean steel" should be "if nonmetallic inclusions or other harmful elements are responsible, either directly or indirectly, for lowering the fabrication capability, in-service properties or requirements of the steel product, the steel is not clean; if there is no such effect, the steel can be considered to be clean, irrespective of the amount, type, size or distribution of nonmetallic inclusions and harmful elements" (World Steel Association, 2006). Clean steel is a concept of material engineering, which means that the cleanliness is adapted to the requirements for processing by customers and its application. The performance and corresponding cleanliness required by different processing and services are different for different grade steels. For customers, different grade steels, including high-quality grade steels and common products, have their own cleanliness requirements, that is, economic cleanliness requirements. Hence, the cleanliness of steels for different grades and different purposes has different meaning.

The term of the high-efficiency and low-cost clean steel production platform derives from the industrial concept of stable and timely supplying clean steel in large quantities. It does not mean to produce a small quantity of sample materials or to pursue cleanliness unconditionally, but means the stable production of quality steels as required by users for processing and services with high efficiency and low cost. A high-efficiency, low-cost production platform for clean steel is concerned with not only results, such as metal quality, product performance, and its function, but also the dynamic operation process for efficiency and cost. In order to achieve comprehensive optimization of results and processes, it should be concerned with establishing a specialized production line by which several kinds of grade steels, not only one steel grade, can be produced with high efficiency and low cost. Specialization is different from simplification. In a steel plant, there could be several specialized production lines for different kinds of clean steel production platforms with mutual consideration.

The high-efficiency and low-cost clean steel production platform is a system group of key common technologies referring to rational factor-structure-function-efficiency. It is not only fit to produce high-quality steels such as IF steel, silicon steel coil, and pipe-line steel, but also fit to produce common steels, such as long products. The high-efficiency and low-cost clean steel production platform is related to the economic cleanliness of steel, product stability in batch, and dynamic integration optimization of process. In other words, the platform is related to the concept of time-space and information control, especially the basic problems such as efficiency and cost in production process (Yin, 2012).

A platform for the effective production of clean steel with low cost is not only concerned with the quality and performance of steel products, but also must consider the efficiency, cost, and comprehensive competitiveness of an enterprise. The platform is related not only to the steel plant operation and its management but also to directing the design of a new steel plant or reconstruction of existing steel plants. It is also related to the R & D problems of process engineering and operation optimization. Meanwhile, it will promote the innovation of theory and method of dynamic-tailored design, as well as the evolution of the new generation steel manufacturing process. Actually, the high-efficiency and low-cost clean steel production platform is a confluence of metallurgical engineering with material engineering. In other words, it is an integration strategic technology system.

Since different steel grades undertake different processing and application situations, as well as require corresponding economic cleanliness, production cost, and batch scale, the clean steel production platform models of different kinds of steels will also be different. For example:

- clean steel production platform for sheet product;
- clean steel production platform for plate product;
- clean steel production platform for common long products;
- clean steel production platform for long product of alloy steels;
- clean steel production platform for seamless tubes;
- clean steel production platform for stainless steel flat products.

Constructing different kinds of clean steel production platforms, the common concepts and theory should be followed, and characteristics and requirements of different kinds of products should be displayed as well.

The clean steel production platform is a dynamic operation system established by optimization and integration of many technical modules. The dynamic operation system will concern engineering design, production control, and management. From the point of optimization and integration of process technology, the connotation of a clean steel production platform should cover a series of basic support technologies and dynamic-orderly logistics operation integration technologies.

1. Analyzed-optimized technology of hot metal pretreatment (metal scrap classification and processing technology);
2. High efficiency and long campaign BOF (EAF) steelmaking technology;
3. Speedy-coordinated technology of secondary metallurgy;
4. High efficiency and constant speed CC technology;
5. Optimized-simplified "process network" technology;
6. Dynamic-orderly mass flow control technology.

It can be considered that the first four technologies in the six mentioned above are supporting technologies, while the optimized-simplified "process network" technologies (eg, reasonable layout) and mass flow control are the integrated technologies.

4.4.3 High-Efficiency and Low-Cost Clean Steel Production Platform and the Dynamic Operation Gantt Chart

The dynamic Gantt chart embodied the selection/adaption relation about the self-growth of mass flow in the manufacturing process and the self-reproduction of mass flow in its procedures/devices. However, the self-reproduction of every procedure at lower level is asynchronous since the conversion rate is different. Therefore, not only the coordination relation of the time factors of mass flow self-reproduction in each procedure should be noticed, but also the matching, buffer, and coordination relationship of the time, temperature, and throughput of the operation of each procedure must be regulated. These problems also include transportation modes, speed, and spatiotemporal relationships (layout, etc.). Based on the information above, it is possible to integrate the self-reproduction of all procedures into continuous self-growth of the entire manufacturing process. Furthermore, the continuous self-growth of a macro-operation mass flow would direct the self-reproduction of every procedure.

Actually, in the system of high-efficiency and low-cost clean steel production platform, especially in the steelmaking workshop, the dynamic operation is the ferruginous mass flow of liquid state at high temperature (from hot metal input to cast products). The object is to achieve sequence casting with long length dynamic-orderly and coordination-continuously. It is the pattern of continuous self-growth with steadiness and efficiency. However, the operation mode of hot metal pretreatment, steelmaking, and secondary metallurgy is batch-type—periodic self-reproduction. The self-reproduction of every reactor must fit the requirements of sequence continuous casting—self-growth. Meanwhile, the program of self-growth will select and direct self-reproduction for the continuous operation of laminar flow. This is just the servo principle as the batch operations obey the continuous ones and continuous operation guides the batches.

The steps for building a dynamic operation Gantt chart in a steel plant are as follows:

1. Establish the time-scheduling plan for the sequence continuous casting of every caster for the steady high-speed casting, and meanwhile determine the time-interval, time-point, time-rhythm, and throughput;
2. Try to determine time-interval, time-point, time-position, time-rhythm, and throughput for every batch procedure/device;
3. Study and analyze the time factors of transportation, waiting, buffering, and temperature fluctuations of procedures/devices, to construct an elastic chain—nonlinear dynamic coupling among procedures/devices;
4. Study the optimization, matching, buffering, and coordination between sequence continuous casting (macro-ferruginous mass flow at upper level) and batch running reactors (partial ferruginous mass flow at lower level). Try to maintain the steady and continuous sequence casting as long as possible and realize the dynamic-orderly operation of the entire steel plant;
5. Improve the logistic routes and try to achieve the steady, laminar, and shortest transportation in mass flow operation.

The working-out and implementation of a dynamic operation Gantt chart can intensify the production rhythm of the procedures, increase the continuation degree of steelmaking production, shorten the auxiliary time, waiting time, and intermittent time, and lower tapping temperature and higher casting rate. The Gantt chart is one of the measures required for low-cost, high-efficiency clean steel production.

For the new generation steel manufacturing process, the production capacity can be dynamic-tailored designed by Gantt chart. For example, a

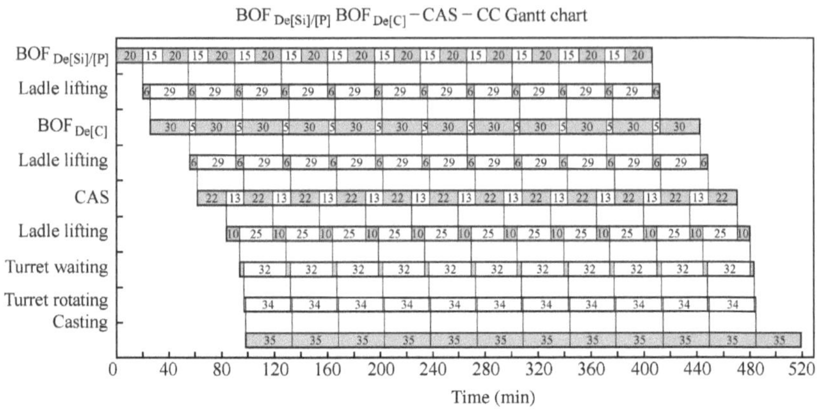

FIGURE 4.5 The dynamic Gantt chart for a large steelmaking workshop.

Gantt chart for $BOF_{De[Si]/[P]}$−$BOF_{De[C]}$−CAS−CC has been shown in Fig. 4.5.

The Gantt chart for $BOF_{De[Si]/[P]}$−$BOF_{De[C]}$−RH−CC or $BOF_{De[Si]/[P]}$−$BOF_{De[C]}$−LF−CC can be formed with a similar arrangement.

4.4.4 Laminar Type or Stochastic Type Running of Mass Flow in Steel Production Processes

The procedures/devices in steel enterprises are connected to each other according to a certain "program" and a layout with certain flow networks has been formed. Because of the different flow networks, the connection modes among procedures/devices are different, too. The connection modes include two types, that is, series connection and series−parallel connection.

1. Series connection—laminar type running

   Series connection mode refers to the fact that the output of a procedure or unit is the input of the next procedure or unit. For a procedure/reactor, mass flow goes through only once. This is laminar type running. The small-scale plant of four one-by-one vessels is the typical case, as shown in Fig. 4.6.

2. Series−parallel connection—chaotic type running

   In the process network of iron and steel enterprises, parallel connection and series connection have often been combined. Series connection is the trunk of the technological process and parallel connection is the replicated and intensified route of the procedure level. For example, the BF-BOF conventional route steel plant adopts the arrangement of two-sinter strands and two BFs in parallel operation. Fig. 4.7 illustrates an example of a series−parallel connection of a sintering machine—BF.

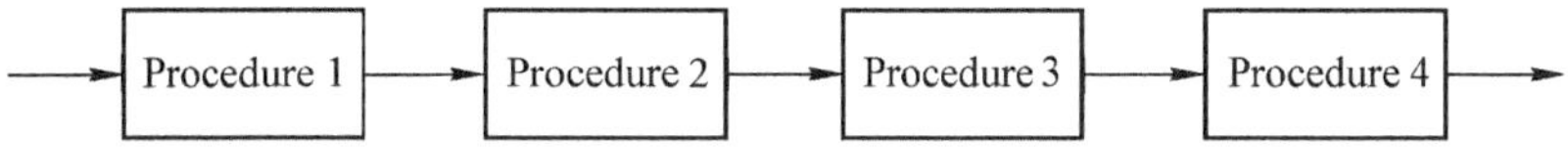

FIGURE 4.6 The series connection for small-scale plant. Procedure 1, a UHP-EAF; Procedure 2, a secondary refining; Procedure 3, a billet caster; and Procedure 4, a wire rod rolling mill.

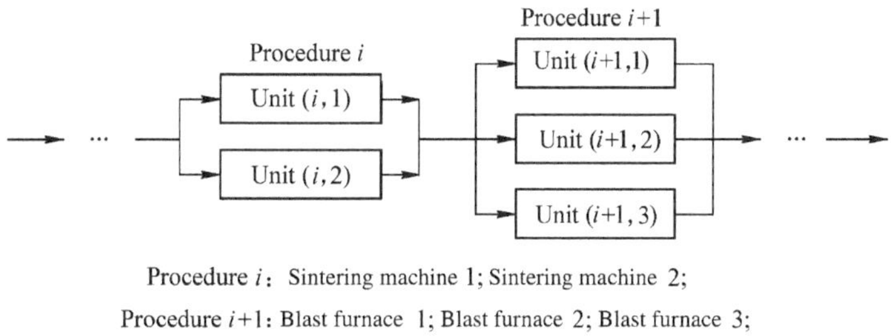

FIGURE 4.7 The scheme of series–parallel connection of sintering-BF route.

Generally, the network of a steel plant takes series connection among upstream and downstream and parallel connection within the same procedure. The network structure of a steel plant is formed on the basis of constructing the static frame of "flow" in the steel manufacturing process. The matching-coordination of "flow" embodies the coordination of upstream with downstream procedures and the synergetic and steady operation among multiprocedures in the steel manufacturing process. From the mass flow relation among procedures/devices with series–parallel connections, the running modes in the steel plant can be summarized as *stochastic type*, *chaotic type*, and *laminar type*, as shown in Figs. 4.8–4.10.

In the stochastic type running mode, the mass flow of neighboring procedures and units are linked randomly, and the input and output of the mass flow are changeable. Hence, its mass flow operation is under extra-disorder. In the chaotic type running mode, the partial mass flow is shown intersecting during undulated production, but the input and output of mass flow remain stationary in reality. The mass flow operation cannot be easily controlled. In the laminar type running mode, the mass flows of neighboring procedures are concurrent and synchronize among procedures/devices, and the input/output of mass flow are steady and controllable. It should be an ideal operation mode with a dynamic-orderly and synergetic-continuous process. In most cases, the idea of laminar type running mode should be established in steel plant design and its operation. The stochastic type running mode must be avoided. Since the external conditions such as market or raw material supply change, the chaotic type running may appear partially or temporally. However, this effect should be minimized. All the above are advanced design concepts to achieve high-efficiency, low-cost, and high-quality

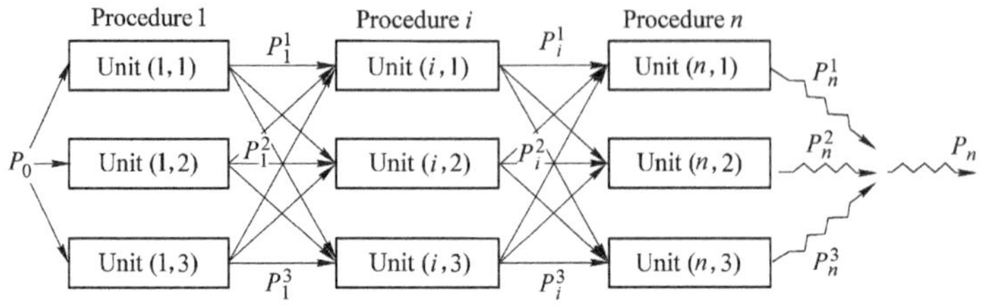

FIGURE 4.8 The scheme of stochastic type running for steel manufacturing process.

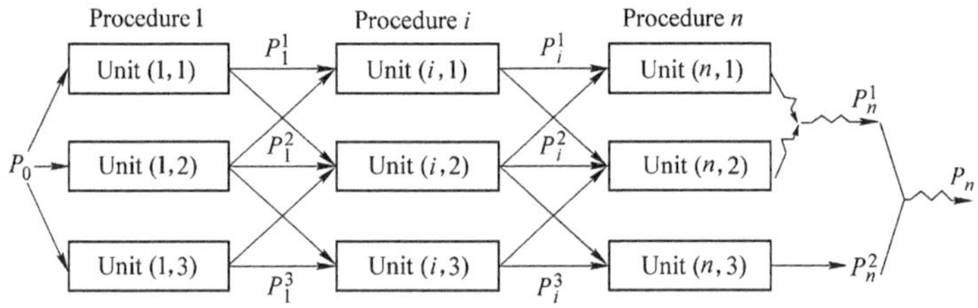

FIGURE 4.9 The scheme of chaotic type running for steel manufacturing process.

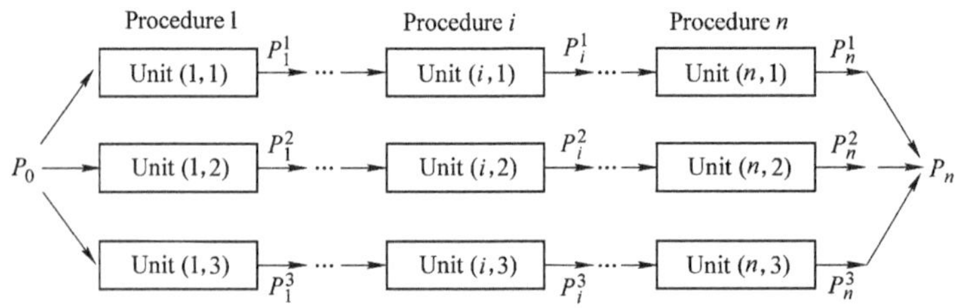

FIGURE 4.10 The scheme of laminar type running for the steel manufacturing process.

production, and are also important measures to solve nonsteady problems in the manufacturing process.

When the layout plan and mass flow mode among procedures are designed, the transport energy consumption for macro-operation will be determined by the operation time and rhythm of mass flow among procedures. This is related to the connection mode in the process network. In series connection mode, production efficiency and transport energy consumption can be optimized easily because of the clear matching relationship among procedures, short waiting time, and stable transport and operation time. In series–parallel connection mode, if the move from procedure to procedure is one-by-one, the production efficiency and transport energy consumption is similar to that of laminar type running. However, in stochastic type running mode, since there is interference, the logistic flows, the transport time, and the waiting time will be increased remarkably. The production efficiency and energy consumption will be deteriorated.

4.5 FUNCTION AND BEHAVIOR OF ENERGY FLOW, AND ENERGY FLOW NETWORK IN THE STEEL MANUFACTURING PROCESS

The physical essence of the dynamic operation of the manufacturing process may be described as: the mass flow (mainly ferruginous flow), driven by the energy flow (mainly carbonaceous flow), operates dynamic-orderly according to a decided program in the specialized network. It is the process of moving forward and mutual conversion of ferruginous mass flow with carbonaceous energy flow in the entire steel manufacturing process.

4.5.1 The Deeper Understanding of Physical Essence and Operation Rules of the Steel Manufacturing Process

The steel manufacturing process is an open system far from equilibrium. The elements of dynamic operation are "*flow*," "*running program*," and "*process network*." The control of ferruginous mass flow and carbonaceous energy flow aims to achieve the dynamic-orderly operation by network concordance and program synergy from a relatively chaos state. The manufacturing process can be highly organized and the multiobjective optimization can be achieved by the hetero-organization related to information flow. Therefore, the optimization problems of "*flow*," "*running program*," and "*process network*" should be studied systematically. Steel industries were only concerned about the ferruginous mass flow during steel production for a long time. Energy was regarded as the external condition for matter state transformation and matter property control. During the investigation on the consumption of energy, it was limited to enthalpy of reaction, phase transformation and deformation, or heat balance of a procedure/reactor. The integration and system of the energy flow behavior and energy flow network were ignored. Actually, problems concerning the "*flow*," the "*running program*," and the "*process network*" of energy also exist in steel production, and in-depth studies of which should be performed.

As far as the dynamic-orderly and synergetic-continuous operation of ferruginous mass flow are concerned, attention to the effective conversion and utilization as well as the recycling of energy flow must be paid in the manufacturing process. In particular, the promotion of the entire process technology evolution and the extension of steel plant functions should be stimulated.

Based on the cognitions above, guidance principles can be proposed for the layout, technical policy, key procedure/reactor, new technology, and energy system of a new steel plant. For existing steel plants, measures and steps may also be proposed for a sustainable improvement of the process and fulfillment of functions of the steel plant.

4.5.2 Research Method and Feature of Energy Flow in the Process

To research energy flow in the steel manufacturing process, instead of static mass balance and energy balance calculation, the dynamic operation model with input/output and fluctuations should be proposed. By analogy to mass flow study, the concepts of "*flow*," "*running program*," and "*process network*" for energy flow ought to be established to study its input and output and the conversion in the open, irreversible, and far-from-equilibrium process. It is necessary to change the mass-energy balance calculation of an isolated spot into the energy flow study in the network of dynamic operation of the entire manufacturing process (Yin, 2010b).

Studies on the input and output character of "*flow*" involve not only the nodes, connectors, and the space arrangement of the nodes and connectors, but also the dynamic operation program, especially the time-characteristic order. The concept of input and output of energy flow involves not only the amount of energy but also the grade of energy and time-space factor as well as the operation program of energy flow. This is helpful for constructing a rational energy network containing secondary energy utilization. This is also conducive to further improvement of the energy efficiency.

4.5.2.1 Input and Output Characteristics and Fluctuation *Phenomena in Energy Flow of Steel Plant*

There are different types of energy carrier and input-output expressions of energy parameters including the type of energy kind (gas, steam, self-generated power, and sensible heat), the grade of energy (gas's kind and calorific value, steam's temperature and pressure, mass flow's temperature), and the amount of energy at procedure nodes in plants, in different spatiotemporal boundaries of the manufacturing process. These energy parameters at different nodes in energy flow possess fluctuation phenomena also. So, it is difficult to describe the dynamic operation by static state separately. It is necessary to establish the rational energy network, which will play to the comprehensive potential of the energy medium including primary energy and secondary energy to achieve the high efficiency and low cost of the entire process. The establishment of an energy flow network, the rational energy recycling scheme with different interaction coupling mechanisms, can be achieved by combining different amounts of gas, steam, sensible heat of matters, potential heat, etc. within a generating spot and time.

4.5.2.2 *The Limitations of Static Mass and Energy Balance Calculation for a Single Procedure*

During the study on the dynamic operation of the entire steel manufacturing process, it is difficult to accurately evaluate the utilization and profit of energy of the entire process by material and heat balance (mass-energy balance) of a single procedure. When calculating static material and heat balance of the single procedure, the result can only be used for one procedure or reactor. In the result, the kind of utilization of the secondary energy is absent, which will affect the energy conversion efficiency. For example, when using BOF gas as a fuel, the effect of its utilization is the thermal efficiency and can be simply calculated. It is also affected by the types of reheating furnace and burner. However, when using BOF gas for the production of quality lime, there are some additional advantages like high activity of lime because of ash-free fuel, and low sulfur in lime to increase desulfurization efficiency, hence, the lime consumption and temperature drop of the hot metal will be lowered. As another example, if a gas/steam pipe network of an enterprise appears to be defective or be in an abnormal state, the emission of gas and steam cannot be avoided. This type of energy loss can hardly be reflected in the material and heat balance of the single procedure.

Therefore, the study of energy flow in the steel manufacturing process should be performed by dynamic modeling of the input/output to the open system. The dynamic modeling should be developed on the fundamentals of "flow," "running program," and "process networks."

4.5.3 Energy Flow and Energy Flow Network in Steel Plants

4.5.3.1 *The Energy Flow Behavior in Dynamic Operation of Steel Manufacturing Process*

Concerning the advanced design, dynamic operation, and information control of the steel manufacturing process, the concept of "flow" running should be established. For the sake of "flow" running dynamic-orderly and synergetic-continuously, it is related to networks including the plane and vertical layout and time-scheduling (eg, operation program). The combination of "flow," "running program," and "process network" leads to the minimizing of the material consumption and the energy dissipation.

1. The roles of mass flow, energy flow, and information flow in the steel manufacturing process.

   In the steel manufacturing process, "flow" expresses itself in three carriers, that is, matter forms—mass flow, energy resources—energy flow, and information media—information flow. The mass flow is the working subject, that is, the processing of material products. Energy flow works as the driving force, chemical agent, heat medium, etc. Information flow is the reflection of the total of mass flow information, energy flow information, external environment information, and artificial control information. Generally, mass flow, energy flow, and information flow interact and accompany each other in the dynamic operation of the manufacturing process.

2. The relationship of mass flow and energy flow.

From the viewpoint of processing, mass flow is the main subject together with energy flow. From the viewpoint of energy conversion, energy flow hardly operates with the accompanying mass flow, and a part of energy flow operates independently. Mass flow and energy flow sometimes accompany each other while sometimes they are separate. When accompanying each other, they interact mutually and a material state transformation, matter property control, and mass flow operation are performed. When separate, they show their own characteristics.

Further analyzing each procedure/reactor, mass flow and energy flow are imported separately at the entrance and interact within each procedure/reactor. Partial energy flow is exported with mass flow at the exit. Meantime, partial energy is exported as residual heat separately because excess energy is necessary for the reaction to proceed spontaneously.

Therefore, not only should the dynamic and arrowed characters of mass flow and energy flow be of importance at the entrance but also at the exit. These are the bases for establishing the energy flow network of the steel plant.

4.5.3.2 The Constituents of Energy Flow Network

The so-called "network" is the system integrating "node" and "link" according to certain graphics systems. For the steel production process, it is a dynamic system including substance-energy-time-space-information. It can be further analyzed as the mass flow network, energy flow network, and related information flow network. Among them, the energy flow network is an operation system composed of "nodes" and "links" and other units according to certain graphics. Namely, the dynamic operation system is constituted by "Energy-Space-Time-Information."

There is primary energy (mainly purchased coal) and secondary energy (such as coke, electricity, oxygen, various types of gases, waste heat, waste energy) inside the steel plant. They are the starting nodes, respectively (such as raw material yard, BF, coke oven, and converter). The energy medium output from the starting nodes along the transportation routes, pipelines, and other connecting pathways reaches the terminal node of energy conversion (such as the various end-users and thermal power plants, steam stations, power stations). Of course, in the delivery and conversion process of energy flow, it inevitably needs to have necessary and effective intermediate buffers (buffer systems), such as a gas tank, boiler, and pipeline—in order, to meet the buffer, coordination, and stability of the energy from the beginning node to the terminal node in number (capacity), time, space, and energy levels and so forth.

Thus, it is not difficult for the energy flow network (energy conversion networks) "energy beginning nodes-coupler-intermediate buffer system-the connector-the energy terminal nodes" to be established according to certain graphics inside the steel plant and achieves some degree of "closed loop." For example, steel plants only buy the coal instead of buying the electricity and fuel oil.

It is to be emphasized that the idea of "only buying the coal instead of buying the electricity and fuel oil" is different from the idea of "power generation with all residual gas." The power generation with residual gas leads to the use of small boilers and small generators, and these types of low efficiency power can hardly enter the state distribution network. The idea of "only buying the coal instead of buying the electricity and fuel oil" means power generation by large-scale boilers and generators with supplementary pulverized coal. The mode of power station achieves higher efficiency and specific energy rate per $kW \cdot h$. If conditions are allowed, a cooperated power station together with power plant could be established on the large-scale capacity for more economic and environmental profits. Therefore, it is important to design a reasonable mode of energy conversion for the network of energy flow in the steel plant. Since electricity is a universal energy form, power generation in the secondary energy recycling is very often in the steel plant. Generators with different capacities have different power generation efficiency, as shown in Table 4.1.

Furthermore, due to the constant progress of energy recovery and conversion technology, the available range of waste energy and waste heat is constantly expanding and the available range of the corresponding nodes of beginning and ending is also expanding, as are the corresponding connectors. This would constitute an "energy flow network" with different levels. For example, if the temperature of the waste heat medium is over 500–600°C, the medium can constitute an "energy flow network" of this level. If the temperature of the waste heat medium is over 300°C, the medium can expand to another level of the "energy flow network." Hence, the "energy flow network" is hierarchical and should be promoted hierarchically. It is necessary to have a clear understanding of the design concept and method of the energy flow network.

Now, in advanced enterprises with the BF-BOF-rolling mill traditional route, approximately 38% of secondary energy is separated from ferruginous mass flow. In order to utilize the primary energy and the secondary energy, the following points should be noted:

- Study the energy flow behavior and the relationship between energy flow and mass flow in the steel plant by input/output model;
- Study the order of the application of primary energy or secondary energy for the object of function-efficiency optimization;

TABLE 4.1 The Energy Consumption and Generation Efficiency of Different Generators With Gas (Heat Generation Excluded)

Generator Capacity (MW)	Boiler With BF Gas		Boiler With Gas and Coal		CCPP Mode	
	Energy Consumption [kg_{ce}/(kW · h)]	Generator Efficiency (%)	Energy Consumption [kg_{ce}/(kW · h)]	Generator Efficiency (%)	Energy Consumption [kg_{ce}/(kW · h)]	Generator Efficiency (%)
3	0.540	22.8	–	–	–	–
6	0.520	23.6	–	–	–	–
12	0.500	24.6	0.480	25.6	–	–
25	0.450	27.3	0.440	27.9	–	–
50	0.425	28.9	0.420	29.3	0.293	42.0
100	–	–	0.400	30.7	0.273	45.0
110	0.410	30.0	–	–	–	–
125	–	–	0.375	32.7	–	–
150	–	–	0.350	35.1	0.267	46.0
300	–	–	0.330	37.2	0.256	48.0
350	–	–	0.320	38.4	–	–

1 kg_{ce} = 29,308 kJ.

- Study the energy flow network by graph theory and information regulation to achieve continuous on-time control and zero emission;
- Develop energy saving and emission reduction technology in ferruginous mass flow operation;
- Study the limit value of energy consumption in the steel manufacturing process with a full utilization of the secondary energy;
- Study the dependence of CO_2 emission per specific amount of steel with structure and efficiency in the manufacturing process.

4.5.4 Macroscopic Operation Dynamics of the Energy Flow in the Steel Manufacturing Process

In the steel production process, the ferruginous mass flow path constructs the mass flow network, which is embodied in the layout of the steel plant. Actually, the layout of a steel plant is the expression of not only the mass flow but also the energy flow and its behavior related to the mass flow. The mass flow and partial energy flow in the steel production interact, but partial energy flow operates independently.

The macro-operation dynamics is observed in the separately running energy flow for dynamic operation. This is one of the main concepts of the general principles for energy flow design and process control. Since the relative instability of energy flow at start nodes (various types of generation reactors of gas or steam, generated amount as well as energy grade) and the high steady energy flow required at end nodes (power generators), the intermediate buffers are necessary for the construction of a reasonable energy flow network. The macro-dynamic operation mechanism with a push force-buffer-pull force in a network is shown in Fig. 4.11.

The dynamic-orderly and coordinated-efficient operation of energy flow is the object of the macro-dynamic system of the energy flow network. It strives for high efficient energy conversion, minimum energy dissipation, and zero gas emission.

In the energy flow network, the buffer is beneficial for efficient utilization and zero emission. In order to achieve the dynamic-orderly and coordinated

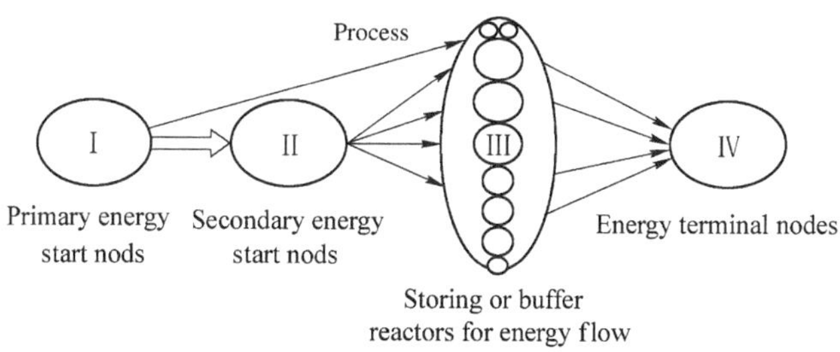

FIGURE 4.11 The scheme of macro-operation dynamics for energy flow of steel manufacturing process.

efficient operation between unsteady start nodes and steady running end nodes, buffer techniques, such as pipe network systems, suitable gas tanks, and auxiliary primary energy, are necessary. Since the main force of the secondary energy application is power generation, the capacity of the generator should be selected correctly. A too small or too large scale of generator would be inappropriate.

4.5.5 Energy Flow Network Control System and Energy Control Center

There were four periods for the energy saving history in China:

1. energy saving of the single reactor and corresponding system in the 1980s;
2. system energy saving based on procedures reconstruction and the entire optimization of the process in the 1990s;
3. energy saving by the full development of the energy conversion function of the manufacturing processes like CDQ, dry dedusting for BF gas, BOF dry dust control, system water saving, power generation, etc. in the early 21st century.

Now the fourth period is coming—the deep energy saving by the promotion of energy control and the management center in the steel manufacturing process. Full application of the energy conversion function will lead to energy saving and emission reduction, and even environmentally cleaner production manufacturing.

4.5.5.1 Energy Control System and Energy Management Center

The concepts of energy flow, the energy conversion program, and the energy flow network are the key theoretical frameworks for the energy control systems of steel plants. The energy control system should be an energy flow network including energy flow, energy flow nodes, energy flow connectors, energy flow buffers, and the corresponding network graph. Based on the information above, through the selection of the reasonable capacity, the number and location of energy conversion reactors, and through integrating a primary circuit, the dynamic physical model and dynamic control model are built. It is now a reality of the energy control system and the management center.

The energy control and management center should be a production unit rather than an administrative unit. This system has equipment, apparatus, a pipeline system, and an information control system. It should operate, monitor, control, maintain, and predict the future all day long. The functions of the energy control center should include the continuous control and coordination of the energy conversion network of the steel plant. Accompanying the mass flow control system, the energy control center can also be applied to predict the future state of the steel plant and to propose corresponding measures.

4.5.5.2 Construct Energy Flow Network and Promote the Energy Control and Management Center of Steel Plant

Constructing the steel plant energy flow network is the basis of applying energy management and the control center. In order to establish the energy flow network, the following concepts, mechanisms, aims, and direction should be understood first:

1. Coordination mechanism of push-buffer-pull for energy flow operation;
2. Idea of high-efficiency and low-cost conversion, selection of large boiler and power generator, as well as a cooperative power station by the steel plant and power plant;
3. Target of zero emission of energy resources, near-zero emission of gas, oxygen, water, etc.;
4. Direction of recycling and reuse of social wastes.

REFERENCES

Haken, H., 2006. Information and Self-Organization: A Macroscopic Approach to Complex Systems. Springer, Berlin, p. 11.

Kerzner, H.R., 2001. Project Management: A Systems Approach to Planning, Scheduling, and Controlling, seventh ed. John Wiley & Sons, New York, NY, p. 728.

Miao Dongsheng, 2006a. Essentials of Systems Science, second ed. China Renmin University Press, Beijing, pp. 22–25 (in Chinese).

Miao Dongsheng, 2006b. Essentials of Systems Science, second ed. China Renmin University Press, Beijing, pp. 30–32 (in Chinese).

Wang Zhongtuo, 2010. Systems Engineering. Peking University Press, Beijing, p. 4. (in Chinese).

World Steel Association, 2006. Clean Steel: Clean Steel Production Technology. Metallurgical Industry Press, Beijing (in Chinese).

Yin Ruiyu, 2009. Metallurgical Process Engineering, second ed. Metallurgical Industry Press, Beijing, p. 236 (in Chinese).

Yin Ruiyu, 2010a. Dynamic operation of process engineering and manufacturing process. Report for the 8th Academic Conference of Chemical Industry, Metallurgy and Material Department of Chinese Academy of Engineering, Harbin (in Chinese).

Yin Ruiyu, 2010b. Comment on behavior of energy flow and construction of energy flow network for steel manufacturing process. Iron Steel 4 (4), 1–9 (in Chinese).

Yin Ruiyu, 2012. Integration technology of high efficiency and low cost clean steel "production platform" and its dynamic operation. Iron Steel 47 (1), 1–8 (in Chinese).

Chapter 5

Dynamic Tailored Design and Integration Theory of Steel Plants

Chapter Outline

Nowadays, there are a lot of challenges faced by the steel industry in China. There are pressures from nature, such as resource shortage and climate change; from the society, such as environmental pollution and population growth; from the economy, such as capital interest rate and financing

Theory and Methods of Metallurgical Process Integration.
DOI: http://dx.doi.org/10.1016/B978-0-12-809568-3.00015-2

channel; from the market, such as supply–demand relation, price factor, and demand of new products; and from process technologies, such as energy efficiency, production efficiency, and stability of product quality. To solve these complex issues in such a complex environment, the elements-structure-functions-efficiency of steel plants should be thought about at strategic levels. This kind of elements-structure-functions-efficiency is inevitably related to the issues on the level of the manufacturing process in the whole steel plant, and relevant issues on the level of various procedures/devices and different products. Obviously, solving these kinds of issues should start from the bottom, such as the structure optimization of the manufacturing process and its design. Proceeding from the needs of the macro-environment and based on the optimization of elements-structure-functions-efficiency, the proper choice and adaptation should be made. The manufacturing process and the industrial production chain construction and extension should be adapting to the constant change and improvement in aspects of nature, society, the economy, the market, and process technology. Optimization of elements-structure-functions-efficiency of the manufacturing process and industrial production chains will inevitably result in the reasonable selection and coordinative integration of procedures/devices by the manufacturing process. Meanwhile, the elements-structure-functions-efficiency of various procedures/devices should also adapt to the optimization of elements-structure-functions-efficiency of the manufacturing process and the reasonable position of the product market. The extension of industrial chains, effective and orderly utilization of mass and energy between different plants are also the development trends of steel plants in the future. Eco-industrial parks centered on the steel plant may be formed in some district.

Such kinds of issues cannot be settled by just solving one problem of the technology or product at the technical science level. It should be solved at the engineering science level of the whole industry and entire process. In particular, the guidance of new design theories and design methods is necessary. People usually know intuitively that market competition refers to the competition of products, especially the competition of product quality and structure. However, the root of competition will be traced to a series of engineering designs, such as process design, procedure function/device design, manufacturing process dynamic design, and design of manufacturing process's dynamic operation rules. Actually, design has become the keynote of market competition. All engineering projects are hardly separated from the design and design is the primary engineering. Design should be an important component of the innovation system. Engineering design is a multilevel and multifactor optimization-integration, a multiobjective optimization system, and a decisive factor that determines the competitiveness and the degree of environmental-friendliness of products.

Production in a steel plant is based on the plant-wide engineering design. In the present view, the design should provide not only the drawing and relevant parameters, but also the dynamic operation rules and programs, which the traditional static design method lacks at present, for the production operation in steel plants.

In this chapter, the tradition and current situation of design and the development progress of design theories and methods are discussed. The necessity of engineering design innovation and the requirements of the new era of dynamic tailored design are explained. Based on the above, the theories, methods, and current situation of the steel plant design in China are discussed. The transformation of the steel plant design theory from an isolated static estimation and drawing to the dynamic tailored design is described. Meanwhile, the core idea and steps of dynamic tailored design are emphasized to guide the dynamic tailored design in the steel plant. This will lead to the steel plant structure optimization being promoted, and the integration innovation of the new generation steel manufacturing process being realized.

5.1 TRADITIONAL DESIGN AND ITS PRESENT STATUS

Design that is closely related to engineering is usually known as engineering design. Moreover, the design itself is a kind of engineering or process, which can be regarded as design engineering.

Design engineering is a primary engineering, that is, any substantial engineering has a design process. Design refers to many types of engineering, such as mechanical engineering, civil and architectural engineering, mine engineering, metallurgical engineering, chemical engineering, textile engineering, hydraulic engineering, and traffic and transportation engineering, communication engineering.

5.1.1 How to Recognize Design

All the engineering projects start from an idea, assumption, plan, and design. And the design has played an important role in the construction and operation process of engineering. The engineering design stage also includes a process which is just like the operation processes in the construction stage and production stage. All these processes have a series of system characteristics, for example, the properties of integrity, hierarchy, openness, orderliness, dynamic, and coordination.

There is often a contradiction between an ideal goal and its realistic feasibility in the designing process, which needs careful weighing, choosing, reciprocating, coordinating, changing, and integrating to determinate the objectives and the construction and operation steps/processes suitable for the engineering under specific conditions. There are challenges and innovation opportunities which require knowledge and wisdom. Therefore, it can be said that the design engineering contains many innovation factors.

From the perspective of the primary engineering or knowledge, design engineering can be understood as follows. Design engineering is an integration of various relevant knowledge and their interrelations and evolutions. Design should be the acquirement, application, organization, and integration that derive from knowledge. Design innovation contains wisdom of the acquirement, application, integration, and discovery of related knowledge, and it will be given a role in the entire life circle of engineering so that it is full of challenges and epochal characters.

There may be two different thoughts about the cognition of design engineering:

1. Design refers to providing the construction program and engineering drawing to create a concrete engineering, and then the constructed engineering entity can be put into operation.
2. Design should provide not only the construction program and engineering drawing for the construction of an engineering project, but also the paths/rules/programs for and the structure/function/efficiency of the engineering's dynamic operation. Namely, design should not be limited to providing the engineering drawings and construction programs, but should start from the engineering requirements, then go through the design, construction, and operation, and finally reach the end to satisfy the engineering requirements. From the perspective of engineering philosophy, design is the combination of engineering essence and engineering process, and the cognition of design on engineering essence is the view on engineering process essence.

The aim of engineering design is to design and construct an engineering integration with optimal elements-structure-functions-efficiency, which is usually shown as an artificial existence. The artificial existence should be able to operate in a dynamic-orderly, coordinative-continuous and effective-steady state and be able to achieve its own value which contains the multiobjective optimization rather than a single-objective optimization.

Engineering design includes thinking activities. The essence of design should be cognized and grasped from the relation between thinking and reality. The keynote of this relation is to provide a reasonable and feasible basis for the construction and the operation of the designed artificial existence. And the connotation of design engineering should be the process of conceptual thinking and the result of problem solving.

Design engineering should be a process of assumption, prediction, rehearsal, and evaluation for the construction and operation of a future-realized engineering system. And this process should solve not only the concrete issues from the technical and tactical angles, but also the global and directional issues from the engineering-wide elements-structure-functions-efficiency aspect and the strategic angle. It is noteworthy that design engineering should not be understood as a simple drawing and an isolated equipment-capacity calculation without respect to environment conditions.

Therefore, engineering design should be *future-oriented*, rather than a simple copy of an existing object. Engineering design should guide the *future*, but should not just hold the *present*. There are some original innovations in the engineering design, but more for integration innovations. The design needs the theoretical direction, so the design method should be continuously updated along with the theory.

5.1.2 Situation of Design Theory and Design Method

Design theories and methods have developed along with the development of the industrial society. According to the differences of objects the design aims at, there are different types of designs such as engineering design, product design, and artistic design. Design that the process manufacturing industry mainly considers is engineering design, while for the equipment manufacturing industry, it is mainly the product design. Most of the equipment is sold in the market as commodities (such as automobiles and machine tools) and in order to improve the competitiveness, the equipment should be upgraded through product design. Culture and art design, which mainly concerns the design of "beauty," belongs to another category of design, and the thought of engineering aesthetics is certainly necessary for engineering and equipment design. There are inseparable relations between engineering design and equipment design, and to realize the engineering objectives, reliable equipment is required as a solid basis. Also, the innovative improvement of certain key engineering equipment will impact on and change the entire manufacturing process.

For a long time, there somehow has been misunderstanding that designing is drawing and the design revolution or modern design is exactly the digital design without any hand drawing. Computer software, such as Auto CAD, Applicon CAD, Pro/ENGINEER, were taken as being the frontier science of design theory. There is also a saying abroad that design is a method, and the main problems that need solving are issues about computer, software, and communication, as stated by the recent research report from Advanced Engineering Environment (AEE) (Zhu et al., 2004). There is no doubt that the rapid development of computer-aided design (CAD) has contributed to the organic connection between equipment manufacturing and design, and also made product upgrading more effective and faster, which, however, is not comprehensive. Without the essential cognition and theoretical research on engineering design system and its dynamic operation rules, and without the scientific description of the activity of design, it is hard to find out what kinds of hardware and software could satisfy the requirements of modern engineering. Calculations and drawing can be rapidly carried out by the design software and it is very convenient to make the comparison between various plans, but depending only on software, the essential characteristics of design objects can be hardly cognized. Take the design of the continuous caster for example. Before the 1980s, in China the machinery design strived to be as handy and pretty as possible. Most of the stands of billet caster, even slab caster, were designed toward lightness. The new caster can operate normally, but temperature distortion of the stands would happen after a period of operation so that the caster could no longer be in a normal and long-time operation. Hence, the thought was brought into most production managers' minds that sequence CC is not reliable and ingot teeming should still be held. That means the design theory is hardly replaced just by the design software. Of course, the design theory should take the successful cases and failure lessons in the practice of engineering design, operation, and management as the basic materials to analyze, study, and finally conclude a new cognition, namely a new theory, by using fundamental science, technical science, and especially the latest achievements in engineering science. Especially for the engineering design of the process manufacturing industry, it should focus not only on the equipment design and function research for key procedures, but also on the relations and the interface technologies among procedures/devices. These relations are usually in nonlinear interactions. A good disposal of these relations depends not only on the traditional decisive theoretic thought suitable for the simple system, but also on the opportunities (random) and possibilities.

The current period has seen the global economic integration, one of whose characteristics is that competitions among enterprises has been brought forward from the from product markets to designs and even to design philosophies. This is because generally, new knowledge that has never been used is adopted in products striving for new properties and rational functions, while all the related knowledge is initially permeated into the products by designing. Thus, under the condition of global economic integration, the design is the starting point of market competition, which requires the optimal integration of multifactors and the comprehensive competition of a multiobjective optimized complex system to be reflected in the design.

It has not been realized that there is a great opportunity for the research and construction of the engineering design theories and new design methods in the design processes of the vast invested domestic engineering projects today. Therefore, for the engineering design, it is unavoidable for there to be occurrences of low-level repetitions or even massive backward copies adversely affecting the domestic manufacturing level in the coming decades, which is worth the concerns. Furthermore, the attempt to build an innovative nation will be heavily discounted or even impossible in some areas without innovative engineering philosophies and engineering design theories. *Attention must be paid to the innovation of engineering design theories and design methods in the coming century.*

5.1.3 Present Status of Design Theory and Methodology for Steel Plants in China

Since the Westernization Movement in the late Qing Dynasty, none of the Chinese, who were eager for a prosperous and powerful China, had expected a strong Chinese iron and steel industry. However, starting from the construction of Han-Ye-Ping Mining Metallurgical Company to the foundation of PRC in 1949, the developing road of the Chinese steel industry was tortuous and slow. The engineering design of steel plants was totally dependent on the foreigners. Since the 1950s, the Chinese learnt steel plant design from the Russians of the former USSR. The first design institute of the iron and steel industry of China was set up in Anshan, Liaoning, in 1953, and the engineering design group for Chinese steel plants came into being. Along with the development of the steel industry, the ability of engineering design for steel plants has also made great progress. Technical process and equipment such as large BF/BOF workshop and bloom/billet caster could be designed gradually. Besides, the Chinese researchers had creatively fulfilled the entire design of the Panzhihua Iron & Steel Co. which specializes in the comprehensive utilization of vanadium-titanium magnetite, the production of quality steel products, and vanadium-titanium recovery. However, the design was made using the static design methods for monomer techniques, that is, first divide the engineering system into several technical units; then the units were designed separately and the capacity of monomer equipment was statically calculated in an isolated state; finally, a superposed manufacturing process would come into being by simple connections of different procedures. The characteristics of this kind of design are as follows. The procedures/devices are designed from their local situations and a surplus capacity is reserved for each of them. Namely, the design, staying at a local procedure, puts forward static requirements rather than requirements on the dynamic, orderly, coordinated, and integrated operation in the upstream and downstream procedures. Disadvantages would occur during the practical operation of the constructed manufacturing process designed by this static method that the capacities of procedures, both front and back, can usually not be dynamically matched, the functions of procedures/devices are not optimized and coordinated, and the spatiotemporal relations in the mass flow operation processes are disorderly, leading to a low material yield, high energy consumption, unsteady product quality, and clogged information.

Since China's reform and opening-up, after depending on the years of summarization of steel plant design experience and digestion of advanced technologies abroad, the design institutions in China had constantly improved the design methods and levels. Most of the engineering designs had been improved, but the general design method still remained at a separate and static level. Therefore, although many designs for single equipment had been updated to an advanced level, the integrated design still referred to the simple superposition of designs of single procedures/devices. The essence of the dynamic-orderly and coordinative-continuous manufacturing process in the steel plant, especially the theoretical conception of "flows" in an open system, was not understood exactly. Meanwhile, the theories and methods for the constructions of mass flow network, energy flow network, and information flow network were also in shortage. Thus, the design theories and methods for the entire process had not been formed, and there was no mature analyzing tool for integrated design. In other words, the traditional design method based on experience still dominated so that there were hardly any breakthrough innovations in the engineering design theories and methods but just some limited partial improvements based on traditional design methods.

The traditional steel plant design method only takes the steel products and their outputs into account, and it focuses on the product grades, scopes and scales, and the make-up of technologies and devices, rather than on the systematic design of structure, function, and efficiency of the dynamic operation of the whole steel plant. Therefore, for the plant-wide engineering design, the knowledge at the manufacturing process level should be concerned with not only the theoretical cognition on the atom/molecule scale, but also the technical issues of certain reactors, devices or fields. These issues are in larger spatiotemporal scales and in more complicated boundary conditions and phenomena, and the optimum relations such as the coordinating-matching between different procedures/devices should be considered. It is found that in the study on the dynamic operation of the multilevel, multiscale, and multifactor phenomena, the optimal conditions for a certain metallurgical reactor are not always the most suitable for the entire metallurgical manufacturing process. Therefore, the research results would involve the structure adjustment and optimization of the entire manufacturing process in the steel plant and instruction on the evolution of the steel plant mode. That is, without the essential cognition of engineering design system, the scientific description of design activities, and the theoretical research on the dynamic operation rules of design engineering system, it would be difficult to figure out what kinds of hardware and software can satisfy the requirements of engineering design for modern steel plants, and it would also be difficult to rationally and accurately distribute functions among various procedures and to put forward the optimal design parameters for key devices.

Since the 1990s, the challenges faced by steel plants in the world have been not just issues of product quality and performance, or even the business scale of enterprise, but the multiobjective comprehensive optimization, such as the product cost, quality and property, the process comprehensive control, process emission, eco-environment, resource and energy. To know and solve the challenges in the multiobjective comprehensive optimization, the essence, structure, and operation characteristics of the overall steel manufacturing process should be researched to realize the top-level design for multiobjective optimization.

In the 21st century, Chinese engineers possessed the modern steel plant design abilities for the process, technology, and equipment of a modern steel plant. Moreover, the iron and steel industry was undergoing the change from a simple pursuit for high output to the precise performance of product, function extension, and harmonious development. The design requirements became more and more exigent in technology innovation, product innovation, energy saving, emission reduction, and environmental friendliness. Meanwhile, competition with large foreign design companies became increasingly fierce in overseas markets. Therefore, the traditional design philosophies, theories, and methods could hardly satisfy the requirements of the industry structure adjustment and upgrading, and could hardly be dominant in the international competition. Thus, it was necessary to put forward new design theories and methods from a macro-idea and study the intrinsic nature of engineering, to promote the industry-wide design levels.

In conclusion, the key topics of engineering design for the steel plant are to solve the multiobjective optimization of process-wide structure, function, and dynamic operation; to solve the issues of the dynamic-orderly, coordinative-continuous/quasicontinuous operation of various procedures/devices; to constitute the mass flow network, energy flow network, and information flow network; to solve the issues on the structure-function-efficiency of equipment and information control units in procedures. Thus, the process engineering science, formed by assimilating the most advanced theory achievements of systematics, dissipative structure theory, synergistics, etc., combined with the achievements of modern information technology with eco-environment ideas should be taken as the theoretical bases of steel plant engineering design. So the modern steel plant engineering design theory is complex. The design theory should take the successful cases and failure lessons in practices of engineering design, operation, and management as the basic materials to analyze, study, and finally conclude a new cognition, namely a new engineering design theory and methodology, by using the fundamental science, technical science, and especially the latest achievements in engineering science.

5.2 ENGINEERING DESIGN

Engineering, which reflects the value orientation of humans, is a human practice activity that creates and constructs substances with various resources and related elements by a planned and organized use of knowledge, and that is characterized by selection, integration, and construction.

In general, modern engineering is integrated to be a technology module group through the selection, assembly, and coordination of related technologies, and is constructed to be an engineering system/integration possessing structure, function, efficiency, and sustainably-reflected value orientation through the optimal allocation of the economic factors (eg, resource, capital, land, labor, market, and environment). The engineering functions should reflect the properties of applicability, economy, efficiency, reliability, safety, and environmental friendliness. Engineering embodies the dynamically integrated operation system of related technologies (especially the advanced technologies) which are the basic connotation of engineering.

5.2.1 Engineering and Design

5.2.1.1 Nature of Engineering

The nature of engineering can be regarded as the unity of the integration process, integration approach, and integration mode of a new substantial reality which is constructed and operated with various resources and related basic economic factors by knowledge application (Yin et al., 2011).

This can be analyzed in three aspects:

Firstly, engineering embodies the integration mode of various factors, and the integration mode is an essential characteristic for distinguishing engineering from the science and technology.

Secondly, the engineering-integrated factors are the unity of technical factors and nontechnical factors (mainly various types of economic factors such as capital, land, resources, labor, market, and environment) which are interrelated, mutually-restricted, and mutually-promoted (Yin, 2008). The technical factors constitute the basic connotation of engineering while the nontechnical ones are the indispensable supplement (Fig. 5.1).

Thirdly, the improvement of engineering depends not only on the progress of scientific-technological elements expressed in basic engineering connotation, but also on the conditions of society, economy, culture, and politics at a certain historical period which are expressed by nontechnical elements.

5.2.1.2 Engineering Activities and Engineering Design

Engineering activities perform as the comprehensive integration and allocation of the natural and artificial factors, as well as the related decision-making, design, construction, operation, and management, etc. Engineering activities, particularly the engineering philosophies, embody the value orientations. Engineering characteristics are the unity of function and value-embodiment of an engineering integration system's dynamic operation process.

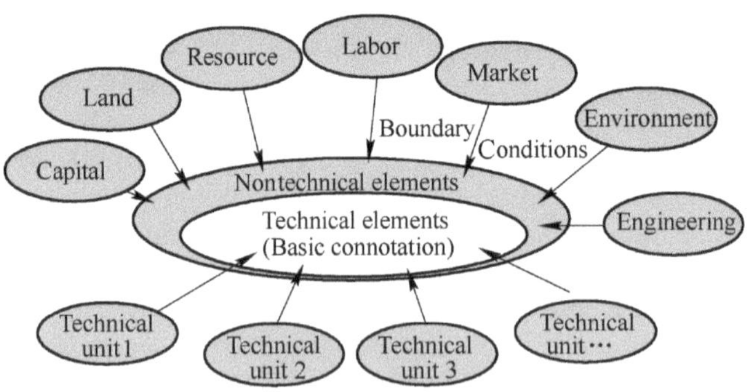

FIGURE 5.1 Engineering activity factors and system structure (Yin, 2008).

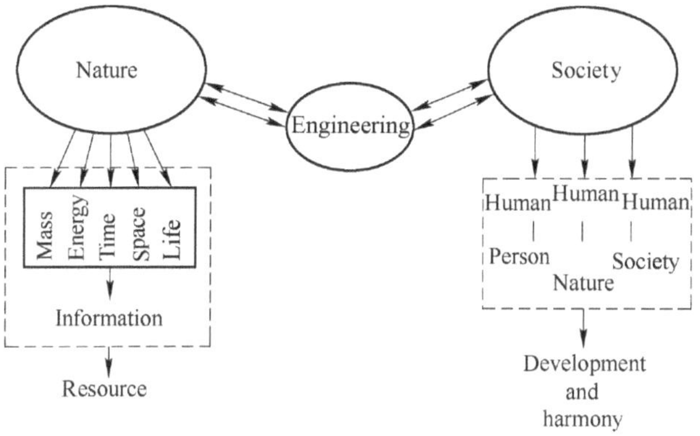

FIGURE 5.2 Relations of engineering to nature and society (Yin et al., 2011).

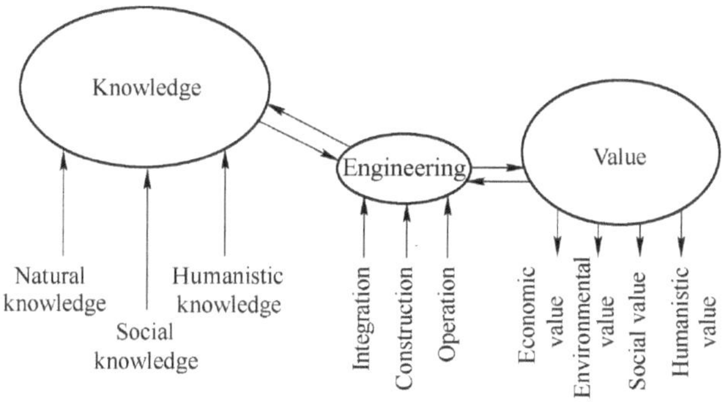

FIGURE 5.3 Value generated by knowledge through engineering (Yin et al., 2011).

Social and economic developments could not be kept apart from substantial engineering activities which are important carriers of economy operation. Engineering activities have two terminals, one is nature (eg, resource) and the related knowledge, the other is the market and society (Fig. 5.2). Engineering, aimed at achieving market value (economic benefit) and social value (harmonious and sustainable development), is a system integrated and constructed by optimized allocations of natural resources and related economic factors.

Engineering equals direct productivity and engineering is a dynamically integrated system of relevant technologies and related economic factors. Generally, scientific discovery and technical innovation should be converted into direct productivity by this dynamically integrated system, and then reflect their values (including additional value, employment, profits, civilization progress, environmental friendliness, etc.) through the market and society, as shown in Fig. 5.3.

For the purpose of science and technology combining with society and the economy, the integration and construction process of transforming sciences and technologies into engineering projects or industries must be paid close attention to. Otherwise, it will take double the effort with half the result, or be too far away to solve the current problems.

In the evolution of engineering and industrialization, engineering design is an important link of the selection, integration, construction, and dynamic operation, and it also can be considered as a "breathtaking jump" of science and technology's conversion into direct productivity. For a long time, engineering design has not been emphasized in scientific and technological work, neither have the design theory and engineering management theory (at least they have not been focused), resulting in the vast majority of engineering designs and constructions being located in a conservative, isolated, superposed, and system-innovation-absent state. In other words, engineering design lacks theories and methods for innovation, or the innovation and theorization of engineering design are still in a scattered state.

In the strategic planning of scientific and technological work, high attention should be paid to engineering and engineering design, and to the transformation of science and technology research results into engineering projects and industries with practical productivities through the organization-and-objective-owned integration and construction. Meanwhile, engineering design theories and methods should be deeply studied at the level of engineering science and engineering philosophy so that new knowledge, concepts, and theory can be acquired to promote the innovation of engineering design.

5.2.2 Innovation View of Engineering Design

Engineering design carries the knowledge integration and construction process for the transformation of achievements, such as scientific discoveries and technology innovation, into practical productivities (Fig. 5.4). During this process, there is not only the rational selection and innovative integration in aspects of process, device, measurement, and control technology but also the creative embodiments of product function and appearance designs.

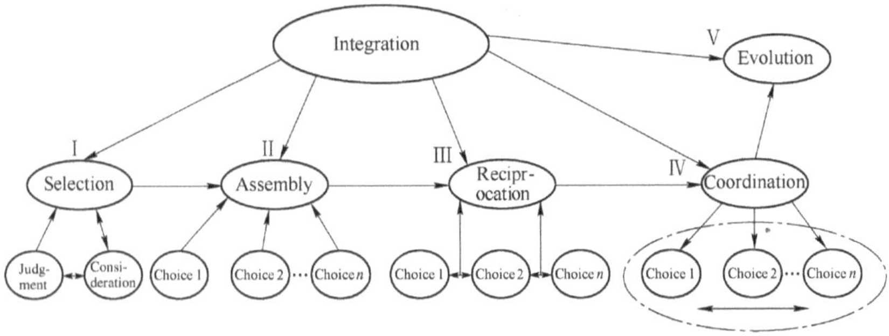

FIGURE 5.4 Integration and evolution in engineering design (Yin et al., 2011).

Engineering design must embody the dynamic operation of various technical factors and technical units (procedures/devices, etc.) to ensure the process-wide orderly, effective, and steady dynamic operation. Meanwhile, engineering design should share with people the "esthetic" enjoyment, namely a harmonious engineering-art combination. Solving problems and discovering new propositions and various requirements, which may guild the consumption and production, should be found throughout the engineering design activities.

The engineering design is the very transformation of research and development achievements into practical and direct productivities through selection, assembly, reciprocation, and coordination. Therefore, engineering design is full of integration innovations and system evolutions.

Selection is a key for engineering design ("selection" includes the concept of "elimination" (Yin et al., 2011), it mainly refers to "choosing something carefully" here), and its connotation should include selections of market, product grade, technical factor, optimal engineering system, and resource rational distribution. "Selection" is based on judging and weighing (such as judgments on market and technology, and weighing up function and value). Therefore, knowing how to make the right judgment in various selections is the foundation stone and the key to engineering design. Judgment, weighing, and selection are the starting point of engineering design and engineering evolution. During the selections, there are phenomena such as continuation (heredity), change (mutation), and evolution (upgrading and transition). Selection should emphasize the reliability, but more importantly, embody the innovation, and it will be eventually reflected in the optimization of the invest-output ratio.

The engineering design in general is used to design an engineering system so that attention must be paid to the system characteristics, such as integrity, hierarchy, relevance, dynamics, and adaptation to external environmental changes. Therefore, on the bases of many selection-judgments, it is very important to effectively integrate the technical factors and basic economic factors in order to achieve the optimal allocation of resources.

The "*assembly*" in engineering design is mainly introduced in terms of technical factors, namely, to achieve the system optimization and innovation by forming the union and intersection relations among these technical factors, and meanwhile take the engineering entirety as the innovation objective which tends increasingly to be the combination of knowledge chains of various technical innovations from many engineering systems. So that the source for creating new value would form via further network assembly (space network, time network, function networks, etc.).

The network assembly based on new technical innovation chains is an innovative view of the dynamically integrated operation which should carry out the network assembly of various technical factors, including the series–parallel connected units and processes both in upstream/downstream procedures and at different levels and in different spatiotemporal scales. Meanwhile, the rationalization and the optimization of different factors will be promoted through the *reciprocation* in dynamic operation.

Views that emphasize the rational and orderly reciprocation of technical factors further promote the coordinative and rational reciprocation of various factors, and the views are taken as the nucleus for the innovation and progression of the entire engineering system. Reciprocations between technical factors, function coordination between constituent units, which boost the effective relevance, compose the basic characteristics of the dynamic innovation view. Further, the engineering entirety is not only determined by the technical factors and their reciprocation, but its elements-structure-functions-efficiency is affected by the whole-part reciprocation.

There needs to be a further discussion on the relation between the unit technology innovation and the engineering-wide innovation. Technical innovation, especially unit technology innovation, is often a local creative response to changes of market demands and surroundings, but it is just a local innovation. To make the overall and comprehensive responses to changes of market demands and surroundings requires a full set of engineering innovation systems to make a strategic action. Only in this way is it possible to achieve overall and lasting results. This relation mentioned above is often neglected in practice, especially in the technical reform in enterprises because the engineering entirety, enterprise structure, and other strategic choices are often invisible and in-depth issues which are difficult to be detected in the short term. The right understanding of these issues should be: Inspiration innovation of unit technology is necessary and important, and is often the "tipping point" of engineering-wide system optimization. However, when the engineering design is considered, especially for well-permitted conditions, only when the evolution attains the engineering-wide system innovation can it be possible to achieve the transition evolution, namely the so-called "*upgrading*" of the enterprise structure or engineering entirety system.

In brief, engineering design must pay attention to dynamic technical innovation and there must be the concept of dynamic integration and the vision of persistent pursuit for engineering entirety evolution.

5.2.3 Engineering Design and Knowledge Innovation

Engineering design should include the various unit technology designs and the related integration process, as well as the rational allocations of the corresponding economic factors. In the engineering design process, the connotation of integration should include judgment, weigh, selection, assembly, reciprocation, coordination, and evolution.

Manufacturing industry is an important part of the national economy and an important embodiment of the substantial economy. The manufacturing industry can be divided into the process manufacturing industry (such as metallurgy, chemical engineering, building materials industry, food processing) and the equipment manufacturing industry (such as machinery and vehicle manufacturing). With the progress of time, it can be clearly seen that the competition in manufacturing is the competition of products in terms of their external form, but actually is the competition deriving from design. If design is the soul of manufacturing, then innovation (especially integrated innovation) is the soul of design.

Design should be a highly concentrated process with lots of knowledge. Knowledge required for design involves complex system problems in multilevels, multiunits, multifactors, multiobjectives, etc. Designers need not only fundamental science knowledge and technology science knowledge, but also new engineering science knowledge. A major characteristic of new knowledge for engineering design in the new century is its dynamic and tailored properties. The new knowledge for engineering design integrity, which stems from the accumulation and advancement of different unit technology knowledge, needs the accumulation of integration knowledge and the related sublimation through the research and comprehension of engineering science and engineering philosophy.

Unit technology knowledge is mainly related to a certain discipline, while the integration knowledge is associated with industries to a large extent (because it refers to products, manufacturing processes, market size, resources, transportation logistics, etc.) and some intersection disciplines (such as system theory, cybernetics, information theory, and synergistics). The knowledge should be put in an appropriate, flexible, and innovative application, accumulation, integration, and evolution in the engineering design practice to form the knowledge bank of design institutes and designers.

New knowledge for engineering design should include new philosophies, theories, concepts, methods, tools (means), and so on. For the theoretical innovation of engineering design, it is important to distinguish different-level knowledge, such as philosophy, theory, method, and tool (means), to get a comprehensive and clear outline.

Engineering philosophy belongs to the general philosophical concept. Briefly, philosophy means the system of thoughts on ideals and totality. There are many specific concepts in philosophy. For example, the "unity of man and nature," "harmonious development," and "conquer nature" philosophies mean different views of the world, systems, and values. Engineering philosophy, which is throughout the engineering activity, is the starting point, destination, and soul of engineering activities. Engineering philosophy will inevitable affect the engineering strategy, decision, planning, design, construction, operation, management, and evaluation (see Fig. 5.5).

Engineering philosophy is not generated without foundation. It is based on human practice, especially on experienced engineering practices. Engineering practices are advancing and developing continuously, as are the

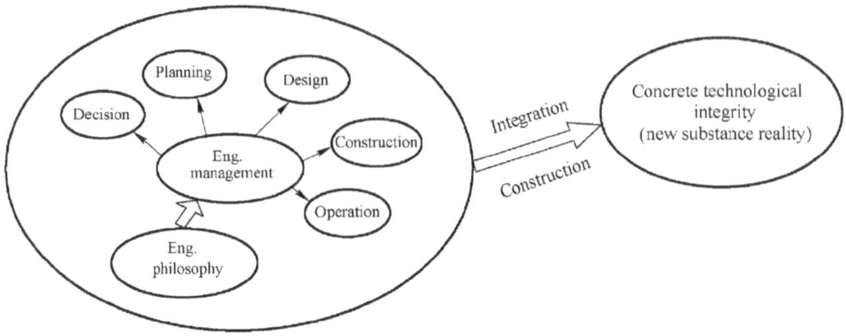

FIGURE 5.5 Relation between engineering philosophy and engineering activity.

engineering activity philosophies. It can be said that engineering philosophy is connected with history and the times.

Development and innovation of engineering philosophy are the dialectical unification of reality and ideality, and the unification of possibility and perfection. There must be constituent and brightness of ideality in engineering philosophy. At the same time, it must carry out the innovative integration and construction based on the knowledge of reality and the basic factors to highlight its advancement.

The concept of engineering design is a concrete reflection of engineering philosophy, which is the key of engineering design innovation and which should be explored in the emerged or emerging dynamic-integrated phenomena (including experience, etc.) in the theoretical progress and engineering operation process of related disciplines. The new engineering design method should be found on the relation-forming process in different unit technical knowledge, such as the formed intersection/union relations among technical factors. Informational control and intelligent control are important methods, tools used during the above "exploring" and "finding."

The competitiveness of a design group lies in the development of times-needed and times-adapted engineering philosophy and relevant new knowledge including new theories, new methods, and new tools. All of them should be correctly applied in order to achieve the engineering design innovation. It can be found that *the core of engineering design knowledge in the new century is to reflect the dynamic-integrated and coordinative-tailored properties of the engineering operation. For the steel plant design, it should be marked with the dynamic tailored engineering design*, and the informational control and intelligent control will increasingly highlight its importance.

5.2.4 Engineering Design and Dynamic Tailored Solution

The task of engineering design is to integrate engineering-system-related technical factors and basic economic factors to realize the objective required by the engineering entirety. Connotations of integration include the structure optimization, the function evolution and the judgment, weigh, selection, assembly, reciprocation, and coordination of these factors.

From the perspective of knowledge, the task of engineering design is, along with the progress of times, to judge, weigh, and select related technologies; then to absorb the advanced, eliminate the backward, and further assemble the advanced elements to conceive and study the reciprocation relations among them; finally, under the requirement of the global coordination-optimization, to adjust and restructure all the reciprocation relations, to determine key parameters of various technical units, and to form the optimized intersection/union in the dynamic-orderly operation and the correlation system of elements-structure-functions-efficiency. The process above should be embodied in concept research and top design, which means that the dynamic-orderly and coordinative-continuous operation mode should be established during the concept research, the concept of dynamic tailored design should be founded during the top-level design, and there should be the engineering design view of integrated dynamic operation and the concept of multiobjective integrated optimization. However, in contrast, the traditional engineering design mainly limits itself to the static or local unit design, namely first dividing an engineering project into several technical units which are then separately designed and finally are put together to finish the engineering design. This design is a static and separate method, and the dynamic weakness would appear after the project is put into operation, resulting in problems in efficiency, effectiveness, and profit. *The conclusion is that during the engineering design, the static weakness should be avoided, and meanwhile, more importance should be attached to the research on dynamic weakness during the operation of the engineering entirety.*

Concept research and top level design of dynamic engineering design should first be done at the engineering science level. Modeling of dynamic engineering design should start with judgment, weighing, and selection of advanced technical units under the principle and multiobjective optimization of dynamic-coordinative operation. Then, these selected technical units should be dynamically assembled to research the coordination and reciprocation relations among them and finally to form a dynamic-orderly, cooperative, and continuous/quasicontinuous integration effect of engineering entirety. The dynamic design modeling assisted by informational and intelligent control is an important way to avoid all kinds of weakness and to realize dynamic-orderly, coordinative, and continuous/quasicontinuous operation.

The three key elements of process system dynamic operation, which are "flow," "process network," and "operation program," should be embodied in the concept research. In the top level design, the optimization of elements-structure-functions-efficiency is required, even the function evolution of engineering system might emerge. The engineering design for process is to make all the related technical units (selected and optimized elements) assembled in network and coordinated in operation program (formation and optimization of structure), which makes the "mass flow," "energy flow," and "information flow" in a dynamic-orderly, coordinative, and continuous/quasi-continuous operation within specified spatiotemporal boundary, realizing the perfect engineering effect and the multiobjective optimization, and forming the source of new values.

From the perspective of engineering science, innovation of engineering design should pay attention to the research on the overall model of multielements, multiscales, multiunits, multilevels, and multifunctions at general or certain circumstances. That means not only the various factors affecting the engineering ought to be considered, but also the reciprocation relations among these factors must be studied. Dynamic engineering design should not only pay attentions to the "1−1" reciprocation relation between units, but also the "2−2" and "many−many" reciprocation relations that are similar to but actually different from the "1−1" relation. Only by the reasonable assembly and effective reciprocation of the engineering system entirety can it realize the dynamic-orderly coordinative operation of the engineering system. Through the reciprocation between procedure-procedure and reciprocation-optimization between procedure-set and procedure-set, it is possible to reflect the coordinative optimization between procedure function−procedure function, and to establish the dynamic-orderly, linked-matched, and continuous-compact relevancies of "mass flow," "energy flow," and "information flow." This is the basis of dynamic tailored design for process engineering and it is necessary to take the selection-assembly-reciprocation-coordination-evolution as the core connotation of the dynamic tailored design. The dynamic tailored design must establish the concepts of "flow," which will certainly involve with "process network" and "operation program." "Process network" is the space framework or even the spatiotemporal framework between various technical units, while the "operation program" includes the function order, space order, time-characteristic order, spatiotemporal order, and related operation information, rules, and strategies. As the design scheme, especially the general layout, is decided (which means the static spatial structure is fixed), the dynamic operation program will be reflected more as the time program and the time allocation. In order to ensure the dynamic-orderly, coordinative, and continuous/quasicontinuous operation, every procedure/device and its parameters or factors (eg, chemical composition factor, physical phase state factor, geometry factor, temperature/energy factor, spatiotemporal factor) should be nonlinearly coupled on the time-axis, which will refer to the information flow, information flow network, and related operation program.

The process manufacturing engineering entirety is not only determined by the constituent elements (units, procedures, etc.) and reciprocation among elements, but also the reciprocation between itself and elements. This means that the reciprocation among elements is different from that between the whole process and elements. Do be attentive that the dynamic-orderly, cooperative-continuous/quasicontinuous reciprocation must express itself on the entire process level, rather than just on the procedure/device level. The process engineering design should ensure the high-efficient, steady, continuous, and safe operation of the whole engineering system.

Sometimes, especially after a breakthrough in a key-technique, key-procedure, or key-parameter and the successful insertion of them into process system, the "stimulation" and "arousal" effects will be generated on the entire coordinative and continuous/quasicontinuous process operation. For example, the sequence CC steelmaking workshop gives full play to the efficiency of the converter and rolling mill. Theoretically, the system capacity can be fully released or a new emergency can arise through the stimulation and arousing effect of certain order parameters within the whole system. Meanwhile, the capacity of other units, elements, procedures, etc. might also be fully released. This is a new form of dynamic engineering design innovation, and factor of engineering integrated innovation that is deserving of attention.

5.3 DESIGN THEORY AND METHODOLOGY FOR STEEL PLANTS

Steel plant design is one of the typical engineering designs. As a theoretical concept of engineering design, the steel plant design cannot be limited to the representation of process phenomena and equipment structure. It is important to deeply, clearly understand and express the physical essence of the steel manufacturing process's dynamic operation. The physical essence of dynamic operation is that, *driven by the energy flow (mainly carbonaceous flow), the mass flow (mainly ferruginous mass) runs dynamically-orderly along a specific established process network according to some established program, to realize multiobjective optimization*. From the viewpoint of thermodynamics, the steel manufacturing process is a kind of open, nonequilibrium, and irreversible complex system that is formed by the nonlinear interactions of the related unit procedures with heterogeneous functions. The nature of its dynamic operation is the self-organization in the dissipative structure. Therefore, in order to make a dynamic tailored design, it is necessary to establish the concepts of *flow*, *process network*, and *operation program* which are called the three elements of the steel plant dynamic operation.

5.3.1 Background for Innovation of Steel Plant Design Theory and Method

The development of the steel industry in China had depended on simply increasing production capacity for a long time. Research and development had been concentrated on the study of local and individual theories about steel products and technologies. All of the work had been limited to a single procedure or unit operation. Although some achievements had been made, they could not be applied into the steel manufacturing process stably and effectively. It is difficult to fundamentally optimize the whole process of the steel industry or steel plant structure and problems still exist, such as disorder structure, unsteady operation, unstable product performance, high energy consumption, high cost, low efficiency, and severe pollution. The original root was the lack of systematic studies on the manufacturing process and its elements-structure-functions-efficiency. The engineering design work had been limited to individual unit procedures/devices. The entire coordination effectiveness among procedures and the structure optimization had been ignored. The metallurgical theory studies mainly focused on micro-problems regarding analyses of chemical reactions, physical changes, or unit devices based on reductionist thought. The intensification of single procedures/devices had been pursued, but the overall optimization and evolution of the steel manufacturing process had not been studied. Now it has been understood that the optimization of single procedures/devices can only solve local problems, but can hardly affect the whole steel plant structure which is a key factor in determining the cost, product, quality, efficiency, investment profit, waste-emission, and environmental effect. Those are the issues affecting the sustainable development of steel plants. So, the optimization of manufacturing process is the root of competitiveness and development of steel plant, of course, it is also the root of engineering design.

Most steel plant design methods in China were learnt from Russian scholars of the former Soviet Union about 50 years ago. The general method is that, by calculation of the static capacity of a single procedure/device and the subsequent simple superposition, an extensive manufacturing process forms accordingly. The characteristics of the traditional design method are that the device capacity is determined based on static calculation of the local procedure's production capacity with a surplus capacity reserved and with the dynamic-orderly, coordinative-integrated operation ignored. However, the surplus capacity that is designed is often different with each designer's subjectivity. The connections between procedures are often static and just simply superposed which lack the spatiotemporal conceptual optimization and coordination, and the coordinative calculation for dynamic operation. The steel manufacturing process and device designed with the above method often have disadvantages in production-capacity matching, function coordination, and information control. So the operating efficiency, product performance, production cost, etc. are difficult to be optimized, which often means that the end-point of the construction of a steel plant manufacturing process is the start-point of the technical reform.

The general layout of the plant is closely related to the dynamic-orderly, coordinative, and continuous operation of the steel manufacturing process.

Actually, the plant layout design is a technology for planning the mass flow, energy flow, personnel flow, and information flow. It is also a technology for organizing workers and substance to evenly, orderly, and efficiently run among various devices. The general layout of a plant, reasonable or not, directly affects the enterprise production efficiency and benefit. The general layout is not only the optimal arrangement of workshops, procedures, or devices, but also the entire optimization of the manufacturing process. It is important for general layout optimization to achieve objectives including the mutual-coordination of mass-energy-space-time-information, and the mass flow-logistics' running in a minimal energy dissipation, material consumption, shortest path, most convenient and rapid way, under the drive of energy flow, and under the control of information flow.

Therefore, the effective combination and optimization of the plant and workshop general layouts is an important guarantee of an enterprise's long-term efficient operation, its development, and even its survival.

Not only the capacity and functions of each procedure/device but also the number, capacities, and reasonable positions of different devices based on the entire manufacturing process's coordinated operation in steel plant design should be taken into account. Namely, the design should think of not only the enlargement and efficiency of devices under the premise that the structure of process network is well optimized, but also the number, positions of, and connecting-lines between nodes in the process network, to make the general layout and the workshop layout for ferruginous mass flow form the figure of the minimum directed tree. This issue has already proposed urgent demands to the design theories and methods in some ways. Meanwhile, in the new century, some questions are delivered, such as whether the function of the steel manufacturing process is only for steel products? The process discharges waste inevitably, is it also be inevitable to cause pollution? Is there a new business growth point for steel plants? How does a steel plant transform itself from environmental pollution to environmental friendliness? How does a steel plant integrate itself into the circular economy? These questions are asking what kinds of roles the steel plant will actually play in future. It is a new topic for steel plant design proposed by the times.

For the process's dynamic operation in steel plant, manufacturing process is intrinsically self-organized. The self-organization can be effectively reflected as the informational hetero-organization's making the flows in a dynamic-orderly, coordinative, and continuous/quasicontinuous operation. It should be noted that the science and technology issues related to the manufacturing process of steel plants present two trends: high differentiation and high integration. Namely, on the one hand, the existing subjects are differentiated constantly and specifically, and the corresponding new theories and new fields are constantly developed. On the other hand, the intersection and combination of different subjects and fields results in several "transversal subjects" like systematic science, cybernetics, synergistics, and dissipative structure theory. These transversal subjects lead science and technology in the direction of comprehensively-integrated global optimization. The two trends complement each other and promote the development of Metallurgical Process Engineering and the innovation of steel plant process engineering design theory.

The so-called advanced manufacturing process engineering should be an effective, orderly, continuous, compact, and dynamically-operated structure. The static layout of procedures and devices cannot represent the whole connotations of the "*process*" and the dynamic-orderly, coordinative, and continuous/quasicontinuous operation, especially the dynamic, coordinative, compact, and continuous properties of process system. Therefore, the design of "*process*" must be based on dynamic operation, that is, the design should realize the highly efficient, orderly, continuous dynamic operation of mass flow and energy flow within a specific spatiotemporal boundary, under the drive of a specifically designed program for information flow. The process pursues the dynamic-orderly operation of ferruginous mass flow under the drive and effect of energy flow in and among procedures. And it also pursues the coordinative, dynamic-orderly operation, according to the coordinative program, of batch-type procedures, such as steelmaking furnace, second refining; quasicontinuous procedures, such as caster, rolling mill; and continuous procedures, such as blast furnaces. Finally, it further pursues the continuity (quasicontinuity) and compactness of process operation.

The steel plant manufacturing process's operation can draw inspiration from dissipative structure theory in an open system, and the nature of the process's dynamic operation is a kind of self-organizing process in a dissipative structure. The dynamic-orderly, coordinative, compact, and continuous operation is the basic goal of the process which also pursues the minimum energy dissipation and material consumption, the optimized material output (high production yield and low waste emission), the reasonable space structure of mass flow, energy flow and time scheduling, and the subsequent multiobjective optimization of economic indexes and environmental loads.

The tasks for the coming years for the Chinese iron and steel industry are to build advanced modern steel plants, to reform existing steel plants, and to eliminate backward production capacities. According to the new situation that steel enterprises are facing in the new century and in order to realize the design and construction of world-class modern steel plants, it is necessary to carry out in-depth and innovative studies on the design theories and methods to avoid the low-level repetitions. The design methods should be established on the design theory for the process's coordinative and continuous operation in the spatiotemporal network in order to describe the rational conversion of mass/energy in a coordinative-continuous manner. And the design methods should also be used to realize the dynamic tailored design of all kinds of technical parameters and information parameters of the mass/energy, and even computer hologram simulation.

5.3.2 Theory, Concept, and Development Trend of Steel Plant Design

Theory seeks the truths and reflects the natures and motion laws of things for people to confirm. Theory also draws a roadmap, aiming at depicting a should-be status (rational status) in people's minds for their practices.

As the old Chinese sayings go: "Investigate things to extend knowledge" and "practice what you have learnt." "Investigate things to extend knowledge" is to know the physical natures and the basic principles of things by an in-depth and systematic research. "Practice what you have learnt" is, according to different specific situations, to reasonably apply the theories and basic principles into practice, such as design and production, to construct new existences (eg, new engineering system), and reform old existences (old plants, technologies, facilities, and products).

As for the steel plant design, what kind of foresight should people have to "investigate things"? It is necessary to think about what the foundation of design theory is. The question is worth asking repeatedly, to obtain scientific meaning and practical values.

Steel plants belong to the process manufacturing industry. The characteristics of steel plants are as follows: The mass flow, composed of kinds of materials, is driven and affected by the energy flow (including fuels and power). The mass flow undergoes heat transfer, mass transfer, momentum transfer, and physical/chemical conversions in specific procedures and devices according to the designed technology process. Therefore, changes take place in mass states, shapes, and properties, and the mass flow is output as the expected product mass flow. In the manufacturing process, the functions of every procedure/device are related and heterogeneous, including chemical and physical conversions, and the processing ways are diverse, including continuous operation, such as blast furnaces; quasicontinuous operation, such as caster and rolling mills; and batch-type procedures, such as convertors and ladle refining.

Superficially, the manufacturing process of steel plants is a series of procedures in simple series—parallel connections. But actually either the process or procedures/devices are dynamic-orderly operations and the operations in or among them belong to an irreversible process which is not aimed at the static equilibrium. Therefore, the manufacturing process will certainly be in a dynamic operation state, and the value of the process will also be embodied in the dynamic operation. From the viewpoint of a dynamic operation, the process composed of a series of procedures/devices is not the simple superposition of procedures but the integration of functions of different procedures. And the integration makes intersections (eg, intersections between mass throughputs, between temperature/energy parameters, and between time-order and time parameters) formed between different procedures/devices. These intersections reflect the correlation and interaction of related procedures, which means a dynamically operating "structure" is formed. The "structure" contains the process networks and the interface technologies. Certainly, the dynamic operation of the process is also influenced by the external environment, such as market, prices, and resource supplies.

The concept of traditional design of steel plant is:

$$\mathrm{F} = \mathrm{I} + \mathrm{II} + \mathrm{III} + \ldots + N \tag{5.1}$$

where, F means the manufacturing process; I, II, III,. . ., N: number of each procedure.

Moreover, the dynamic capacities of procedures were unequal in most cases.

$$\mathrm{I} \neq \mathrm{II} \neq \mathrm{III} \neq \ldots \neq N$$

In the traditional design, the capacity of a procedure was often determined from itself, namely an estimated static capacity plus a surplus capacity. The designed capacities vary with different designers so that these designed capacities can be hardly given full plays, making the dynamic-orderly and coordinative operation impossible and leading to high material consumption, energy dissipation, long process time, large occupied space, obstructed information, low operation efficiency, low investment efficiency, and more environmental loads. The root of these problems is the design being based on a static capacity calculation and the improper assumption of the surplus capacity.

Clearly, it is necessary to establish steel plant design theory on the physical essence of a process's dynamic operation, namely the operation dynamics theory of a process's dynamic-orderly operation. This conforms to the objective laws of a process's dynamic operation and is beneficial for improving all kinds of technical and economic indexes, reducing the investment amount, increasing investment profit, and environmental effectiveness.

In the new century, it is inevitable that the steel plant design theory advances from the simple estimation and superposition of a procedure's static capacity to the integration theory in process-wide dynamic-orderly, coordinative, and continuous/quasicontinuous operation, which, moreover, is possible in method. The conception is as follows:

Eq. (5.1) becomes to:

$$\mathrm{F} = \mathrm{I} \underset{\cap}{\overset{\cup}{\quad}} \mathrm{II} \underset{\cap}{\overset{\cup}{\quad}} \mathrm{III} \underset{\cap}{\overset{\cup}{\quad}} \mathrm{IV} \cdots\cdots (\mathrm{N}-1) \underset{\cap}{\overset{\cup}{\quad}} \mathrm{N}$$

where the capacities of procedures in dynamic operation should be $\mathrm{I} = \mathrm{II} = \mathrm{III} = \ldots = (N-1) = N$, and F is manufacturing process; I, II, III, . . ., N, the number of each procedure; $\cup$, the union of sets of procedure's functions; $\cap$, the intersection of sets of procedure's functions.

The core idea of steel plant design, which is guided by the integration theory in a process's dynamic-orderly, coordinative, and continuous/quasi-continuous operation, is, under the condition that the dynamic capacities in upstream and downstream procedures are well matched, to carry out the analysis-optimization of sets of procedures' functions, the coordination-optimization of sets of procedure relations (including relations of neighbor procedures and long-range relations among several procedures), and the reconstruction-optimization of sets of procedures (eg, the elimination of backward devices and the effective embedment of advanced procedures/devices) in the whole process.

The physical essence of the steel manufacturing process is a kind of open, far from equilibrium, and irreversible complex process formed by the nonlinear interaction and mutual nesting of related heterogeneous units with different structure and functions. In the process, the ferruginous flow (including iron ore, scrap, hot metal, liquid steel, cast strand, finished steel products, etc.) runs dynamic-orderly, driven by the energy flow (including coal, coke, electricity, vapor, etc.) according to designed networks (eg, general layout and plane layout of workshop), and programs (including function order, time-characteristic order, space order, spatiotemporal order, information flow control program, etc.). This kind of process operation contains the optimal integration and the multiobjective optimization of operation factors.

Concerning the steel production behavior, the process contains three levels of scientific issues. They are fundamental science (mainly reactions on the atom/molecule scale), technical science (mainly on the reactor/device scale), and engineering science (mainly "flows" on the manufacturing process scale). In the steel manufacturing process, three kinds of issues are mutually-nested and become a coupled integration. Therefore, for steel plant design theory, it is necessary to have a deep and historical understanding of the three scientific issues, especially to make the three levels of dynamic mechanisms (atom/molecule level, procedure/device level, and process level) dynamically-orderly mutually-nested. Meanwhile, for steel plant design theory, it should also make the continuous, quasicontinuous, and batch-type procedures run in a coordinative manner to achieve the entire integrated continuous/quasicontinuous operation of the process. It can be seen that the theoretical problems in steel plant design raise scientific propositions at different levels reasonably arranged in the design in order to properly solve the engineering scientific proposition of the dynamic-orderly, coordinative, and continuous/quasicontinuous operation (see Fig. 5.6).

The so-called reasonable arrangement on the atom level involves the optimization of unit operation, procedure arrangement, and equipment design. In other words, the reactions at the atom/molecule level should be embedded in procedures/devices properly. For example, the desulfurization reaction should be reasonably arranged and distributed among sintering, ironmaking, hot metal pretreatment, steelmaking, and secondary refining. Comparatively,

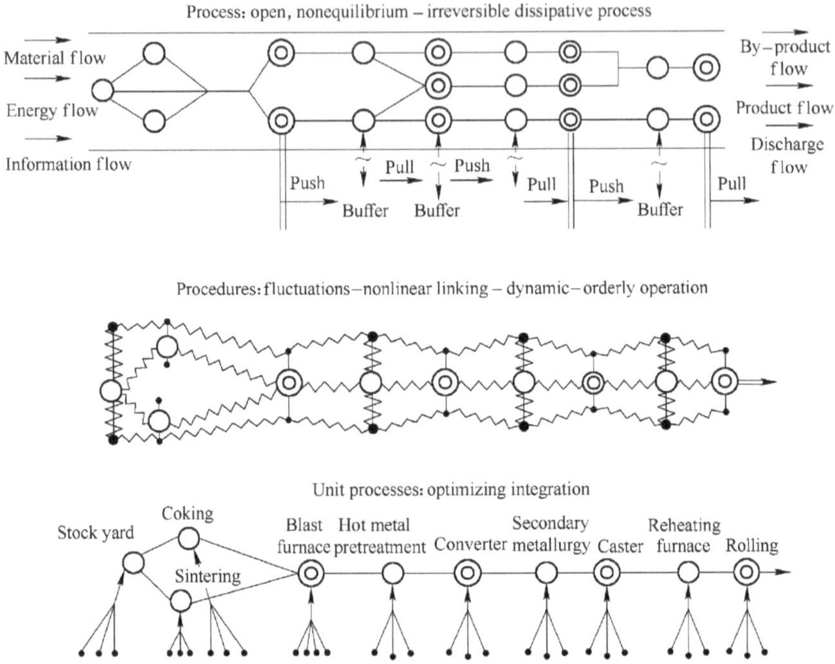

FIGURE 5.6 The integration-analysis relation among manufacturing process, procedures, and unit operations (Yin, 2007).

hot metal pretreatment is most rational for desulfurization considering the reaction rate, cost and efficiency stability, and ignoring the control of sulfide inclusion morphology. For another example, the thermodynamic condition for desulfurization is contradictive with that for dephosphorization in the steelmaking process, and so is the relation between decarburization and dephosphorization. These contradictions are inherent in the metallurgical reaction thermodynamics. But in the process, these problems would be solved by the analysis-integration with De[S]-De[Si]/[P] pretreatment. That is to say, for the metallurgical process, if only the issues at the atom/molecule level are considered or only the intensification of one reaction is pursued, it will lead to the discordant and unreasonable time-rhythm and temperature on the process scale. However, if these issues are considered from the process level, the reactions can be well coordinated and optimized, namely the matching of time, temperature, composition, and throughput is essentially the analysis-optimization of the procedures' functions.

The so-called reasonable arrangement on the procedure/device level involves the instability (the fluctuations of dynamic operation) in the procedure/device's dynamic operation process and the corresponding nonlinear interaction and dynamic coupling. For example, in order to optimize the operation of BOF, secondary refining and CC and to realize the sequence CC, there should be nonlinear interaction and dynamic coupling (matching and coordination) among them. Similarly, EAF's being successfully embedded into the sequence CC process is due to the reasonable arrangement of its functions at the procedure level. Namely, divide the EAF smelting functions into the three traditional stages (melting stage, oxidizing stage, and reducing stage) and the reducing stage is accomplished by the LF. The tap-to-tap time of EAF is reduced from 180 min to 45–60 min so that the EAF can be well matched with the sequence CC. Conversely, the open hearth has been eliminated because it cannot be embedded in the sequence CC because of its long smelting time. In order to meet the nonlinear interaction and dynamic coupling in process dynamic operation, some devices, such as the open hearth furnace, have been eliminated, and some functions have been rearranged so that some new procedures/devices arise, such as the new added hot metal pretreatment and secondary refining.

The so-called reasonable arrangement on the "process" level firstly involves the rational arrangement of function orders. For example, in the De[S]-De[Si]/[P] pretreatment process, the order of desulfurization → desiliconization → dephosphorization or desulfurization → desiliconization → dephosphorization should be clearly considered. For another example, when low carbon Al-killed steel is produced, one of the secondary refining procedures, RH and LF, etc., should be selected. The function-order arrangement is inevitably connected with the reasonable arrangement of space order, such as general layout, elevation of workshop. However, seen from the requirement of the process's dynamic-orderly, coordinative, and continuous operation, it is not enough to just consider the function order and space order, and the time factors such as time-characteristic order and time-rhythm, even the spatiotemporal order must be arranged properly. Then the information flow could run fluently and meanwhile, it could properly control the operation of mass flow and energy flow to realize a quasicontinuous and compact process with minimum process dissipation.

From the analysis of factors for the dynamic operation of a steel plant's manufacturing process, the process has three elements: "flow," "process network," and "program." Therefore, in the dynamic tailored design, both the dominant design idea and the design method should be based on the cognition of the above three elements.

"Flow" refers to the various resources (including mass and energy) or events (eg, oxidation, reduction, heat transfer, mass transfer, momentum transfer, deformation, phase transition) running in the open process system. Actually, "Process network" is a kind of mass-energy-time-space structure for integrating "resource flow" through "nodes" (procedures, devices) and "connectors" (transport means, method, and routes) in the open system. This "process network" should adapt to the operation rules of "flow," especially the rules for the dynamic-orderly and continuous-compact operation of mass flow. "Program" can be regarded as the set of all kinds of orders, rules, strategies, and operation ways, and it also reflects the optimized information flow programs and the dynamic operation rules.

By researching the operation of the steel manufacturing process, it can be seen that the dynamic operation of ferruginous mass flow is reflected as the steel product manufacturing function; the dynamic operation of energy flow (mainly carbonaceous energy flow) embodies the function of energy conversion; the interaction between ferruginous mass flow and carbonaceous energy flow could realize not only the multiobjective optimization of the steel manufacturing process, but also the function of treatment and recovery of large bulk wastes. Namely, the steel plant is supposed to and also able to have the following three functions: (1) steel product manufacturing; (2) energy conversion; and (3) waste treatment and recovery. It is necessary to expand the No. 1 function to the three functions in steel plant design theory and design contents. When the three functions come into play, a new economic growth point may arise and be gradually integrated into the future circular economy (see Fig. 5.7).

5.3.3 Innovation Roadmap of Steel Plant Design Method

The characteristics of steel plant engineering design are as follows:

1. The steel manufacturing processes have complex characteristics concerning time, space, mass, energy, self-organization, and hetero-organization. Those are reflected as the multiscale, multilevel, multiunit, multifactor, and multiobjective optimization;
2. Engineering design of steel plants is a process to realize the selection, assembly, reciprocation, coordination, and meanwhile, the optimization and evolution of performance, cost, investment, efficiency, resource, environment, etc.;
3. Engineering design of steel plants is a process to realize the system optimization of the entire process based on the integration of optimized unit procedures;
4. Engineering design of steel plants puts forward the integration and optimization of each procedure/device for realizing dynamic-accurate and continuous-effective operation;
5. Engineering design of steel plants should adapt the innovation trend to realize the three functions of steel plants instead of just the steel product manufacturing function.

Therefore, steel plant design should not only be for the engineering construction drawings, but should also focus on the dynamic operation of the steel manufacturing process, which further requires procedures/devices to achieve the dynamic-orderly, coordinative-continuous/quasicontinuous operation, and also requires the design method to achieve a corresponding level.

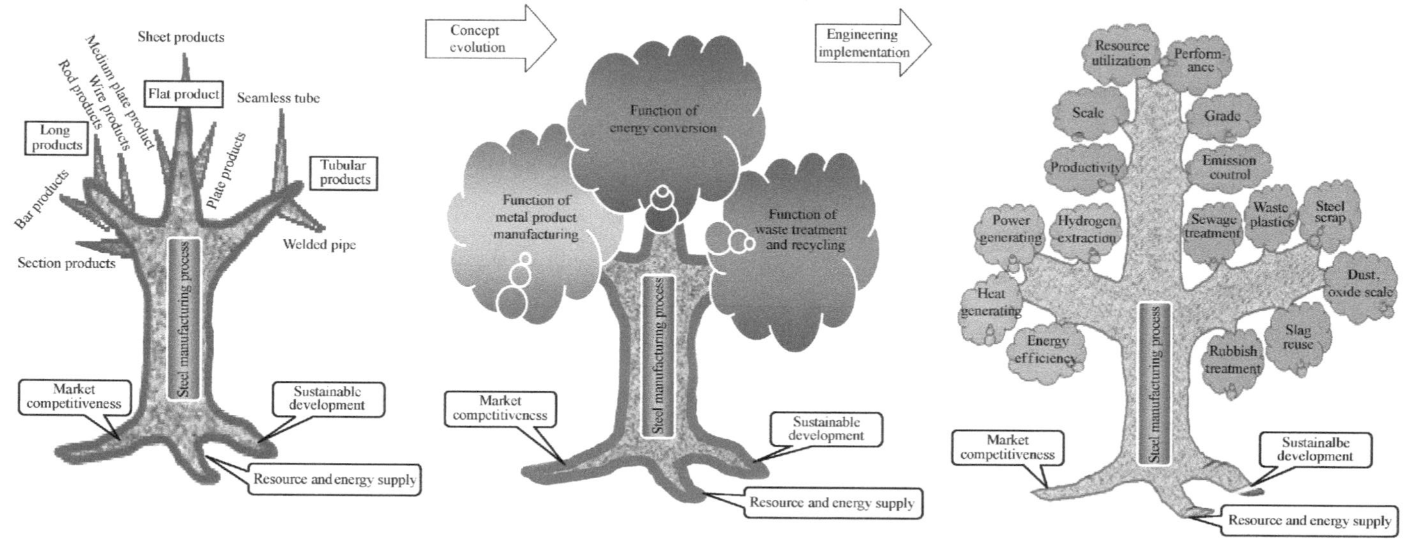

FIGURE 5.7 The coming trend of the functions of the steel manufacturing process (Yin, 2006).

That is to say, the design method should be based on the process design theory that could describe the rational conversion of mass/energy and describe the dynamic-orderly, coordinative, and continuous/quasicontinuous operation. Meanwhile, the design method should strive to make all kinds of information parameters in the mass/energy flows' dynamic operation be dynamic, accurate, steady, and controllable, and strive to further develop a computer virtual reality.

Currently, the negative "stereotypes" about design methods should be eliminated. Some people think that "design has always been confined to the old methods. It accords with the tradition and it is good." Some people think that "we must obey the user's demands that we can do nothing but in this way." These thoughts are seemingly reasonable. However, aren't quite a few normal design institutes in China called design and research institutes? It should be said that this kind of positioning of these institutes is correct and strategic. But only by strengthening the research and development work while conducting designs can these institutes adapt to the requirements of the times, obtain core competitiveness of self-dependent innovation, and satisfy the user's demands.

Compared with the traditional static capacity estimation of procedures and devices, studying the dynamic-orderly, coordinative, and continuous/quasicontinuous operation laws and design methods is a practical and accurate cognition method. Only by the incessant and active study of the accurate design method knowledge is it possible to truly acquire the engineering design knowledge and promote the design method. Conversely, in the current engineering design methods, there are a lot of vague, uncertain, inaccurate contents which should be transformed into engineering science knowledge by researching the dynamic-orderly, coordinative, and continuous/quasicontinuous operation laws. In this way, the innovative dynamic tailored design method probably could be developed.

5.3.4 Dynamic Coupling in Steel Manufacturing Process's Dynamic-Orderly Operation

The steel manufacturing process is a complex and open process system. The thermodynamic openness and nonequilibrium, the dynamic behaviors on different levels, and the spatiotemporal scales in the process's dynamic operation are usually realized by nonlinear interaction. The relations among units on different levels and scales are complex, and here lies the requirement of self-organization. The self-organized system of function, space, and time, namely the dynamic-orderly operation system, is realized by nonlinear interaction among units on different levels and scales, and by the adjustment of artificially input hetero-organizing means.

The nonlinear coupling in the steel manufacturing process's dynamic-orderly operation is reflected in the following three aspects: the dynamic-orderly sectional operation, the coordinative interface technologies, and the rational process network.

5.3.4.1 Dynamic Ordering of Section Operation

As is well known, the manufacturing process of the steel plant can be generally divided into three sections.

The first section is ironmaking where sintering and BF ironmaking procedures are continuously operated. The ironmaking in BF is a smelting reduction process in a moving bed of a shaft furnace. This process requires continuous and steady operation rather than an undulatory and intermittent one. To ensure the continuous operation in BF, raw materials and cokes should be supplied timely and steadily. The ironmaking section should center on the continuous and steady operation of BF. That means the stock yard, coke ovens, and sintering machines should adapt to the BF's continuous operation which, in return, puts forward requirements of proper material input-output rhythms and stable product qualities on the sinter, coke oven, and stock yard and related transporting system.

The second section is steelmaking where only the continuous caster is quasicontinuously operated. Procedures, starting from the BF tapping to the hot metal pretreatment, BOF or EAF steelmaking and secondary refining, should adapt to the operation of sequence CC. The essence of the CC operation is a quasicontinuous/continuous heat exchange-solidification-cooling process. Its efficiency and benefit are intensively reflected in the realization of a long-time operation of sequence CC and the improvement in continuous caster utilization. Generally, the operation period of sequence CC is expected to be as long as possible. So this section centers on the sequence CC procedure which must be adapted or obeyed by batch-type procedures, such as hot metal tapping, hot metal ladle transporting, hot metal pretreatment, BOF/EAF steelmaking, and secondary refining. These procedures, in return, are required to operate in proper time-rhythms.

The third section is hot rolling where the operation of hot rolling mills in a roll-change to roll-change period can be regarded as a quasicontinuous operation. From the cast product output to the operation in a rolling cycle, the rolling mill is in an alternating continuous and batch-type operation, but they can be regarded as quasicontinuous in a roll-changing cycle. The rolled pieces passing the rolling machine in the manufacturing process or in one rolling cycle are expected to be as often as possible, which is beneficial to energy-saving, productivity, and rolling yield. So, the batch-type output of metal pieces should obey the requirement of continuous rolling. Additionally, continuous rolling puts forward proper requirements of time-point, time-interval, and time-rhythm on the casting blank output, transport, and buffering, and its going into/out of the reheating furnace.

Moreover, the mass flows should keep laminar-type operation to ensure the orderly and steady operation in the three sections, and the interference between different logistic flows should be avoided as far as possible.

In order to keep the procedures/devices in a dynamic-orderly, coordinative, and quasicontinuous/continuous operation in the entire process, the following regulations should be made and executed:

1. Batch-type procedures/devices should obey the quasicontinuous/continuous procedures/devices. For example, steelmaking and ladle refining should adapt to the requirements of sequence CC on temperature, composition, and especially time-rhythm.
2. Quasicontinuous/continuous procedures/devices should guide and regulate the behaviors of batch-type procedures/devices. For example, high efficient-constant speed CC should put forward requirements of molten steel throughput, temperature, cleanness, and time factors on related procedures/devices on hot metal pretreatment, steelmaking, and ladle refining.
3. Low-temperature continuously operated procedures/devices should obey high-temperature continuously operated ones. For example, ore sintering or pellet production should adapt to the requirements of BF operation.
4. In a series–parallel connection structure, the process should be in a laminar-type operation rather than a stochastic one. For example, the relatively stable specialized production lines for different products form along the steelmaking-secondary refining-CC procedure.
5. Matched capacities and compact layout among upstream and downstream procedures/devices are the premise of a laminar-type operation. For example, slab hot charging and rolling relies on the matched capacities and coordinative operation among continuous caster, reheating furnace, and hot rolling mill.
6. The manufacturing process entirety should establish the macroscopic operation dynamic mechanism on the dynamic-orderly, coordinative, and continuous/quasicontinuous operation of the push-source-buffer-pull-source.

5.3.4.2 Coordinating by Interface Technology

It is essential to establish the following concepts in order to well understand the interface technology and the coordination with it.

Relation between process and procedures: Procedures/devices are the unit elements of a process; a process is a dynamic operation system integrated by various factors including all the procedures/devices.

Procedures/devices: Optimized procedures/devices should reflect the analysis-optimization of sets of procedures' functions. Several different processes with various functions may run in one procedure/device. When a certain procedure/device is chosen and integrated in a manufacturing process, its function sets should be analyzed and optimized according to the requirements of the manufacturing process. Some necessary functions should be intensified and the useless functions should be weakened or even abandoned. It should emphasize that operation modes of procedures in the process system are different, for example, the continuous, quasicontinuous, and batch-type operations. In the engineering design of a steel plant, attention should be paid to the operation modes of a procedure/device operation and their function.

Process: A process reflects the engineering system in a dynamic-integrated operation. A process integrates correlative procedures/devices owning different functions and operation modes to become an optimized dynamic operation system. A process system reflects the analysis-optimization of sets of procedures' functions, the coordination-optimization of sets of procedures' relations, and even the restructuring-optimization of sets of procedures in the process. So people should study the physical nature of the process system whose fundamental elements are "flow," "process network," and "operation program."

Interface technology: From the analyses on the three key elements of dynamic operation, in order to make the procedures/devices be a dynamic-orderly, coordinative, and continuous/quasicontinuous "flow" through the process's dynamic operation, the interface technologies must give play to the functions of linkage-matching-buffering-coordination.

1. The meaning and role of interface technology. The interface technologies, being relative to the main procedures, such as ironmaking, steelmaking, casting, blooming, and hot rolling, are the linkage-matching-buffering-coordination technologies among the technically reformed procedures and the corresponding devices. It must be said that interface technologies consist of not only technologies and devices, but also a series of engineering technologies including spatiotemporal rational allocation and device number matching in steel plants (see Fig. 5.8).

   From the viewpoint of engineering science, the interface technologies in the steel manufacturing process are mainly applied for linking, matching, coordinating, and stabilizing the process parameters like temperature, time, space, mass flow (including throughput, composition, phase structure, and shape), and energy flow (including primary energy, secondary energy, or energy terminal using). Therefore, to optimize the manufacturing process and especially to realize the engineering design innovation, the research and development of interface technologies must be paid attention to. All of the above is applied to solve the dynamic "weakness" problems in the manufacturing process and to promote the steadiness, coordination, efficiency, and continuity of dynamic operation.
2. The development of interface technology. Before the mid-20th century, the main procedures, such as ironmaking, steelmaking, ingot casting,

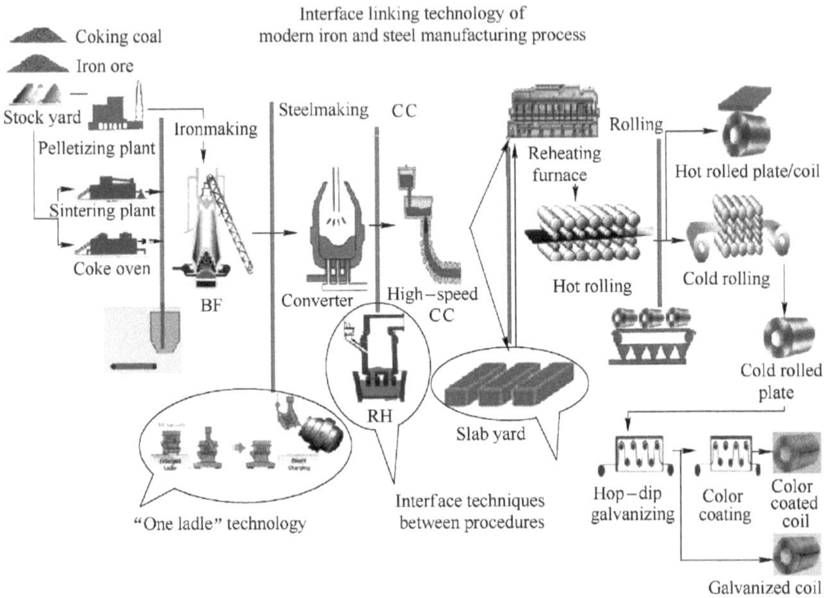

FIGURE 5.8 Interface linking technology of modern iron and steel manufacturing process.

blooming, and hot rolling, ran independently according to their own rhythms. The dynamic relations among procedures were not of enough concern. Procedures/devices were often in a mutual-waiting, randomly connected state. They were connected as simply as input, wait, storage, conversion, and output. During the above process, the technological steps, including repeated heating-up, unnecessary cooling, and repeated moving and shifting for storage and transport, led to long production times, high energy consumption, low yield, low efficiency, unstable product quality, and other problems, such as large land coverage, poor profit, and heavy environment load.

After the 1950s, a series of key-common technologies, such as BOF, CC, wide strip tandem mill, large volume BF, UHP-EAF, was put into operation. These key-common technologies had huge effects on process structure and operation macro-dynamics. For example, the BOF, CC, and wide strip tandem mill had accelerated production rhythm and expanded the continuation degree of the process, the large volume BF increased the throughput of mass flow and energy flow as well as efficiency. Based on the above, the development of these key-common technologies and their applications in process integration caused the evolution of sets of procedures' functions, caused the reallocation of sets of procedures/devices' functions, and further changed the sets of procedures' relations. The analysis-optimization of sets of procedures' functions and the coordination-optimization of set of procedures' relation provided a technical platform for the re-orderliness and higher efficiency of procedures/devices, and then promoted the restructuring-optimization of sets of procedures. The evolution and optimization of sets of procedures' functions and sets of procedures' relations led to a series of variations of interface technologies, and even the appearance of several new interface technologies that had formed some efficient combinations in different production sections. These new interface technologies and their effective combinations had directly affected the structure of the steel manufacturing process, including the technical structure, equipment structure, plane layout (space structure), operation time structure, product mixes, etc.

In short, interface technologies are a group of technologies for coordinating and optimizing relations among procedures developed on the basis of process design innovation, including unit procedure function optimization, operation program optimization, and process network optimization. The interface technologies are also established on the rational procedure relations between adjacent procedures and long-range multiprocedures.

The general layout of a modern steel plant has gained some new characteristics due to the evolution and progress of interface technologies among procedures, such as interfaces of ironmaking-steelmaking, steelmaking-secondary refining-CC, CC-reheating-hot rolling, hot rolling-cold rolling, and especially the thin slab casting-rolling. The tendency of the layout between BF and BOF is to be shorter distance, with fewer hot metal ladles and faster hot metal ladle turnover. Besides, the mixer should be eliminated and the torpedo car questioned. Due to the worldwide establishment of sequence CC system, the continuation degree increased between CC and hot rolling. Therefore, the long-range high temperature connection from hot metal pretreatment to hot rolling has been developed in various ways, leading to the changes of general layout (space structure):

a. The distance between the steel plant and the hot rolling plant is getting shorter and shorter. The linkage between each other should mainly depend on the roller bed, while railway transportation should not be adopted anymore;

b. The capacity of the continuous caster and hot rolling mill should be matched to establish the high speed and high temperature linkage, even the one-to-one or integer correspondence between continuous caster and hot rolling mill;

c. Optimized and specialized production lines, concerning aspects such as steel grades and specifications, should be formed. However, the "universal products" manufacturing lines should be gradually eliminated;

3. The establishment of interface technologies for the dynamic-orderly continuous and compact operation of process system. It is necessary to connect those parameters with linkage-matching-buffering-coordination functions among procedures for the engineering effect of elastic chain resonance (Yin, 2009). Due to the differences between procedures' functions and between devices' operation modes, interface technologies of different types are necessary. Interface technologies can be briefly divided into the following types: the spatiotemporal interface technologies for mass flow; the interface technologies for material property change; and the interface technologies for energy/temperature conversion.
   a. The spatiotemporal interface technologies for mass flow
      The details of spatiotemporal interface technologies for mass flow include:
      i. The optimization of the plane and vertical views—this is the embodiment of optimized networks (including the mass flow network, energy flow network, and information flow network);
      ii. The rationalization of capacities, numbers, and space positions of procedures/devices;
      iii. The rational routes of mass/energy flows among procedures/devices and proper operation modes of procedures/devices ("laminar type" or "stochastic type," continuous operation or bath operation);
      iv. The throughput among procedures/devices should be equal, and the laminar-type operation should be attained;
      v. Selection, optimization, and coordination of transport means and paths;
      vi. Time scheduling of operations among procedures/devices.
   b. The interface technologies for property change of materials
      The details of interface technologies on property change of materials include:
      i. Selection of procedure/device functions, including the analysis-optimization of sets of procedures' functions and coordinative-optimization of sets of procedures' relations;
      ii. Operation modes of procedures/devices and optimization of their spatiotemporal orders;
      iii. The coordination and matching of relevant parameters such as throughput, temperature, and time in the dynamic operation.
   c. The interface technologies for energy/temperature conversion
      The details of interface technologies for energy/temperature conversion include:
      i. Rationality analysis of input-output vectors of energy flows in different procedures/devices;
      ii. Optimization of efficiency and distribution of energy conversion in different procedures/devices;

iii. Rational control of temperature parameters at different time-cycles and time-points;
iv. Optimization and regulation of energy flow input-output models in unit procedures;
v. Optimization and regulation of energy flow networks in a process system, etc.

Interface technologies should connect and integrate all the factors, such as physical phase state factor, chemical composition factor, energy-temperature factor, geometry factor, surface feature factor, space-position factor, and time-time sequence factor in manufacturing process, in order to obtain multiobjective optimization (including high efficiency, minimum energy dissipation, less material consumption, stable product quality, proper product performance, and environment friendliness).

Interface technologies embody the integration by selection/assembly, reciprocation/optimization, and coordination/design. They also embody the multiobjective, multiscale, and multilevel optimization of multifactor flow through the "intersection-union" formed by the "reciprocation-coordination" of the above-mentioned factors. From the viewpoint of irreversible thermodynamics in an open system, in order to form an organized structure with optimized efficiency and minimum process dissipation, it is important to recognize the coordination-continuity/quasicontinuity-compactness of multifactors. All the coordination and continuity should be expressed by process time, and the dynamic-orderly and continuous-compact operation should also finally express itself in process time (time-axis). Therefore, for the continuity, the time-axis is the axis for coupling every parameter in the steel manufacturing process (Fig. 5.9).

Fig. 5.9 shows the coordination and coupling with the time-axis of multifactor flows (phase state factor, chemical composition factor, temperature (energy) factor, geometry factor, surface feature factor, space-position factor, and time-time sequence factor) in the continuous manufacturing process. It can also be said that only when the process factors are coordinated at some optimized time-points on the time-axis, can the engineering design of metallurgical process be called optimum and perfect.

4. Interface technologies in engineering design. In a design for the whole steel plant, a workshop, or a procedure/device, in order to realize the dynamic-orderly, continuous, and compact operation and the high efficiency, high quality, competitive cost, and environmental friendliness, the following topics should be considered:
   a. The function, number, capacity, and space position of procedures/devices should pay attention to the dynamic adaption, buffering, and mutual-complementation among procedures/devices in the upstream and downstream, to achieve continuous-compact and stable-effective production.

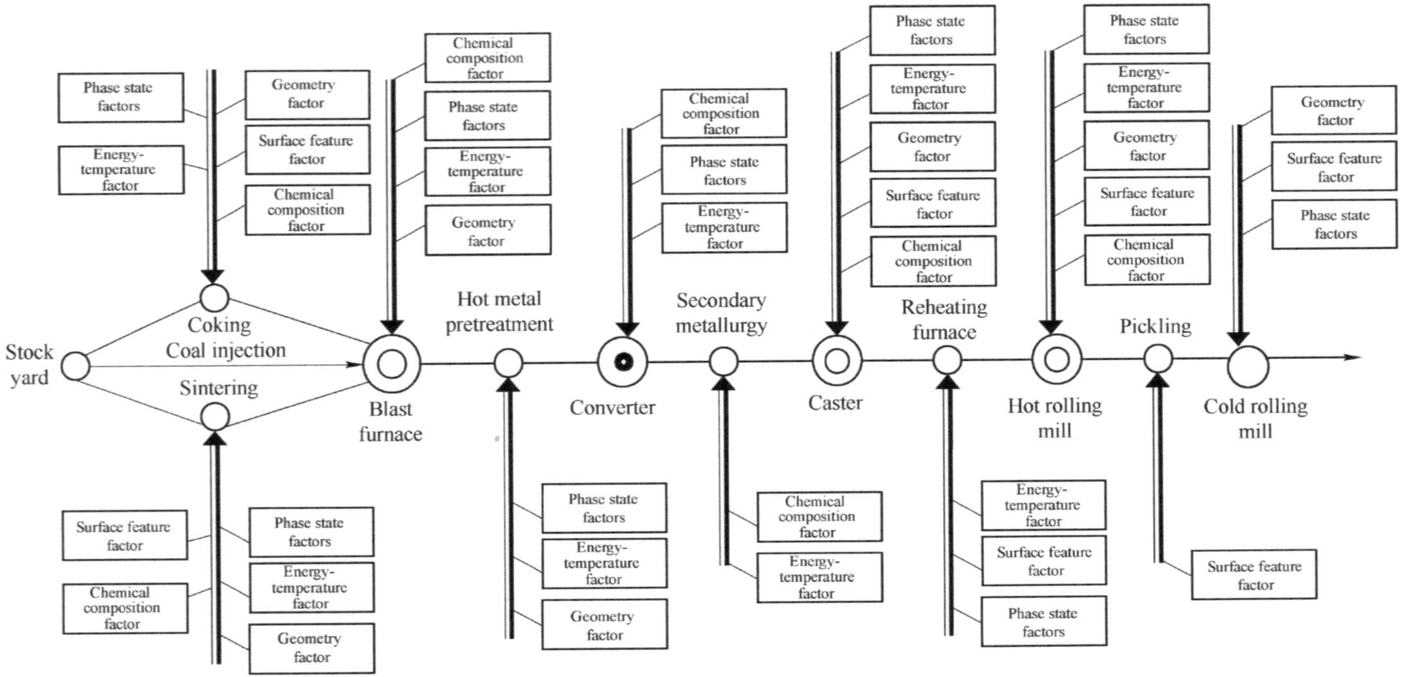

FIGURE 5.9 Diagram of relevant factors in the time-axis under dynamic operation in the steel manufacturing process (Yin, 2009).

b. Decrease the transiting-linking time between procedures/devices, be compact in the space layout, and keep the operations of material flows and energy flows away from interference by the optimization of spatiotemporal factors.
c. Given the operation time, operation rhythm, and operation cycle, the rational matching and necessary buffering and transition are of great importance to steady and flexibly realize the dynamic-orderly operation.
d. If there are several production lines in the manufacturing process, the capacities of procedures in the upstream and downstream and in different production lines should be matched in one-to-one or integer correspondence. The laminar-type operation, in principle, should be kept in different production lines to avoid interference.

Therefore, in the engineering design process, the interface technologies can be further manifested as follows:

1. Simple and direct paths for mass flows and energy flows (eg, plane layout);
2. Buffering-steadiness-coordination among procedures/devices (eg, dynamic Gantt chart);
3. Optimization of network nodes and connectors in the manufacturing process (eg, device number, device capacity, rational location, transport path, transport distance, and transport rules);
4. Optimization of mass flow efficiency and throughput;
5. Optimization of energy flow efficiency and energy saving and emission reduction;
6. Coordinative optimization of mass flows, energy flows, and information flows.

In this kind of engineering design, the rational interface technologies in the upstream and downstream procedures/devices will be beneficial to coordinating the manufacturing process (self-organization enhanced by the effect of informational hetero-organization), to realizing the informatization in the manufacturing process and management regulation, to promoting the throughput, efficiency, and productivity, and to saving investment.

5.3.4.3 Rationalization of a Process Network

The motion of flow is practically a spatiotemporal dynamic operation process, and the input-output of flow in its dynamic operation process is vectorial. To reduce process dissipation, a simple-direct, compact, and optimized network is necessary; otherwise it may lead to disorder or chaos in the flow operation. This is one of the important guiding principles for the construction of new steel plants or renovation of existing plants. Moreover, the rational process network will guide the equipment enlargement and product specialization under the premise that the process structure is optimized.

When the relations among procedures and the transport ways are decided in a plane layout, the transportation energy consumption per unit of steel products mainly depends on the time-length and time-rhythm of the mass flow/logistics operation among procedures, which is closely related to the connecting mode in the process network.

That is to say, with the fluctuations of moving "nodes" and the coordinative relation among the fluctuations of nodes in the network, a process-wide nonlinear interaction field (structure) may be formed by building a reasonable and optimized process network (eg, the general plane layout). At the same time, a self-organizing or hetero-organizing control program that reflects the physical essentials of the dynamic operation of the process can be developed to implement the nonlinear coupling among procedures/devices in an open system. So the minimum energy dissipation and the lowest material loss in the dynamic-orderly operation of flows will be reached. And then a reasonable dissipative structure in an open system will be formed.

5.3.5 Energy Flow Network of Steel Manufacturing Process

Since the beginning of 21st century, due to the rapid development of the economy, there have been acute shortages of resource/energy supplies and environmental challenges, such as global warming, have become increasingly serious. Therefore, steel plants should not only make full use of the steel product manufacturing function, but also pay high attention to the improvement of the energy conversion function and give full play to the function of waste treatment and recovery.

In the steel manufacturing process, there are three kinds of "flow": mass flow, energy flow, and information flow. The carrier of mass flow is material, for energy flow it is energy, and for information flow it is information. The mass flow is the main force in the manufacturing of steel products. The energy flow acts as a driving force, chemical agent, and thermal medium. And the information flow is the sum of the behaviors of mass and energy flows, and reflections of the environment and artificially controlled information. In the dynamic operation of the manufacturing process, the mass flow, energy flow, and information flow sometimes run together, sometimes run separately but they are always interacting. However, in the traditional engineering design and production, the energy flow and its operation system were only considered to be supporting the steel production. The energy/power system was always in a dependent position as a public auxiliary department supporting the steel production. The importance of energy being used efficiently and developed as a valuable resource had been ignored. Because of the advancement of function expansion in steel manufacturing, the energy flow should be regarded as important as the mass flow.

Considering the coupling of energy flow with mass flow, it is necessary to study the energy flow behavior and the energy flow network in engineering design.

5.3.5.1 The Behavior and Operation Laws of the Steel Manufacturing Process

In the steel manufacturing process, sometimes the energy flow operates together with mass flow, sometimes separately (see Fig. 3.2). Flows embody their respective characters when they are separate, but they interact and are mutually affected when they accompany each other.

For the local procedure/device, the mass flow and energy flow enter independently at the entrance. Whilst inside the device, the mass flow and energy flow interact. At the outlet terminal, they are expressed as the mass flow output with part of energy, and meanwhile, secondary energy in different form may be outputted separately to the mass flow. It is necessary to keep an energy surplus to get a high efficiency of production in procedures/devices, thus inevitably there would be some extra energy flow output. For example, for the ironmaking process, iron ore (sinter or pellet), coke, coal, and air blast are charged separately, but inside the BF they are mixed together, and reactions occur, such as combustion, heating-up, carbon solution loss, ore reduction, decarburization, and desulfurization of molten iron. Finally, the hot metal and molten slag are formed and tapped out periodically. The mass flow of hot metal and molten slag at tapping holes carries most of the energy output. Meanwhile, a lot of blast furnace gas is output independently as kinetic energy, thermal energy, and chemical energy at BF tapping. Similarly, there are similar phenomena and processes in the sintering, coking, steelmaking, and reheating furnace and all of these are phenomena of an open system.

Research on energy flow behavior in the manufacturing process should be based on the characteristics of dynamic operation in an open system. The method of static calculation of material balance and heat balance must be abandoned, because this kind of calculation only focuses on the static calculation of a certain state and lacks the concept of dynamic operation and the concept of the vector between upstream and downstream. To analyze the behavior of the manufacturing process, it is necessary to establish the concept of input/output flows' dynamic operation in an open system. The establishment of the concept of energy flow's input/output refers to not only the energy, energy grade, time, and space, but also the energy flow operation program which is conducive to constructing the energy flow network (energy conversion network), improving the energy utilization efficiency, and building the corresponding information control system. The establishment of the energy network is in order to build a dynamic operation model for the input/output. Therefore, the concept of "flow," "process network," and "program" should be built in to studies on the input/output of energy flow in an open, nonequilibrium, and irreversible complex process. It requires a change in concept from the static concept focusing on a certain location to a dynamic concept focusing on the dynamic operation of an energy flow network. The concept should include the time-space-information of energy flow instead of the single calculation of material-energy balance. For the engineering design and reform of steel plants, people should focus not only on the designs of the mass flow network, the reaction and transformation in mass flow, and their operation programs, but also on the design of the energy flow network, the energy conversion, and its "program." Moreover, by integrating these designs with information technologies, an energy-environment control center could be constructed in the steel plant.

5.3.5.2 The Constitution and Design Principle of an Energy Flow Network

The "network" is the system which consists of "nodes" and "connectors" according to certain graphics. For the steel production, the network is a dynamic system including mass-energy-space-time-information. Further classified, it can be analyzed as a mass flow network, an energy flow network, and a related information flow network, among which the energy flow network is an operation system composed of "nodes" and "connectors," namely a dynamic operation system composed of "energy-space-time-information."

In steel plants, there is primary energy (mainly purchased coal) and secondary energy (eg, coke, electricity, oxygen, gases, waste heat, and waste energy) which form their own energy flow nodes (eg, every procedure of steel production, every conversion procedure and waste energy recovery device). As shown in Fig. 3.5, the energy flow network consists of these nodes along with connectors (eg, pipelines, conveyers, circuits). Of course, in the energy flow transport and conversion procedures, there is an inevitable need for necessary and effective buffers, such as gas tank, boiler, and tubes, to satisfy the buffering, coordination, and steadiness of number (capacity), time, space, and energy grades between start and end-points. Thus, it is not difficult to see that the energy flow network (energy conversion network) constructed according to certain graphics inside the steel plant has the form of a "closed loop" to some extent.

So, in the design and the reform of a steel plant, people should pay attention not only to the mass flow, the mass conversion, and its operation program, and the mass flow network design, but also to the energy flow, the energy conversion program, and the energy flow network design.

To design the energy flow network structure and its operation control system in the iron and steel enterprise, people must break the thought shackles of static material balance calculation and heat balance calculation theoretically, and in engineering design, should break with the design limitation on common-auxiliary facilities which are related to gas application, steam distribution, electricity supply, distribution, etc. The thermal processes, such as energy conversion, utilization, transport and storage (eg, boiler, generator), should be integrated with the material processing processes (eg, blast furnace and caster) to form an entire energy flow network for steel manufacturing and its operation program. Therefore, the design of the energy flow network should pay attention to the following principles:

1. Establish the input-output concept of energy flow, not be confined to the concept of the static material balance and heat balance;
2. The reasonable selections of capacity/number/efficiency of primary energy conversion devices (eg, the enlargement of coke oven, the optimization of number and location of oven chamber);
3. The reasonable selections of priorities for secondary energy application (eg, the BOF gas is applied for lime calcinations as a priority);
4. The optimization of the capacity, function, and efficiency of the terminal energy conversion device (eg, gas and steam, as the main forms of the waste energy, are converted into electricity, so that the capacity, number, and efficiency of the generator need be emphasized);
5. The energy flow network should be designed according to the concept of the primary circuit in graph theory (eg, the secondary energy and waste heat must be timely and fully recovered so that the energy flow network should be connected with a simple primary circuit);
6. Solution of optimized means and capacities of storage and buffering devices for sustained and steady operation of energy flow in the network (eg, the number, volume, and location of the gas tanks);
7. The design of "energy flow network" should be designed and constructed hierarchically (eg, the energy flow network for 300°C recovered waste heat, or that for 150°C recovered waste heat);
8. Construct the energy-environment control center through the energy flow network design and move toward the goal of near-zero emission.

5.4 DYNAMIC TAILORED DESIGN FOR STEEL PLANT

The steel production is one of the manufacturing processes. The so-called "manufacturing process" should have an effective, orderly, continuous dynamically-operated structure. The static layout of procedures/devices, which is just a static space structure, cannot embody the whole innovation of the process especially the innovation of the process's dynamic-orderly, coordinative-continuous/quasicontinuous operation. Therefore, the design of the process must be founded on the dynamic operation. The dynamic operation will realize the high efficient, orderly and continuous operation of mass flow and energy flow in a specific spatiotemporal boundary under the drive of a specifically designed operation program for information flow. The process design seeks the dynamic-orderly operation of mass flow (mainly ferruginous mass flow) under the drive and effect of energy flow within and among procedures. It also seeks the coordinative and dynamic-orderly operation of batch-type procedures (such as BOF, secondary refining), quasicontinuous procedures (such as CC, rolling mill), and continuous procedures (such as BF) according to "programs" in coordinative operation. Further, the process design would seek the continuation and compactness of process operation.

Therefore, the theory of the dynamic tailored design for a steel plant should be established on the basis of the physical essence conforming to the dynamic operation, especially on the basis of dynamics theory in the manufacturing process's dynamic-orderly, coordinative, and continuous/quasicontinuous operation. That is, the dynamic tailored design would realize the design of mass flow/logistics in the dynamic-orderly, coordinative-continuous/quasicontinuous operation, the design of timely recycled and efficiently converted energy flow, and the design of the open system centered on energy-saving and emission-reduction (the steel plant is an important link in the circular economy society) under the direction of the advanced conceptual study and the top-level design, and with the help of the Gantt chart and the effective interface technologies.

In the design of steel plants in the 21st century, it is inevitable that steel plant design theory advances from the simple estimation and superposition of the procedure's static capacity to integration theory in process-wide dynamic-orderly, coordinative, and continuous/quasicontinuous operation, which, moreover, becomes possible in practice due to the support of information technologies. So it is necessary to strive to develop the steel plant dynamic tailored design theory and methods through in-depth study.

5.4.1 Difference Between Traditional Static Design and Dynamic Tailored Design for Steel Plant

In the traditional steel plant design and manufacturing process, problems were solved by dismantling them into parts habitually. In the mechanical dismantling method, not only was the design or the production entirety dismantled into details (parts) but also the details (parts) were dismantled from their surroundings or the upstream and downstream sections by "classical" methods, after which, it seemed that the complex interactions between the researched issues and their surroundings could just be neglected and the issues could be easily solved. However, the entire steel plant design and manufacturing process is not an isolated system, but an open system inserted in a certain surroundings which keeps an incessant input and output of mass, energy, and information and lies in a nonequilibrium open state practically.

The steel manufacturing process can be divided into several technical sections, such as the ironmaking, steelmaking, and hot rolling. They operate separately and the optimal isolated operation rules might be found for themselves. For the manufacture process, the sum of partial optimums of isolated sections is not the optimum of the whole dynamic operated process. Therefore, it is improper by using the mechanistic disassembling method to accomplish the integration of related procedures that are idiosyncratic and hardly in a synchronous operation. It is important to study the dynamic operation process engineering in a multifactor, multiscale, and multilevel open system and it is also of great importance for the design and the operation of the manufacturing process to distinguish the outside-inside, cause-effect, and coupling relations between technical phenomena and physical essence of the dynamic operation. When the engineering design and operation process is considered, the following aspects should be taken into account:

1. The "substantiality" has been designed, but the "flows" (virtuality) should not be ignored;
2. The operation superficially is a "substantiality", however, it is in fact a part of moving "virtuality" ("flow");
3. The "virtuality" ("flows") integrates the "substantiality" (procedures/devices), but the "substantiality" embodies the "virtuality"; "virtuality" and "substantiality" rely on each other;
4. The "flows" reflected the combination between processes and engineering (system); input and output of flow must be considered with the issues at the atom/molecule scale or the procedure/device scale;
5. Superficially, the operation or design seemed to be the "substantiality," but practically they are the embodiments of the running of the "flow" in essence. The main purpose and soul are to realize the dynamic-orderly, coordinative, and continuous/quasicontinuous operation of the "flows."
6. Both the production operation and engineering design need to combine the "virtuality" with "substantiality" ("flow" with "matters"). It should first determine the engineering philosophy (the "virtuality") then put it into practice (the "substantiality") through the engineering design and dynamic operation.

A distinguished engineer should master not only the intensive knowledge in his profession, but also the extensive knowledge in economy, society, and humanity, as well as having a comprehensively integrated high-level and holistic innovation ability. This proposition appeals to Chinese engineers to strive for new concepts, new styles, new knowledge, and new abilities.

In conclusion, the difference between the dynamic tailored design and the traditional static design is that the dynamic tailored design has considered the dynamic and practical operation from the beginning, which is different from the static ideas that "design is just for manufacturing devices and building plants." The dynamic tailored design covers not only the mentioned ideas but also the dynamic-orderly and coordinative-efficient manufacturing process (Fig. 5.10). The objectives and tasks of the dynamic tailored design must be embodied in the multiple-object optimizations from the production-beginning to the daily production.

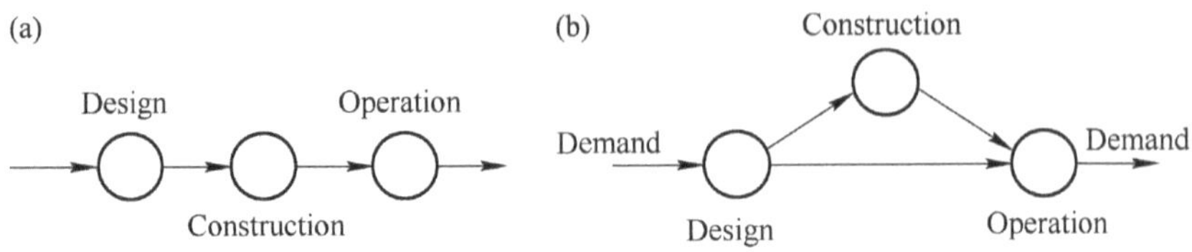

FIGURE 5.10 The difference between the dynamic tailored design and the static design: (a) static design; (b) dynamic tailored design.

Thus, the design of the steel manufacturing process not only provides the project and engineering drawings for construction, but also prepares the rational routes, rules, and programs for the future operation, making the manufacturing process into a dynamic-orderly, coordinative, and continuous/quasicontinuous operation and gives play to its excellent function and efficiency, to achieve the multiple-object optimization.

Therefore, it is necessary to develop a series of technical modules for the dynamic-orderly, coordinative-continuous/quasicontinuous operation. The method and its routes are as follows:

1. Three sorts of dynamic Gantt charts are as follows:
   a. The dynamically operating Gantt chart for the ironmaking system centered on the blast furnaces' continuous operation. And expand it as the base of the global energy flow network;
   b. The dynamically operating Gantt chart for the steelmaking system centered on the sequence CC. And extend it as the high efficient and low cost clean steel production platform;
   c. The dynamically operating Gantt chart for the hot rolling system centered on the optimal plastic deformation in a roll-change to roll-change cycle of the hot rolling mill. And coordinate it with the sequence CC cycle for the formation of CC-HCR or CC-DR dynamic-time control chart.
2. The development of three interface technologies is as follows:
   a. The rapid-effective hot metal transportation system from blast furnace tapping to BOF charging. The system includes the adjustment-control of hot metal ladle transport distance/pattern, rational number, accurate tapping ratio, bottom-left hot metal amount, and ladle turnover speed, etc.;
   b. The dynamic operation system of the high temperature slabs/billets to different reheating furnaces. The system includes the caster strand run-out time and distance to reheating furnace, slab/billet transportation distance/pattern, and storage (buffer) location, as well as the coordinated relation between the rolling rate per minute and the export rate from reheating furnace (including bar split rolling);
   c. The temperature-time control of the rolling metal in and after the hot rolling process. It should form an efficient and rational dynamic control system for rolling and cooling.

3. The construction of series–parallel mix-connected and simple-efficient process networks (the mass flow network, energy flow network, and information flow network). The concept of a simple-smooth-efficient "minimum directed tree" of the mass flow is the foundation. At the same time, high attentions must be paid to the design and research of energy flow network. On the basis of the simplicity-efficiency of the mass and energy flow networks, the information flow network and its operation programs can be designed easily and controlled effectively and steadily.

5.4.2 Process Model for the Dynamic Tailored Design

A model is the accumulation of knowledge. However, it must be noted that without constantly refreshed professional knowledge and an accurate mathematical simulation built on reliable physical mechanisms, the so-called model is just a mathematical amusement. To deeply understand some new technologies and their new applicable conditions, it is essential to theoretically analyze and experimentally verify the rationality and reliability of the new structured physical model to acquire new knowledge.

The mass flow operation is the foundation of the dynamic operation of the steel manufacturing process so that the dynamic tailored design should be based on the mass flow design. From the viewpoint of the model, the dynamic tailored design of mass flow is hierarchical. Different kinds of design methods can be compared as follows:

1. The separate design method which depends on the static structure design and static capacity estimation of procedures/devices is shown in Fig. 5.11.

   This kind of separate design method emphasizes the structure design drawing of procedures/devices and the static estimation of their capacity. This design does not include coordinative operation among procedures/devices, or informational self-adjustment. The connection mode between the upstream and downstream sections, based on waiting for each other and stochastic linkage, belongs to the simple and rough design method.
2. The design method which depends on the static structure and its inner semidynamic operation of procedures/devices (the separate design

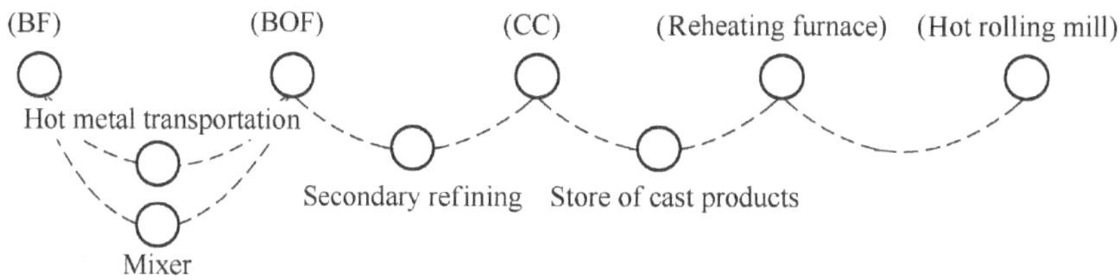

FIGURE 5.11 Process design by the structure design and static capacity estimation of procedures/devices.

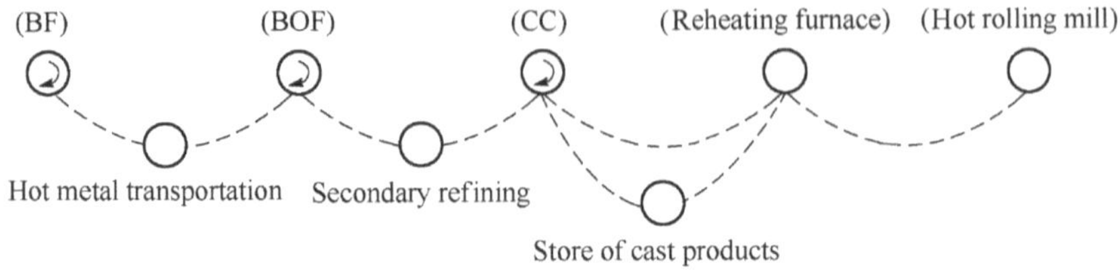

FIGURE 5.12 The design method by the static structure and semidynamic operation.

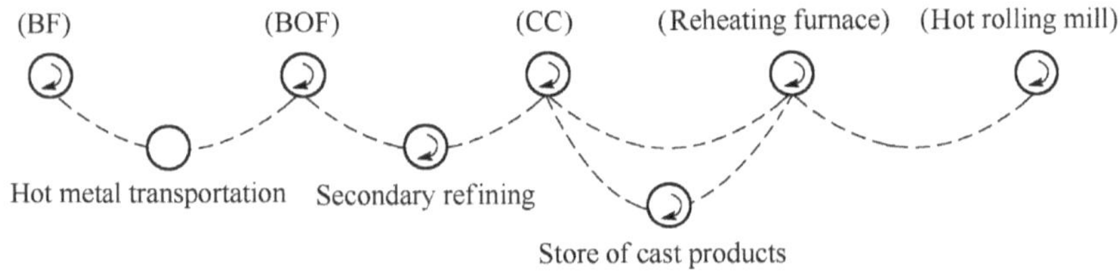

FIGURE 5.13 The design method by inner semidynamic operation of procedures/devices and coordinative regulation of a part of procedures.

method of the unit procedure/device plus a simple expert system) is shown in Fig. 5.12.

Actually, this design method is still a separate design method. Some elementary automatic control measures or simple expert systems are attached to the devices for semidynamic adjustment on the basis of the static structure design. This method does not include the dynamic-orderly adjustment of relations among procedures, and its connection modes among procedures still depend on waiting for each other and random assortment.

3. The design method which depends on the inner semidynamic operation of procedures/devices and the dynamic-orderly operation of a part of procedures is shown in Fig. 5.13.

   This method concerns the design of local dynamic-orderly substantiality (hardware) and virtuality (software). Namely it includes the dynamic and coordination design concept and methods among local procedures, for example, the dynamically-orderly operated steel plant with sequence CC.

4. The dynamic tailored design method which is based on the dynamic-orderly, coordinative, and continuous/quasicontinuous operation of the entire process is shown in Fig. 5.14.

   This method, which is actually the timely self-adapting dynamic tailored design method under the adjustment of informational hetero-organization, belongs to the integration of substantiality (hardware) and virtuality (software) in the dynamic-orderly, coordinative, and continuous/quasicontinuous system.

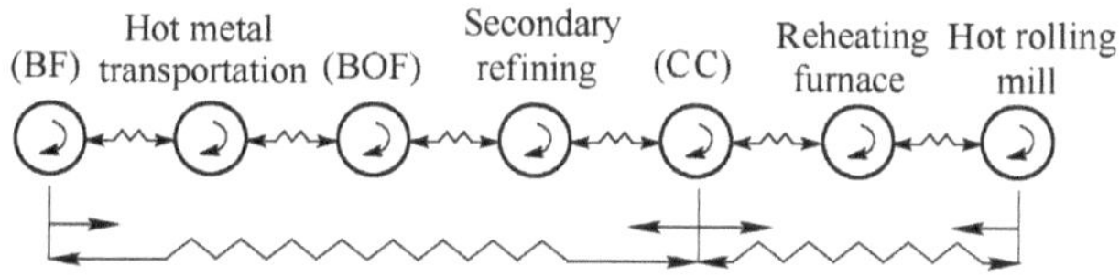

FIGURE 5.14 The dynamic tailored design method of entire process.

From the above schematic diagrams of different design methods, it is clear that the steel plant design method in China almost belongs to stage I (Fig. 5.11) and II (Fig. 5.12). In order to reach stage III (Fig. 5.13), and especially stage IV (Fig. 5.14), the design method needs the guidance of new design theories, methods, and means. Namely, the traditional separate method with static structure design and static capacity estimation needs to be developed into the informational dynamic tailored design. The new analysis, integration-optimization, and operation dynamics of the process should be studied to formulate the dynamic-orderly, coordinative, and continuous operation model under the support of informational and intelligent technologies. This work would involve the following aspects:

1. The reasonable arrangement of function orders and space orders of the process and procedures/devices;
2. The program coordination of the time-characteristic order of the process, procedures/devices;
3. The time-space orders in continuous, compact, and laminar-type operation of the process, procedures/devices, and the related informational/intelligent control;

   To solve the problems mentioned above, it needs

4. The analysis-optimization of sets of procedures' functions;
5. The coordination-optimization of sets of procedures' relations;
6. The restructuring-optimization of sets of procedures in process.

Consequently, a special emphasis for concept research should be placed on the integration optimization for the dynamic-orderly and coordinative-continuous/quasicontinuous operation process system. It should first research and determine the analysis-optimization of the function orders and the sets of procedures' functions; then research and determine the effective spatiotemporal coupling of all the optimized procedure/device functions in process operation, and the optimization of spatiotemporal orders (embodiment of the process network) based mainly on the plane layout, making the "flow" run within the specified spatiotemporal network boundary to ensure the dynamic-orderly, continuous, and laminar-type operation of the mass/energy flow. Finally, based on the above, a computer hologram simulation, namely a mathematic emulation system and the matched-software (tool) with reliable physical mechanism, could be developed for the steel plant design. Thereby the steel plant design method can be upgraded and refreshed, the combination can be more effective for the informational and intelligent technologies with the physical model of the process's dynamic operation. And moreover, the designed steel plant can obtain new competitiveness.

The dynamic tailored design for the new generation steel plant is not only the accurate calculation of the technology parameters, such as single device (equipment), workshop, and energy medium, but also the tailored design of basic parameters and derivative parameters related to the rational energy/material conversion and the entire process's dynamic-orderly, coordinative, and continuous/quasicontinuous operation which contains these basic parameters of mass flow rate/mass throughput, time, and temperature, etc. The dynamic tailored design begins with the perfect connection of all the parameters above, which make mass flow rate (throughput) per minute (per second or per millisecond for tandem rolling) well-matched, to achieve an efficient-continuous process. The mass flow rate will affect the capacity/number/operation-time of a single device and the space-distance/connection-mode between devices, etc. Eq. (5.3) expresses the principles that the dynamic tailored design should obey from aspects of the matching of mass low rate per minute among procedures, the expected continuation degree of the production, and the comprehensive energy consumption per tonne steel.

$$\left.\begin{array}{ll}(1) & Q_{PF}=Q_{FN}, Q_F=Q_{IN}, Q_I=Q_{cc}=Q_{rh}=Q_{ro} \\ (2) & \Sigma t_1^{\mathrm{d}}+\Sigma t_2^{\mathrm{d}}+\Sigma t_3^{\mathrm{d}}+\Sigma t_4^{\mathrm{d}}+\Sigma t_5^{\mathrm{d}} \rightarrow min \\ (3) & \Sigma E \rightarrow min\end{array}\right\} \quad (5.3)$$

where Q_{PF} is the feeding rate of charge materials before ironmaking, $\mathrm{t_{(iron)}}$/min;

Q_{FN}, the demand charging rate of raw materials for the ironmaking system, $\mathrm{t_{(iron)}}$/min;

Q_F, the average hot metal tapping rate of the ironmaking system, $\mathrm{t_{(iron)}}$/min;

Q_{IN}, average demand hot metal charging rate of the steelmaking system, t/min;

Q_I, the average molten steel tapping rate of the steelmaking system, t/min;

Q_{cc}, the average output of cast products per minute from casters, t/min;

Q_{rh}, the average output of cast products per minute of reheating furnaces, t/min;

Q_{ro}, the average rolled pieces per minute of rolling cycle, t/min;

$\sum t_1^{\mathrm{d}}$, the sum of the designed operation time of mass flow through every procedure and device in the process;

$\sum t_2^{\mathrm{d}}$, the sum of the designed transport time of mass flow streaming along the process networks;

$\sum t_3^{\mathrm{d}}$, the sum of the designed waiting (buffering) time of mass flow operating in the process network;

$\sum t_4^d$, the sum of the different maintenance time that retard the process operation;
$\sum t_5^d$, the sum of the repair time of breakdowns in operation and unplanned downtimes;
$\sum E$, the comprehensive energy consumption (coal equivalent) per tonne steel, kg/$t_{(steel)}$.

It should be noted that Eq. (5.3) must be considered as simultaneous equations and it cannot be understood and applied individually. Eq. (5.3) expresses the relations of the coordinative operation of mass/energy flow. Dynamic tailored design can never satisfy just one of the principles. Only when three principles are simultaneously satisfied would the continuity, compactness, and dynamic-orderliness be reflected. Eq. (5.3)(1) mainly reflects the dynamic-coordination and matching of mass flow within a long time range. Eq. (5.3)(2) mainly reflects the principles of compactness and dynamic-orderliness, that is, the minimal operation time is based on the compactness, ordering, and rationalization of the process network. The minimization does not mean the time must tend to be zero or that the work time of a certain procedure is always found to be the shorter the better. There is a special explanation that the compactness involves both the space concept and the operation time. The short operation time does not depend purely on the local process intensification but mainly on the compact, reasonable network, the steady laminar-type operation of mass flow, the rationally arranged, and the quality-ensured repairing or maintenance. Eq. (5.3)(3) mainly reflects the purpose of the dynamic-orderly operation: to reduce dissipation for the minimization of energy consumption. This is an important mark of the continuous, compact, and dynamic-orderly operation.

5.4.3 Core Idea and Step of the Dynamic Tailored Design

The dynamic tailored design is established on the dynamic-orderly, coordinative, and continuous/quasicontinuous tracing of reasonable material change and energy conversion as well as their operation process. Afterwards, it will design information parameters for the dynamic-orderly, coordinative, and continuous/quasicontinuous operation of the whole process, and then it may extend to the computer virtual reality.

5.4.3.1 The Core Idea of the Dynamic Tailored Design

The dynamic tailored design derives from the evolution of engineering design. The following core ideas of the dynamic tailored design must be emphasized: time-point, time-position, time-interval, time-rhythm, and time-cycle based on the continuity and irreversibility of the time arrow. So the importance of the dynamic time scheduling management will be outstanding.

1. To establish the spatiotemporal coordination relation. Time shows the continuity of process, the coordination among procedures, the coupling of technical factors on the time-axis, the energy loss with temperature drop during transporting and waiting. However, the static space structure of the steel plant is fixed, namely the spatiotemporal boundary of the steel plant is defined, after the determination of equipment type/number, process flow sheet, and general layout. Consequently, to state the definition of the "flows," "process network," and "operation program," the node number (node stands for procedure/device), the link path, and the arc/line length between node positions should be fully considered in the design of the process flow sheet and general layout. Moreover, the coordinative and matched operation time and the time factor manifestation forms among the upstream and downstream procedures must be carefully calculated.
2. To structure the process network, the fundamental difference of the dynamic tailored design from the traditional static design is that the former structures the process network under conception of dynamic operation of "flows." The process network is both a carrier and the framework for the spatiotemporal coordination. It should establish the concept of the process network from the viewpoint of dynamic tailored design, dynamic operation, and informational control of the steel manufacturing process. A simple, compact, and smooth process flow sheet and general layout could be taken as the physical framework of the process network to make the flow in a dynamic-orderly and continuous-compact operation, so that the minimization of the energy dissipation and material loss may be realized. It must be noted that the process network should be first reflected in the ferruginous mass flow and then attention should be paid to the development and research of the energy flow network and the information flow network.
3. To highlight the integration innovation in top-level design. The steel plant design, which is essential for solving the multiobjective optimization in design, is a procedure/device-based creative behavior that cooperates with several specialties. Each engineering design project varies according to the site, resource, environment, climate, terrain, transport condition, product market, etc. At the same time, the designer will introduce new technologies in design according to the relevant technical progress. So the combination (or the effective "insertion") of a new technology with the existing advanced technology becomes the integration innovation which is one of the important constituents and modes of independent innovation. The integration innovation requires not only the optimum innovations of unit technologies but also the organic and orderly combination of the optimized unit technologies, which are highlighted as the process-wide integration innovation of top-level design, and the formation of the dynamic-orderly, coordinative, and continuous guiding ideology for top-level design. It should be pointed out that some "frontier" technologies in exploration just belong to local

attempts (and may be immature or failures), which do not have to be enclosed in the top-level design. Reserving or deleting these individual "frontier" technologies depends on whether they can be inserted into the process network and whether it is mature or not. Meanwhile, integration innovation never means to simply gather the different individual "frontier" technologies.

4. To pay attention to the relations among procedures and the development of interface technologies. One of the important thoughts of the dynamic tailored design is to emphasize the optimizations of the relevant procedures, especially the relations among procedures, and the development and application of interface technologies. Meanwhile, different means, such as a Gantt chart, should be applied to carefully predesign the arrangement of devices and their operation modes in manufacturing process.
5. To emphasize the steadiness, reliability, and efficiency of the dynamic operation of entire process. The dynamic tailored design should establish the dynamic-orderly, coordinative, and continuous/quasicontinuous operation rules and programs. And at the same time, it emphasizes the dynamic operation effectiveness, especially the steadiness, reliability, and high efficiency, of both procedures and the whole process. This is the objective of the dynamic tailored design.

5.4.3.2 The Detailed Steps of the Dynamic Tailored Design

The engineering design for a steel plant should conform to the following steps on the condition that the product mix and production scale have been demonstrated and confirmed.

1. Concept research and top-level design. In this step, the concept of the dynamic tailored design and the engineering design view of the integrated dynamic operation should be set up. The dynamic concept of "flow" in the manufacturing process should be established and the dynamic-orderly, coordinative, and continuous/quasicontinuous operation should be emphasized as the basic concept of the process operation. It should highlight the integrity, hierarchy, dynamic characteristic, correlation, and environment adaptability in the top-level design, emphasize the element selection, structure optimization, function expansion, and efficiency improvement as the principles of top-level design, and emphasize thinking logic in methodology that the top level decides the basic level and the lower level is observed from the upper level. The integration optimization of the entire process system should also be particularly stressed in this part, with the steps as follows:
   a. Firstly, to study and determine the function orders and the analysis-optimization of sets of procedures' functions;

b. Secondly, to study and determine the effective time-coupling of all optimized functions of procedures/devices in process operation;
c. Thirdly, to study and determine the spatiotemporal orders based mainly on the general layout and then fix the "flows" in a proper spatiotemporal boundary of network to make the mass flow and energy flow in a dynamic-orderly, continuous, and laminar-type operation.

2. The construction of the static structure of the steel manufacturing process. The preliminary calculation of capacities, number, and rational positions of devices is carried out backwards from the rolling line, CC, BOF (EAF), BF to the ironmaking system according to the determined product mix, production scales, and the obtained metal yield at each procedure, after which, the steel plant manufacturing process framework design would be accomplished.
3. The rational function orientation of the manufacturing process and the rational selection of device capacities based on the analysis-optimization of sets of procedures' functions. The dynamic-orderly operation of the steel manufacturing process is the premise for the division of the procedure/devices' functions. For several devices in the same procedure, the division of the procedure's functions means the division of work for each device. For instance, when two sets of rolling mills are adopted, the product grade/specification/market division needs to be considered comprehensively. The function division of upstream and downstream procedures means the reliable arrangement of space order and time-characteristic order in the process for realizing one or some functions, for instance, the rational selection and control of the slab-width-adjustment function of caster or hot rolling mill and the rational selection and order arrangement of desulfurization, desilication, and dephosphorization for hot metal pretreatment.
4. The design of the interface technology and the expression of coordination and complementation among procedures based on the coordination optimization of sets of procedures' relations. From the viewpoint of engineering science, the interface technologies in the steel manufacturing process are mainly expressed in the linkage, matching, harmony, and steadiness of the mass flow, energy flow, temperature, time, etc. in the manufacturing process. Hence, the dynamic tailored design should perfectly deal with the interface technologies which will be placed at interfaces of coking/sintering (pelletizing) to ironmaking, ironmaking to hot metal pretreatment, hot metal pretreatment to BOF, BOF to secondary refining, secondary refining to CC, CC to reheating furnace/hot rolling, etc. The proper interface technology is essential for the steady, continuous, and compact dynamic operation of process.
5. The establishment of the process network and operation program based on restructuring-optimization of sets of procedures. The main tasks of this step are to establish the diagrams of the ferruginous mass flow network,

the energy flow network, and the information flow network, in order to realize the rational conversion/transformation and the efficient utilization of energy/resources, as well as the recovery of secondary energy and waste heat, in order to make the information flow run effectively through the whole mass flow network and energy flow network, and to adjust the optimum dynamic operations of mass flow and energy flow.

6. The check on the productivity and equipment utilization rate. The productivity calculation of every procedure/device is the concentrated reflection of the available operation times in technical documents and drawings. To solve this problem, it is necessary to make a dynamic-output diagram for material-hot metal, a time-characteristic order diagram for hot metal transporting, a production operation Gantt chart for the steelmaking workshop, a rolling plan diagram and rolling schedule for rolling mill, etc., depending on the product mix, equipment technology parameters, process flow sheet, and general layout. The calculation of productivity and the equipment utilization rate is carried out in the condition of the process's dynamic-orderly, coordinative, and continuous/quasicontinuous operation. Meanwhile, the balance check of material throughput per minute in the manufacturing process should be done according to the production mode.
7. The evaluation of the process's dynamic operation efficiency.

The dynamic tailored engineering design method, which starts with the general goal of dynamic-coordinative operation, is to judge, weigh, and select the technical units the process needs, then to dynamically assemble the technical units and study their reciprocal and coordinative relations to generate a dynamic-orderly, coordinative, and continuous/quasicontinuous engineering integration effect which could be used to estimate the efficiencies of mass flow, energy flow, and information flow.

5.5 INTEGRATION AND STRUCTURE OPTIMIZATION

The steel manufacturing process's dynamic operation itself is exactly the integration of fundamental science, technological science, and engineering science, and also the integration of process technology, equipment technology, and information technology, and even includes the integration of resource, capital investment, process efficiency, and process emission. Thus, the structure optimization, function optimization, and efficiency optimization of steel production can hardly be kept apart from the integration, and neither can the engineering design for the steel manufacturing process, especially the dynamic tailored design for the steel plant.

Integration is closely, relatively related to analysis. Although integration and analysis seem contradictory, they are actually a dialectical unity which is that integration-optimization should be based on analysis-optimization.

However, the effect of the analysis-optimization and its efficiency will be limited since they remain at the level of analysis-optimization. Only when both analysis-optimization and integration-optimization, especially the integration-optimization at the macro-system level are emphasized, will the effects of the optimum factors, optimum structure, optimum functions, and high efficiency be fully reflected. Intuitively, the integration-optimization manifests as the optimization of the system elements (eg, procedures/ devices). But actually, the optimization of elements will inevitably lead to the optimization of the nonlinear interaction relations among elements, and then lead to the structure optimization of the process system.

5.5.1 Integration and Engineering Integration

5.5.1.1 What Is Integration

Integration is not a simple superpositioning of constituent elements, or a simple combination of analysis-optimization units, but an artificial integration system with a dynamic-coordinative structure, optimum function, optimum efficiency, and the related evolution and advancement of this system. The connotation of integration involves aspects of theory, element, structure, function, and efficiency. *The essence of integration is the comprehensive optimization of the structural incidence-relation formed in the multisubjects' interaction and mutual-restriction.*

5.5.1.2 The Meaning of Integration and Integration Technology in Steel Plant Design

Integration in the steel manufacturing process is the solution to problems at the process scale and at the engineering science level. In steel plant design, there are different constituents of different-level connotations. Fig. 5.15 lists the specific constituents involved in engineering science, production technology, engineering application, decision-making, and investment.

In the 21st century, the function of the steel plant will expand to steel product manufacturing, energy conversion, and waste treatment and recovery, so that the new generation steel plant design should focus on developing the following integrated technologies:

1. Integration technologies for the dynamic-orderly, coordinative, and continuous/quasicontinuous operation, including various interface technologies and technologies for sectional dynamic-orderly operation and process network optimization;
2. Technologies for high-efficiency and low-cost clean steel production platform;
3. Technologies for efficient energy conversion, full utilization, and system energy-saving in the process;

(a)

Engineering science

Analysis and integration of steel manufacturing process—Metallurgical Process Engineering

- Cognition science;
- Engineering logic: steel plant structure analysis and evolution under system integration;
- Analysis and optimization of procedure functions;
- Synergy-optimization of procedure function sets and procedure relation sets in process;
- Reconstruction and optimization of procedure sets;
- Union of process science with information technology

Production technology

Metallurgical manufacturing process global optimization technology

- Core issues: the multi-factormass flow control system, that is, the harmonization, interlinking, coordination, and management of three items: mass state transformation, mass property control, and mass flow control;
- Basic parameters and derivative parameters in the manufacturing process;
- Comprehensive control of matter flow, energy flow, and information flow;
- "Critical-compact-continuous" effect;
- "Generalized loop engineering"—"flexible loop engineering";
- Operation mode in manufacturing process

Engineering application

Engineering design for achieving new generation productivity

- Selection of plant mode and the system optimization of process: the transformation of steel plants from "universal" product–mix to "specialized and serialized" product mode.
  - —For new steel plant selection of proper plant mode and innovation of entire process;
  - —For old plant renovation orientation of plant structure optimization, relation between sectional optimization and global optimization;
- Great change of process technology under existing conversion principles:
  - —Ironmaking: combination of charge preparation and ore reduction with energy conversion;
  - —Steelmaking: high efficiency and low cost clean steel continuous production platform;
  - —Solidification processing: high speed solidification (no equilibrium phase change) and near-net-shape casting (intensified and compact shaping technique);
  - —Metal rolling: compression ratio decreasing, precise cross-section and feature, microstructure and property control (including amorphous metal), continuous and speedy rolling;
  - —Product mix, process flow sheet, proper economy scale.
- Function expanding and eco-industrial chains
  - —Mass flow network, design, and operation;
  - —Energy flow network, design, and operation;
  - —Information network, regulation;
  - —Different circular economy areas

FIGURE 5.15 The constituents of analysis and integration of the steel manufacturing process at different levels (Yin, 2000).

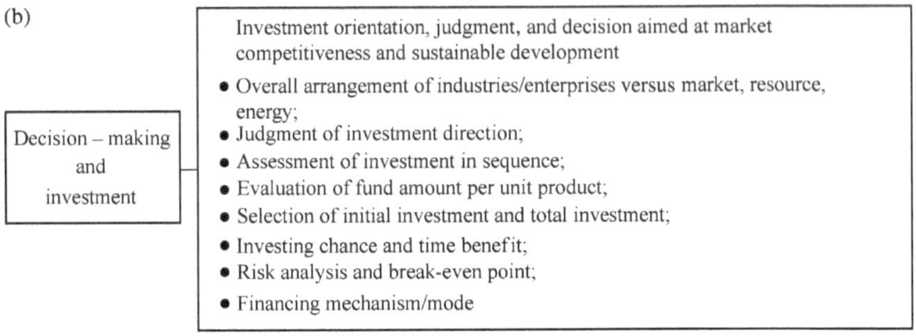

FIGURE 5.15 Continued

4. Technologies for information control and management of mass flow, energy flow, and emission flow;
5. Technologies for waste recovery and treatment;
6. Technologies for clean production and environmental information monitoring;
7. Technologies for industry chaining.

5.5.1.3 Engineering Design and Integration Innovation

In the construction of an innovative country, engineering innovation is important. It is the main field for the implementation of independent innovation strategy. Design innovation is the basis and keynote of engineering innovation. To renovate design methods and establish new design theories are issues that design innovation has to deal with.

Engineering design innovation should be thought of as a manufacturing process. The engineering design and its innovation could be understood as the following actions: (1) the integration of elements, that is, the basic characteristic and action; (2) the construction of networks, that is, the tracking and boundary of event's motion; (3) the formation of structure, that is, the static framework and dynamic-orderly operation of process, as well as the effect and efficiency; (4) the expanding of function and highlighting of efficiency.

One of the engineering design features is that the integration focuses on the selection and improvement of elements (units). The essence of integration is the comprehensive optimization of the structural incidence-relation formed in the multisubjects' interaction and mutual-restriction. So the incidence-relations constitute the structure of the network. In this integration process, there is a need to match various elements (including their parameters), connect and coordinate various units, and also adjust various demands. All of these should be done after complicated selection and consideration.

Engineering design innovation is often expressed as the combination of innovative technologies and existing advanced technologies. And it is almost impossible to update all the technologies. Therefore, the interface technologies and "network" technologies, which are developed on the basis of the combination of new and existing technologies, are also important parts of the engineering integration innovation, which could reflect the "beauty" of the whole coordinative innovative engineering system. The design often pays attention to "modeling" which, actually, advocates the esthetic feeling of reasonable structure and artistry, and pursues the harmony between the engineering subject and nature.

The fundamental engineering design innovation is the activity of breaking old structures and making new ones. Its implementation will reconstruct the spatiotemporal concept and greatly improve the efficiency of conversion and utilization of matter and energy. The fundamental engineering design innovation in evidence becomes the composition of the innovation of the systems of industry and society, namely the formation of new production modes, new lifestyles, new spatiotemporal boundaries, even triggering new languages and new economic-social structures. Generally, engineering and design innovations mean the restructuring of industrial activities, social lives, and social civilizations.

Design derives from engineering ideas. The changing and improving of engineering ideas directly affect the planning and developing strategies of engineering, and then further accelerate the creation of new generation design theories and design methods. Here, people should valorously question the traditional ways, incessantly deliberate the existing incomplete engineering design theories, and the rationality of existing design methods and have the courage to challenge the technical difficulties to acquire independent innovation abilities.

In modern industrial production, the correlations among industries are gradually promoted and the correlative, interdependent and interacting relations between technologies become increasingly apparent. The breakthrough or improvement of a single technology can hardly change the global situations, and there is no engineering or manufacturing process in existence that uses just a single technology. Productivity and market competitiveness can be achieved only through the independent innovation of professional core technologies, the effective integration of related technologies, and the establishment of innovative process systems. Therefore, for the design innovation of the process industry, it is of great importance that there is optimization and the integration innovation of processes and technologies at different steps and levels.

Integration innovation is an important constituent and expression of independent innovation. It requires not only the optimal innovation of a single technology, but also the perfect combination of various optimal new technologies to integrate the new generation manufacturing process with new technical equipment under a thorough understanding, to produce competitive new generation products. Meanwhile, the industry chain would be extended through the expansion of function of manufacturing processes, which will consequently create new business fields and new management modes, as well as new economic growth points.

The main force of integration innovation is industrial enterprises who, however, should not conduct integration in isolation during the implementation of their integration innovation strategy. Integration innovation should never be limited to enterprises and people should note that: "how farsighted your vision determines how big a thing you can see and how great your ambition determines how big a thing you can do." In the implementation of innovation strategy, cooperation is necessary between enterprises and the engineering design institutes, professional research institutes, and universities. Colleges and universities should not be constrained in the theory of the pure science field, but should link with applied science, and should pay attention to disciplinary intersection and the combination of subject knowledge and engineering practice. Engineering design institutes and professional research institutes should not confine themselves to research on some regular problems, but should study the natural connotation, essence, and operation laws of engineering systems for the establishment of new design theories and new design methods. Only in this way will it be possible to effectively promote the implementation of independent innovation strategies.

5.5.2 Structure of Steel Plant

During the exploration of engineering activities, especially engineering design issues, the construction and perfection of the "structure" is an important part of design and an important demonstration of the quality of engineering design. The design innovation is often closely related to the structure optimization and structure improvement. Therefore, it is important to discuss and understand the "structure."

5.5.2.1 Engineering Analysis of Steel Plant Structural Optimization

Steel plant structural optimization is conspicuously reflected in the product structure, which is closely associated with market demand and rational economic scale. Steel plants have always valued the importance of rational economic scale, which means an economic scale with a rational structure. It is hard to judge whether or not a steel plant's scale is economic and rational by merely looking at its quantity. One of the decisive assessing factors is whether the structure is optimized or not.

For the selection of the product mix, two types of product and two kinds of consumer should be paid attention to. Of the products, one type is general products at the large scale (such as constructional long product, common plate and sheet); the other is special steel products (such as oriented silicon steel, deep oil well pipe). The differences of the two types are obvious in investment, sales volume, sales region, production technology, equipment level, and production scale. In general, the large-scale general products are for the regional market with a reasonable sales radius because of their low investments, short construction periods, and low additional values, while the special steel products, whose total demand is limited, are for a big market region with special objectives because of their large investment amounts, severe technical conditions, and obvious additional values. Market analysis, technical analysis, product analysis, and customer analysis are the bases of steel plant structure optimization.

Fig. 5.16 (Yin, 2002) shows the logic conception of steel structure optimization. The figure shows the connotations of steel enterprise structure. The first thing to do in optimizing the steel enterprise structure is to select products series that meet the market demand and then select the optimized process, equipment level, and capabilities, starting from the requirements of manufacturing these products. What should be pointed out is that the selections of process and equipment not only depend on the selection of products, but also are influenced by other factors, such as resources, energy, transportation, environment, human resource quality, and capital. Therefore, the optimization of the steel plant structure is in fact an optimization with constraints.

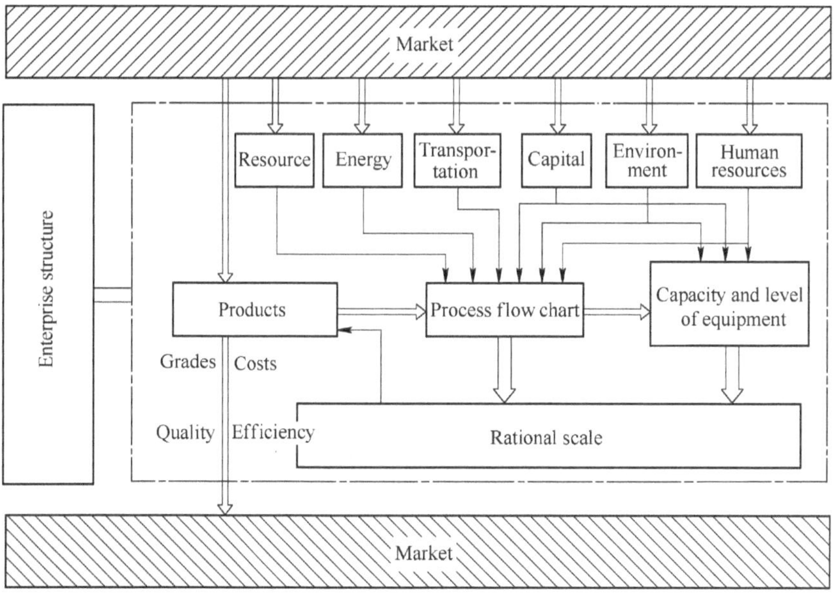

FIGURE 5.16 Logic block diagram of steel plant structure (Yin, 2002).

To describe the structure of steel manufacturing process from the viewpoint of metallurgical process engineering, it is bound to involve the procedures' functions and the relations among procedures. This is of great significance for steel plant structure optimization which is characterized by many batch/quasicontinuous process patterns.

Taking the comprehensive view of the technological progress in steel manufacturing process, the thinking of structure optimization of steel plant is principally based on the following:

1. Analysis-optimization of sets of procedures' functions;
2. Coordination-optimization of sets of procedures' relations;
3. Restructuring-optimization of sets of procedures.

The optimization of the above three sets has been boosting the progress of the steel manufacturing process and the optimization of steel plant structure, thus forming a modern steel enterprise based on the analysis-optimization of procedures' functions and on the coordination-optimization of relations among procedures.

5.5.2.2 Principles of Engineering Design in Optimizing Steel Plant Process Structure

During the 1980s almost all of the world's steel plants began to use CC to connect and coordinate the chemical or physical metallurgical process during the restructuring of the manufacturing process. In addition, steel plants generally tried their best to apply sequence CC to coordinate the entire manufacturing process (including the BF–BOF process and EAF process). Therefore, during the restructuring, special attention should be paid to the following aspects:

1. To determine the outline of the product mix of each steel plant (not steel company group) according to the assessment on market distribution, consumer demand, and investment benefit, and then to analyze the rational capacity of modern rolling mills related with these products. It is necessary to pay attention to the rational scale size of every hot rolling mill and the compatibility between different types of mills, which would influence the product mix and process structure and economic scale (for instance, bar mills and wire mills are of better compatibility, so are the thin strip mills of different widths, but there is little compatibility between hot strip mills and bar/wire mills or between strip mills and seamless tube mills).
2. To analyze the capacity of a continuous caster (including caster type, strand section, strand numbers) and to match the capacities of the continuous caster and rolling mill in a one-to-one or integer correspondence. This part includes the designs of the connection-matching relations and transport patterns/routes for mass flow, coordination-buffering relation

for temperature and time-process (rhythm) between continuous casters and hot rolling mills; the full use of heat energy of cast products based on the compactness between steelmaking workshop and rolling mill. Generally, the billet-export roller bed (or the billet-export cold bed) of a continuous caster ought to be directly connected to the billet-import roller bed of a reheating furnace rather than transport billets with cranes or other devices.

3. To analyze and select the optimum-coordinated interface technologies among steelmaking furnace—secondary metallurgy—continuous caster for the quasicontinuous and compact linkage. Concerning the coordination-buffering of technological process and equipment, people should notice the transverse compatibility and coordination when different types and various scales of casters are used in a synchronous operation. Meanwhile, the rational layout (particularly the locations of steelmaking furnaces-main refining devices-continuous casters) for improving the smoothness and harmony of mass flows in a steelmaking workshop is necessary.
4. To determine the volume, number, and rational location of blast furnace based on the scale and process structure of the steel plant (generally two BF if possible, sometimes one or three). Interface technology between BF and BOF should be determined by the requirements of the product mix. The layout of area between BF–hot metal pretreatment–BOF and the transport route of the hot metal ladle would be improved to promote the dynamic-orderliness, quasicontinuation, and compactness of the high-temperature mass flow.
5. On the basis of the above thoughts, to analyze and calculate the longitudinal coordination and the transverse compatibility among ironmaking–hot metal pretreatment–steelmaking–secondary refining–CC–reheating hot rolling. That is, to rationalize the manufacturing processes of different steel products and properly group the different but compatible steel products together in a comprehensively competitive structure system for the rational production. This is a complex design system integrating market demand, technical progress, economic benefit, and environmental burden assessment into a whole.
6. When the dynamic structure of the mass flow is determined, the energy flow network should be designed for primary energy, secondary energy, residual heat energy, and waste heat. The rational utilization of various energy media and the promotion of the energy conversion efficiency should be enhanced. The residual energy and waste heat should be recycled timely and fully. The plant-wide energy flow network (Energy Control Center) and its operation control program should be formed.
7. For the constitution of the eco-industrial chains in a moderate scope, a regional circular economy society should be gradually formed according to the relations among the processing chain, energy utilization chain,

logistic chain, fund appreciation chain, and the knowledge extending chain.

The topics mentioned above, if implemented as the technical guidance in manufacturing process design, would lead to the following strategic principles being formed:

1. Equalization of mass flow throughput between upstream and downstream procedures/devices;
2. Stability of temperature with time of mass flow in the process ("convergence") principle;
3. Continuous/quasicontinuous operation of mass flow in the process;
4. Highly efficient conversion and full recovery of material and energy in the process;
5. Compactness principle about space and time in the process;
6. Simplicity and smoothness principle for process network structure ("node number," "connector type," general layout);
7. Design and adjustment of the dynamic operation process with informational and intelligent controls based on the optimization of mass flow and energy flow.

REFERENCES

Yin Ruiyu, 2000. Structural analysis on steel manufacturing process and some problems of engineering effectiveness. Iron Steel 35 (10), 1–7 (in Chinese).

Yin Ruiyu, 2002. Energy-saving, clean production, green manufacturing and sustainable development of steel industry. Iron Steel 37 (8), 1–8 (in Chinese).

Yin Ruiyu, 2006. Topic of new generation steel manufacturing process. Shanghai Met. 28 (4), 1–6 (in Chinese).

Yin Ruiyu, 2007. Some science problems about steel manufacturing process. Acta Metall. Sin. 28 (4), 1121–1128 (in Chinese).

Yin Ruiyu, 2008. Engineering in the field of vision of philosophy. Eng. Sci. 3, 3–5 (in Chinese).

Yin Ruiyu, 2009. Metallurgical Process Engineering, second ed. Metallurgical Industry Press, Beijing, p. 169, 199 (in Chinese).

Yin Ruiyu, Li Bocong, Wang Yingluo, et al., 2011. Theory of Engineering Evolution. Higher Education Press, Beijing, p. 26, 27 (in Chinese).

Zhu Aibin, Mao Hongjun, Xie Youbo, 2004. Research on the advanced engineering environments. Chinese J. Mech. Eng. 40 (8), 1–6 (in Chinese).

Chapter 6

Case Study

Chapter Outline

Theory and Methods of Metallurgical Process Integration.
DOI: http://dx.doi.org/10.1016/B978-0-12-809568-3.00016-4

Combining theory with practice is the gist of this book. In the study of the integration theory and method of the metallurgical process, case studies play a significant role. As a bridge between theory and practice, case studies not only facilitate the "implementation" process of theories, but also serve as a solid foundation for the ongoing theories to "take off."

6.1 PROCESS STRUCTURE OPTIMIZATION IN STEEL PLANT AND BF ENLARGEMENT

In recent years, Shougang Jingtang United Iron & Steel Co., Ltd. (hereafter referred to as Shougang Jingtang Steel) in China has been seeking to achieve three functions (Yin, 2004; Yin, 2010): steel product manufacturing, energy conversion, and waste treatment and recycling. Therefore, a new-generation recycling iron and steel process flow has been established, which is the foundation for a dynamic and orderly, continuous and compact, high efficiency and harmonious production process.

For the modern steel plants, apart from getting high-quality pig iron for steelmaking, the production process of BF also involves a significant amount of energy conversion and input/output of information (Yin, 2004). Therefore, the theory of metallurgical process engineering (Yin, 2004) should be applied to evaluate process structure optimization of steel plants as well as the issue of BF enlargement, in order to analyze the different levels of the entire production process in steel plants, namely, the consideration of aspects of ferruginous mass flow, energy flow, information flow, etc. in the whole plant. The relationship between BF and process structure optimization in a steel plant should be analyzed comprehensively from the four fundamental functions of BF, namely, oxidation reduction and carburetor, hot metal generator and continuous supply, energy converter, and metallurgical quality regulator (Yin, 2009a). The ironmaking process is a critical control link for production cost and energy utilization efficiency in steel plants, thus BF ironmaking technology should be developed with the goals of high efficiency, low consumption, high quality, long campaign life, cleanness, etc. High efficiency is not simply the intensification of production; more attention should be paid to economic, environmental, and social benefits. Long campaign life is not limited to extending BF working life; more attention should be paid to its technical advancement and the viability of sustainable development. With the fierce market competitiveness at home and abroad, managers of steel plants are confronted with issues on how to optimize the process structure of steel enterprises in China, as well as how to select the rational process flow for steel plants in the future. In this chapter, focusing on BF ironmaking—the main critical process in steel plants—a systematic analysis and study is performed on how to determine scientifically and rationally the BF capacity, number, and volume by the theory of metallurgical process engineering. Economic efficiency and feasibility of different proposals on BF configuration are studied under the precondition of the same production scale, which intends to be a reference in the process structure optimization and BF enlargement for steel enterprises in China.

6.1.1 Development Trend of BF Ironmaking

In recent two decades, BF ironmaking technologies at home and abroad, have made substantial progress in the aspects of enlargement, high efficiency, long campaign life, low consumption and environmental protection. In particular, there have been notable achievements in individual techniques like BF long campaign life, high blast temperature, coal injection with enriched oxygen, and fully dry dedusting of BF gas (hereafter referred to as BFG), and top-pressure recovery turbine (hereafter referred to as TRT), etc. These advances are mostly in two important aspects—lower energy consumption and higher production efficiency—and they are inextricably connected with the development of BF enlargement. On the premise of process structure optimization of the whole steel plant, enlargement and a few numbers of BFs have become the development trend for current BF ironmaking technology.

With consideration of network structure optimization on the flow of mass, energy, and information, it is an appropriate choice to equip two or three BFs at a steel plant. In determining the capacity, number, rational volume, reasonable location, and layout of BF, the process structure optimization of the whole steel plant should be taken as the precondition. An over-emphasis on size, based on the one-sided view of "the larger the BF, the better," or the excessive pursuit of "the largest," without consideration of the market and other factors, should be avoided. BF enlargement and the number of BF are the essence of production process structure optimization, and they are related to high-efficiency optimization of dynamic operation on mass flow, energy flow, and information flow at steel plants.

6.1.1.1 Development of BF Enlargement Abroad

1. Development of BF enlargement in Japan

   The number of BFs on service in Japan has decreased from 65 in 1990 to 28 in 2016, a decrease of 56.9%, while the average volume of BF has increased from 1558 m^3 to 4157 m^3, an increase of 166.8%. Furthermore, the fuel ratio of BF in Japan has dropped below 500 kg/t generally, pulverized coal injection (hereafter referred to PCI) has reached more than 120 kg/t, and coke ratio has dropped below 380 kg/t. Fig. 6.1 shows the changes in BF number and volume over the past 30 years in Japan. Fig. 6.2 shows changes in the fuel ratio of BF ironmaking over the past 30 years in Japan. Figs. 6.1 and 6.2 basically reflect the development of BF ironmaking technology in Japan. At present, there are generally two or three BFs at a steel plant in Japan, and there are cases where only one BF is equipped. In 2010, the annual production capacity of Oita Plant of Nippon Steel Corporation (NSC) was 9,634,000 t with only two 5775 m^3 BFs in operation.

2. Development of BF enlargement in Europe

   The number of BF in service in Europe has decreased from 92 in 1990 to 58 in 2016, a decrease of 37%. The average working volume of BF has increased from 1690 m^3 (effective volume[1] being approximately 2150 m^3)

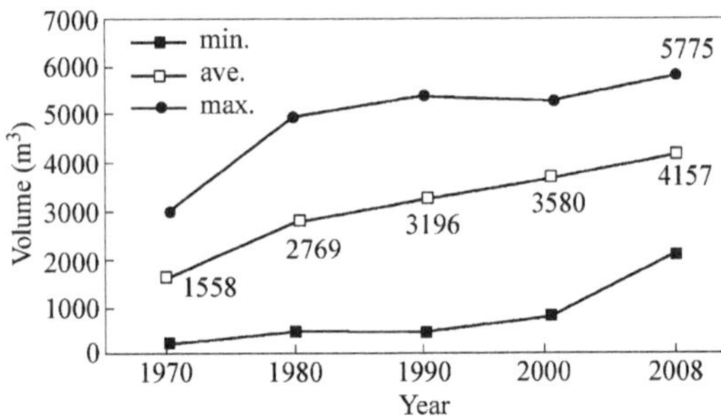

FIGURE 6.1 Changes in BF volume in Japan (Miwa, 2009).

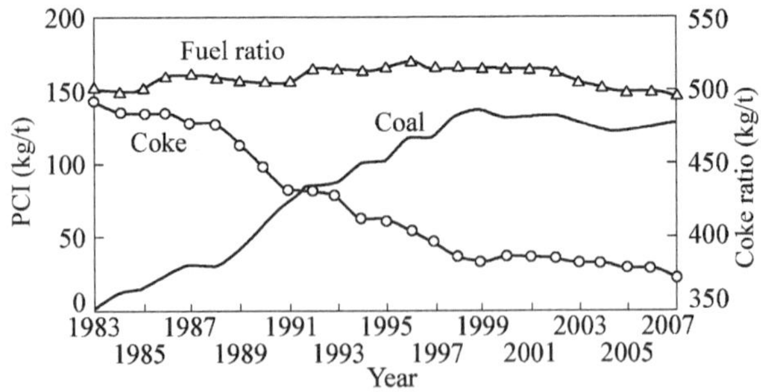

FIGURE 6.2 Changes in fuel ratio of BF ironmaking in Japan (Ariyama et al., 2009).

1. BF effective volume refers to the volume in BF between the bottom plane of bell and the center line of taphole.

to 2063 m^3 (effective volume being approximately 2480 m^3), an increase of 22%. The fuel ratio of BF ironmaking in Europe has been reduced to 496 kg/t, coke ratio reduced to 351.8 kg/t, PCI has reached above 123.9 kg/t, and heavy oil/natural gas injection has reached 20.3 kg/t. Fig. 6.3 shows changes in BF number and volume over the past 20 years in Europe. Fig. 6.4 shows changes in the fuel ratio of BF ironmaking over the past 20 years in Europe. Figs. 6.3 and 6.4 represent the development of European BF ironmaking technology. In Europe a single steel plant generally is equipped with two or three BFs. For instance, ThyssenKrupp Steel AG (TKS)'s Schwelgem Plant has a production capacity of approximately 7,800,000 t/a, with only two BFs (1×4407 m^3 + 1×5513 m^3) in operation.

6.1.1.2 Development of BF Enlargement in China

The development trend of BF enlargement in Japan and Europe started in the 1980s, and has accelerated since 1990. Under process structure optimization, the number has reduced and volume has increased by the increasing output of individual BFs. In the 1990s, the Chinese steel industry was developing rapidly and the steel output continued to increase. With major breakthroughs in key and common technologies, such as high-efficiency CC, BF, PCI, long campaign life of BF, continuous rolling, process structure optimization and BF enlargement in steel plants developed promptly.

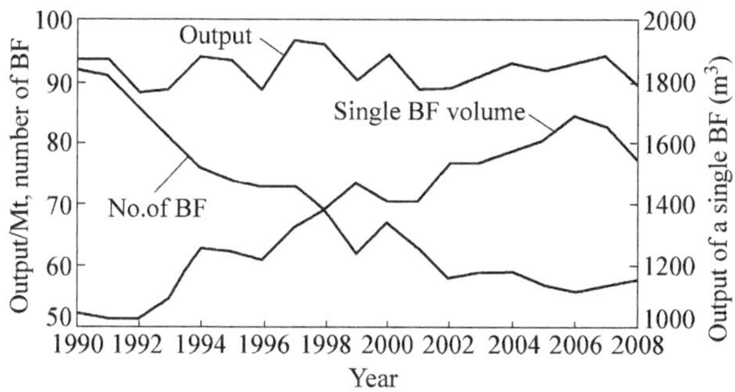

FIGURE 6.3 Changes in the number of BF and output in 15 European countries (Michael and Bodo, 2009).

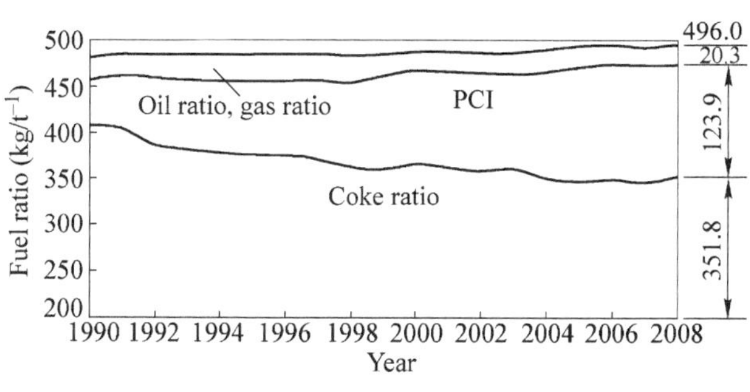

FIGURE 6.4 Changes in fuel ratio of BF ironmaking in 15 European countries (Michael and Bodo, 2009).

In September 1985, as an important milestone of BF enlargement development in China, Baosteel 1# BF (4063 m^3) was put into production at Baosteel. However, the BF enlargement that pushed process structure optimization in steel plants was promoted in the early 21st century. Incomplete statistics show that the number of BFs in service or under construction, with a volume larger than 1080 m^3, was approximately 227 by 2010; the total volume was 429,420 m^3. There were 119 BFs with a volume of 1080–1780 m^3, 61 BFs with a volume of 2500–4080 m^3, and up to 20 of them have a volume larger than 4063 m^3. Fig. 6.5 shows the structure distribution of the BFs with a volume larger than 1000 m^3 in China in 2010.

It has been found that BF enlargement has driven the technical progress of BF ironmaking in China. In 2010, the fuel ratio of BF ironmaking in Chinese key steel enterprises was reduced below 520 kg/t, coke ratio reduced to 370 kg/t, and PCI reached above 150 kg/t. The energy consumption of BF was reduced to less than 410 kg_{ce}/t. Fig. 6.6 shows the change in BF fuel consumption and blast temperature in Chinese key steel enterprises in the 21st century. Fig. 6.6 shows that with the increase in the quantity of large-scale BFs (larger than 1080 m^3) since 2005, the fuel ratio of BF ironmaking and the coke ratio of BF have been significantly decreased.

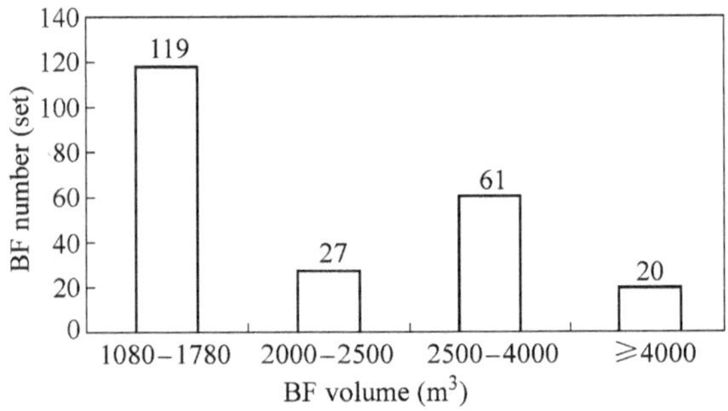

FIGURE 6.5 Quantity distribution of different volume of BF above 1000 m^3 in China in 2010.

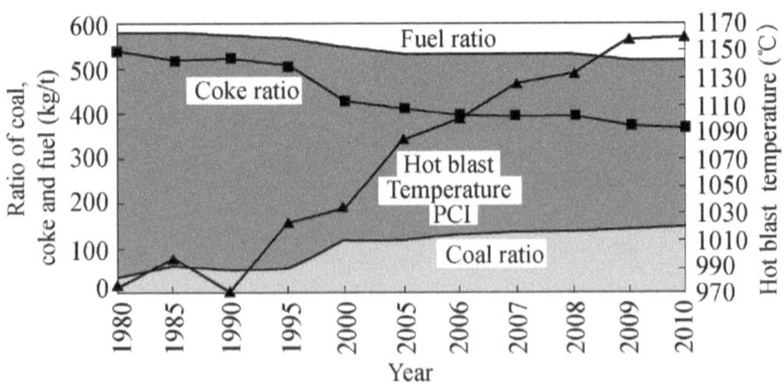

FIGURE 6.6 Change in fuel ratio of BF ironmaking in Chinese key steel enterprises.

6.1.2 BF Enlargement with the Premise of the Optimization of Process Structure in Steel Plants

Production capacity of steel plants is determined with social development and market demand. In other words, optimal selection of production positioning and production scale in a steel plant should be carried out in accordance with regional market demand and changes of product mix requirement, by considering adjustable measures to the local conditions and the market capacity of relevant zones. Top level design should be considered, based on an evaluation of the rationality, efficiency, and economy of the whole process structure in steel plants. Then rolling mill composition is engineered comprehensively and productivity is reasonably evaluated to determine the quantity of BFs and their volumes. Meanwhile, investment orientation and the prospective target of the enterprise development must be considered.

6.1.2.1 Criterion for Determination of BF Number and Volume

In determining the number and volume of BF, the following factors should be considered: the demand of hot metal output, the influence of the number and location of BF on the mass flow network, energy flow network, and information flow network in the plant, and the corresponding dynamic operation procedure. A brief analysis is given on the mass flow network and energy flow network of Shougang Qian'an Iron and Steel Co., Ltd. (hereinafter referred to as "Shougang Qian'an Steel") and Shougang Jingtang Steel. (Refer to Figs. 6.7–6.10.)

The steel output scale of Shougang Qian'an Steel and that of Shougang Jingtang Steel are quite similar (within the range of 8.00–9.00 Mt/a). Each plant has two rolling mills with similar production capacity, namely one with 1580 mm rolling mill and another with 2160 mm (2250 mm) rolling mill. However, due to differences in their process structure, the number and

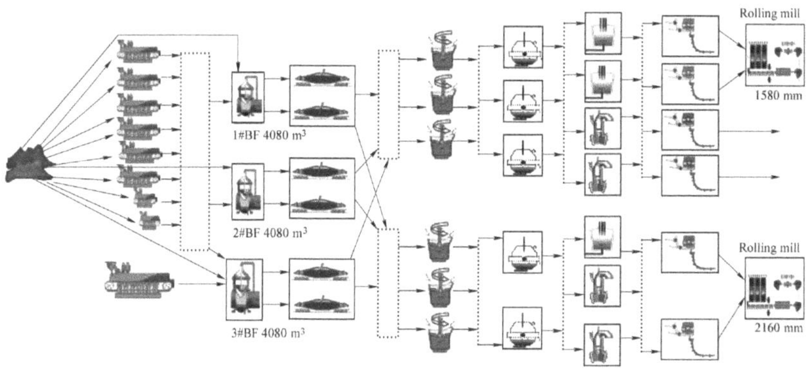

FIGURE 6.7 Mass flow network at Shougang Qian'an Steel (scale 8–9 Mt/a).

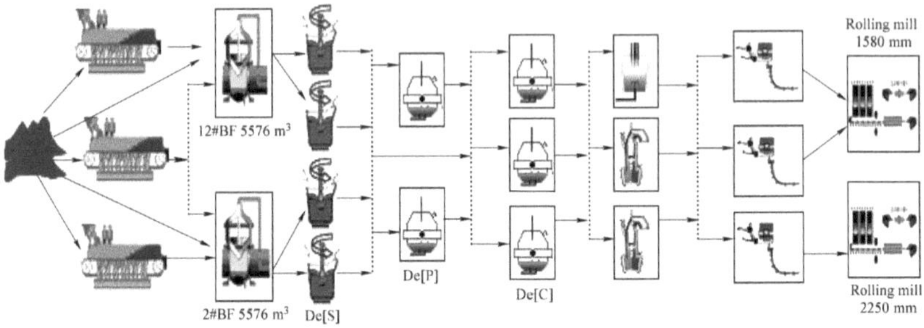

FIGURE 6.8 Mass flow network (8–9 Mt/a) at Shougang Jingtang Steel.

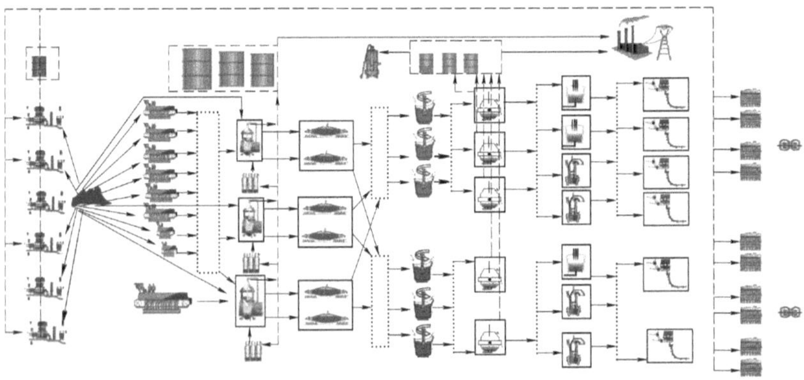

FIGURE 6.9 Energy flow network at Shougang Qian'an Steel.

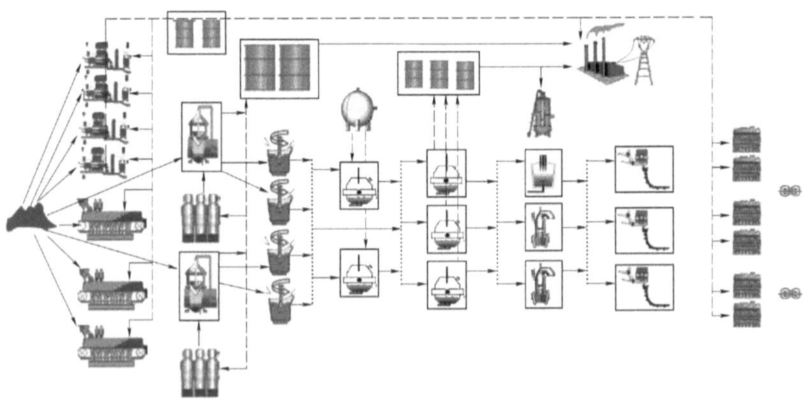

FIGURE 6.10 Energy flow network at Shougang Jingtang Steel.

volume of BF and the mass flow network (Figs. 6.7 and 6.8) and energy flow network (Figs. 6.7 and 6.8) of the two plants differ greatly.

In terms of ironmaking equipment, there are three BFs with volumes of 4000 m^3 at Shougang Qian'an Steel, while two BFs with volumes of 5000 m^3 at Shougang Jingtang Steel. Shougang Qian'an Steel has two steelmaking plants (five decarburization converters), seven CCs, and two rolling mills, in a configuration of 3-5-7-2 (namely 3-2-2 structure). There are many pieces of equipment at Shougang Qian'an Steel (eg, 6 KR devices, 7 refining furnaces). Two steelmaking plants at Shougang Qian'an Steel are confronted with considerable complexity and difficulty in production scheduling and system management. There are two BFs with volumes of 5000 m^3 at Shougang Jingtang Steel, corresponding to one steelmaking plant (three decarburization converters), three CCs, and two rolling mills, forming a configuration of 2-3-3-2 (namely 2-1-2 structure). Obviously, equipment number at Shougang Jingtang Steel is lower (eg, 2 KR, 4 LF) than those at Shougang Qian'an Steel. In comparison with Shougang Qian'an Steel, Shougang Jingtang Steel adopts a single steelmaking plant system with characteristics of simple process flow, fewer connections (number of equipment) of the process network and a more concise process network, facilitating easier scheduling and management.

The above comparisons show that from the viewpoint of metallurgical process engineering, the total throughput of Shougang Qian'an Steel and that of Shougang Jingtang Steel are basically similar (similar production scale), but Shougang Jingtang Steel adopts two BFs (instead of three BFs), reducing the nodes greatly in the integrated process network (see comparison between Figs. 6.8 and 6.7). The mass flow network and energy flow network of Shougang Jingtang Steel are simpler and smoother, which has apparent advantages in information control, joining and matching of process operations, as well as production scheduling and management. Obviously, there are also natural advantages in high-efficiency utilization of energy flow. Hence, BF enlargement, the design of BF number, volume, position in general layout, etc., not only influence BF ironmaking itself (increased output and decreased energy consumption), but also have decisive effects on dynamic operation, structure optimization, and efficiency of mass flow of steel plants.

Therefore, BF enlargement in China is adopted and promoted with the integrated process structure optimization in steel plants. It advocates neither an excessive number of BFs in a steel plant, nor blind pursuit of larger volumes of BF. The number and capacity of BFs should be determined on the basis of process optimization of steel plants with necessary attention to their location. It should be noted that the determination of capacity and volume of BFs should consider the rationality of process structure in the steel plant—blindly pursuing the so-called BF enlargement is inadvisable. At the same time, it is not advocated to follow the convention and build an excessive number of small BFs. Those unrealistic and blind pursuits of "comparison of volume" or "number one" in engineering should be avoided.

6.1.2.2 Capacity and Productivity of BF

The current development of ironmaking technologies at home and abroad shows that the production capacity of BFs in different scales has increased by a large margin. In keeping with the technical equipment level, operation level, and material/fuel condition of BFs of different levels, BF production capacity has been kept basically within a reasonable range. Fig. 6.11 shows the change of average productivity of BF at Chinese key steel enterprises over the past 30 years. It can be seen that the average productivity of BF has increased by 1.15 t/(m^3 · day) over this time-frame.

In the past, the assessment of production efficiency of BF was based on two technical indexes, that is, "smelting intensity" and "productivity." As the BF hearth reaction is a fundamental metallurgical process, it is more scientifically convincing to employ the "production rate of BF hearth area" to evaluate the production efficiency of BF (Tang, 2005; Zhang and Yin, 2002). This coefficient reflects the fundamental characteristics of the BF smelting process.

BFs of different sizes have different BF volume and hearth area. For instance, productivity of a 1260 m^3 BF is 2.5–2.7 t/(m^3 · day), and the rate per hearth area is 61.16–66.05 t/(m^2 · day). For a 2500 m^3 BF, the BF productivity is 2.4–2.6 t/(m^3 · day), and the rate per hearth area is 60.93–66.01 t/(m^2 · day). For a 3200 m^3 BF, the BF productivity is 2.3–2.5 t/(m^3 · day), and the rate per hearth area is 60.98–66.28 t/(m^2 · day). For a 4080 m^3 BF, the BF productivity is 2.2–2.4 t/(m^3 · day), and the rate per hearth area is 61.51–67.10 t/(m^2 · day). For a 5500 m^3 BF, the BF productivity is 2.1–2.3 t/(m^3 · day), and the rate per hearth area is 62.04–67.95 t/(m^2 · day). Under the same smelting conditions, BF productivities between small-scale BFs and larger-scale ones cannot be simply compared. Fig. 6.12 shows the annual average BF productivity and annual average hearth area utilization coefficients of BFs at different volume levels in

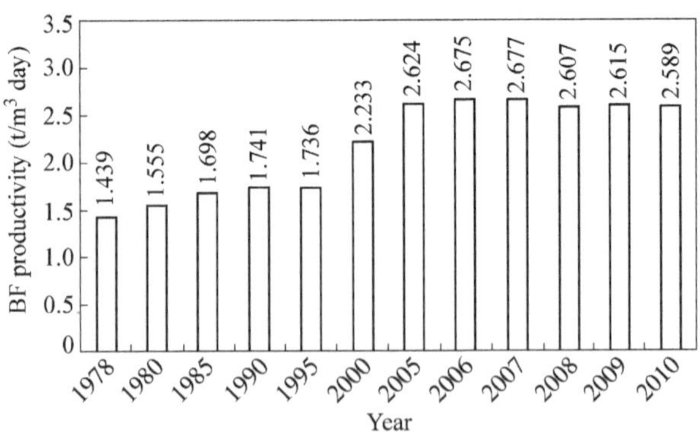

FIGURE 6.11 Change on average BF productivities in Chinese key steel enterprises.

China in 2010. It indicates that with an increase of BF volume, BF productivity and production rate per hearth area present different trends of change.

6.1.2.3 Rational BF Number in Steel Plants

According to the analysis of production efficiency and production capacity of BFs, the expected annual output of BFs with different volumes could be determined, and the reasonable number and volume of BFs for steel plants of different scales could be designed. Table 6.1 shows the production capacity and production efficiency of BFs with different volumes. For large-scale steel plants with production of steel sheets, whose capacity is normally within a range of 8–9 Mt/a, an optimal choice is to build 2–3 BFs for process structure optimization. Simultaneous operation of more than three BFs may cause scattered logistics, transportation road jams, long hot metal transfer time, and remarkable dropping of hot metal temperature, which are disadvantageous to the hot metal desulfurization pretreatment. Furthermore, it scatters and complicates the energy flow network, creates a disorder in operation, and reduces the utilization efficiency of secondary energy such as BFG. Therefore, the volume and number of BFs directly affect the production efficiency of steel plants, which is the optimization of the mass flow network, energy flow network, and information flow network in a steel plant. BF enlargement could be beneficial in order to reduce BF number, in order to build a simpler and smoother process for improving energy efficiency and emission reduction, which is also conducive to information control.

6.1.3 A Comparison of Technological Equipment of BFs with Different Volumes

BF enlargement has technical advantages in high efficiency and intensiveness, energy saving and emission reduction, informationization control, etc. Specifically, BF enlargement is beneficial to achieve the three functions, namely production of hot metal with high quality, high energy efficiency, and effective waste treatment and recycling. At the same time, BF enlargement

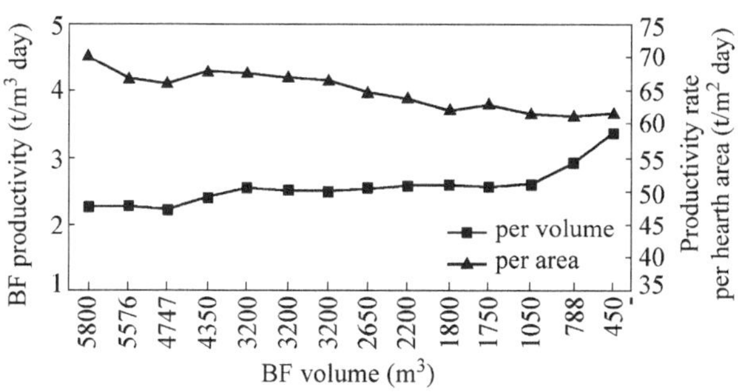

FIGURE 6.12 Situation of typical BF utilization in China in 2010.

TABLE 6.1 Production Capacity and Expected Annual Output of BFs at Different Levels (Zhang and Yin, 2002; Qian et al., 2011)

BF Volume (m^3)	1260	1800	2500	3200	4080	4350	5000	5500
BF productivity ($t/(m^3 . day)$)	2.5–2.7	2.4–2.6	2.4–2.6	2.3–2.5	2.2–2.4	2.2–2.4	2.1–2.3	2.1–2.3
Production per area ($t/(m^2 \cdot day)$	61.16–66.05	60.98–66.06	60.93–66.01	60.98–66.28	61.51–67.10	61.32–66.89	61.90–67.79	62.04–67.95
Annual operating days (days)	350	350	350	355	355	355	355	355
Desired annual output (Mt/a)	1.10–1.19	1.51–1.63	2.10–2.27	2.61–2.84	3.12–3.40	3.39–3.70	3.72–4.08	4.10–4.49

has been driving the development of ironmaking technical equipment, the development of large-scale metallurgical equipment and refractory technology, as well as the synergetic development of many other techniques, such as informationization technology, long campaign life of BF, concentrate, bell-less top, cast house equipment, high blast temperature, PCI with enriched oxygen, dry dedusting of BFG, and TRT.

Therefore, for steel plants with different production scales and product mix, there are various technical proposals for the determination of BF production capacity, number, and volume. For steel plants with different production scales and different product mixes, such as 2–3 Mt/a construction bar/wire rod mill or plate/strip mill, 4–6 Mt/a strip mill (sheet or sheet + plate), 8–9 Mt/a large-scale strip mill, the determination of BF number and volume must be made on the basis of a comprehensive consideration of the different levels of the top level design of the entire plant in order to realize mass flow, energy flow, and information flow at their own reasonable process flow network in coordinated and high-efficiency operation and carrying BF enlargement with a premise of integrated production process structure optimization of steel plants.

For a steel plant with an annual output of 900 Mt/a, three 4000 m^3 BFs or two 5000 m^3 BFs are the available choices. The above two technical proposals are compared, analyzed, and studied in this book. Shougang Qian'an Steel 4080 m^3 BF and Shougang Jingtang Steel 5576 m^3 BF are compared by technical and economical indexes, crude fuel adaptability, energy and power consumption, electric installed capacity, project investment, production management and operation cost, occupied land of general layout, energy saving, and environmental protection.

6.1.3.1 Main Technical and Economical Index of Large-Scale BF

Table 6.2 lists the main technical and economical indexes of three newly built large-scale BFs in China, indicating that the 4000 m^3 BF falls considerably behind the 5500 m^3 BF in fuel ratio and other critical indexes.

6.1.3.2 Adaptability to Crude Fuel for BFs Larger Than 4000 m^3

After investigation and survey of major Japanese steel plants, it is revealed that the BF enlargement and construction of BFs over 5000 m^3 in Japan were achieved mainly by means of volume expansion, overhaul, and revamping of existing BFs. The volumes of many BFs has been expanded from the original 4000 m^3 to over 5000 m^3, while neither the procedures like coking, sintering, etc., nor crude fuel conditions were fundamentally changed. The practices in Japan show that the crude fuel conditions for 4000 m^3 BFs could meet the production requirement of 5000 m^3 BFs.

Fig. 6.13 is a comparison of the profile and effective height between a 4350 m^3 BF and a 5576 m^3 BF. The two BFs differ in volume by 1226 m^3,

TABLE 6.2 Main Technical and Economical Indexes of Three Large-Scale BFs

Items	4350 m^3 BF		4080 m^3 BF		5576 m^3 BF	
	Design value	Actual value in 2009	Design value	Actual value in 2010	Design value	Actual value in 2010
BF productivity (t/(m^3 · day))	2.1	2.2	2.4	2.387	2.3	2.37
Production per hearth area (t/m · day)	57.71	60.46	68.44	68.07	67.07	69.12
Coke ratio (kg/t)	320	307	305	331.2	290	305
PCI (kg/t)	200	197	190	167.8	200	175
Fuel ratio (kg/t)	520	504	495	499	490	480
Blast temperature (°C)	>1250	1242	1280	1280	1300	1300
Top pressure (MPa)	0.25	0.25	0.25	0.25	0.28	0.28
Oxygen enrichment ratio (%)	≤3	3	3.5–5	4.17	3.5–5	4.0
Agglomeration ratio (%)	95	-	90	90	90	90
Occupied land area of single BF (10^4 m^2)	28.200	-	-	-	31.925	-

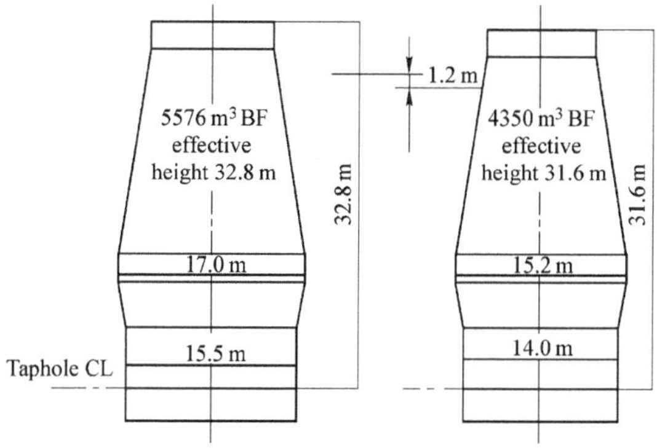

FIGURE 6.13 Comparison of inner profile between 4350 m^3 BF and 5576 m^3 BF.

but the difference of effective length is only 1.2 m due to the difference of hearth volume. There is little difference in the height of material columns. Therefore, compared with the 4000 m^3 BF, the requirements of the 5000 m^3 BF on coke mechanical strength (M40, M10), Coke Reactivity Index (CRI), and Coke Strength after Reaction (CSR) are not remarkably rigorous.

The requirements of crude fuel conditions for BFs are stipulated in the code for design of BF ironmaking technology (GB50427-2008) (China Metallurgical Construction Association, 2008). It can be seen from Tables 6.3 and 6.4 that there is no remarkable difference between the requirements of crude fuel conditions for a 4000 m^3 level BF and a 5000 m^3 level BF.

6.1.3.3 Comparison of Technological Equipment

Table 6.5 shows the comparison of technological equipment between the 2×5576 m^3 BFs and the 3×4080 m^3 BFs. Compared with the 3×4080 m^3 BFs, it is the 2×5576 m^3 BFs that apparently have lower quantities and items of equipment: the quantity of equipment is 15% lower; total equipment weight is 14,820 t, 22.6% lower; and the main charging belt conveyor is shorter by 118.8 m. With a decrease of equipment quantity, the equipment investment for operation, maintenance, and consumption of spare parts is reduced. Power consumption, job-site manpower, pollutant emissions, etc. are also lowered, reducing the cost of production and operation.

6.1.3.4 Energy Consumption, Energy Conservation, and Environmental Protection

Table 6.6 shows a comparison of energy consumption, energy conservation, and environmental protection and installed capacity between 4350 m^3 BF, 4080 m^3 BF, and 5576 m^3 BF.

It is shown that the 5576 m^3 BF has advantages in water consumption, fuel consumption, installed capacity of motor, power consumption, cleanness,

TABLE 6.3 Requirements of Material Quality for 4000 m^3 Level BF and 5000 m^3 Level BF (China Metallurgical Construction Association, 2008)

Items		4000 m^3 BF	5000 m^3 BF
Ore grade at BF entry (%)		≥59	≥60
Agglomeration ratio (%)		≥85	≥85
Sinter	Grade fluctuation (%)	≤±0.5	≤±0.5
	Fluctuation of basicity (%)	≤±0.08	≤±0.08
	FeO content (%)	≤8.0	≤8.0
	Fluctuation of FeO content (%)	≤±1.0	≤±1.0
	Tumbler index +6.3 mm (%)	≥78	≥78
Pellet	Iron content (%)	≥64	≥64
	Tumbler index +6.3 mm (%)	≥92	≥92
	Abrasion index −0.5 mm (%)	≤4	≤4
	Normal temperature compressive strength (N/pellet)	≥2500	≥2500
	RDI +3.15 mm (%)	≥89	≥89
	Swelling index (%)	≤15	≤15
Lump	Grade (%)	≥64	≥64
	Hot decrepitation (%)	<1	<1
	Fe variation (%)	≤±0.5	≤±0.5

TABLE 6.4 Quality Requirement of Fuel for 4000 m^3 Level BF and 5000 m^3 Level BF

Items	4000 m^3 BF	5000 m^3 BF
Crushing strength M_{40} (%)	≥85	≥86
Abrasion index M_{10} (%)	≤6.5	≤6.0
CSR (%)	≥65	≥66
CRI (%)	≤25	≤25
Ash content of coke (%)	≤12	≤12
Sulfur content of coke (%)	≤0.6	≤0.6
Grain size range of coke (mm)	75-25	75-30
More than the upper limit (%)	≤10	≤10
Less than the lower limit (%)	≤8	≤8

TABLE 6.5 Comparison of Technological Equipment Between $2 \times 5576\ m^3$ BFs and $3 \times 4080\ m^3$ BFs

Items	$3 \times 4080\ m^3$ BF	$2 \times 5576\ m^3$ BF
Charging system/set	3	2
Length of main charging belt conveyor (m)	1014.0	895.2
Top charging equipment/set	3	2
BF proper/set	3	2
Cast house system/set	3	2
Dust catcher system/set	9	6
Hot blast stove system/set	12	8
Granulated slag treatment system/set	12	8
BFG dedusting system/set	39	30
Top pressure power generation (TRT)/set	3	2
Blower/set	4	3
BF control system/set	3	2
BFG holder/set	3	2
Total weight of equipment (t)	65,433	50,613

and environment protection, and especially in energy conservation and environmental protection, compared with the 4000 m^3 BFs (Table 6.6).

6.1.3.5 A Comparison of Project Investment and Land Occupation in General Layout

The 5576 m^3 BF of Shougang Jingtang Steel and the 4080 m^3 BF of Shougang Qian'an Steel all belong to Shougang Group. They were designed by the same engineering company and were completed and put into operation in roughly the same period, enabling them to be compared reasonably. A comparison in project investment and land occupation in general layout is performed between the technical proposal of establishing $2 \times 5576\ m^3$ BFs and that of establishing $3 \times 4080\ m^3$ BFs. The results show that under the same conditions, the construction of $2 \times 5576\ m^3$ BFs could reduce the investment by about 100 million USD, which means a decrease of 12–14%. Meanwhile, because of the application of large-scale equipment with compact and intensive process flow, the occupied land area of the BF zone was reduced by 158,100 m^2, which represents a decrease of 20–24%. Reduction of occupied land not only saved land resources, but also greatly shortened

TABLE 6.6 A Comparison of Consumption of Energy/Power, Energy Conservation, and Environmental Protection Between 4350 m³ BF, 4080 m³ BF, and 5576 m³ BF

Items	4350 m^3BF	4080 m^3 BF	5576 m^3 BF
Technological Equipment			
BFG dedusting	Dry/wet method simultaneously	Fully-dry method	Fully-dry method
Installed capacity of TRT (MW)	24.3	30.0	36.5
Average hot blast temperature (°C)	1242	1280	1300
Dust catcher facilities	Dedusting system is completely provided with 100% operation rate		
Indexes of Energy and Resource			
Energy consumption of BF (kg_{ce}/t)	391.5		373
Coke ratio (kg/t)	307	331.2	269
Coke nut ratio (kg/t)			36
PCI (kg/t)	197	167.8	175
Fuel ratio (kg/t)	504	499	480
Fresh water (m^3/t)	9	0.6	0.49
Water reusing rate (%)	98	98	98
Product Index			
Yield of pig iron (%)	100	100	100
Indexes of Pollutant Emission			
Dust (kg/t)			≤0.10
SO_2 (kg/t)			≤0.02
Waste water (m^3/t)	0	0	0
Slag (kg/t)	280	299.6	250–270
Utilization Index of Wastes Recovery			
Coke nut recovery	Recovery devices for coke nut and iron ore nut are provided		
Recovery rate of BF slag (%)	100	100	100
Recovery rate of BFG dust (%)	100	100	100
Installed capacity of single BF motor (kW)		26,205	26,476

TABLE 6.7 Comparison of Project Investment Between 5576 m^3 BF and 4080 m^3 BF

Items	Shougang Qian'an Steel 3#BF	Shougang Jingtang Steel 1# BF
BF volume (m^3)	4080	5576
Investment (billion USD)	0.298	0.367
Unit BF volume investment (10^4 USD/m^3)	7.31	6.67
Occupied land area of single BF (10^4 m^2)	26.500	31.925

TABLE 6.8 Occupied Land of General Layout and Project Investment Between 2 × 5576 m^3 BFs and 3 × 4080 m^3 BFs

Items	2 × 5576 m^3 BFs	3 × 4080 m^3 BFs
Length of occupied land (m)	1243	1506
Width of occupied land (m)	695	626
Area of occupied land (m^2)	63.85×10^4	79.50×10^4
Area of occupied land per ton of hot metal (m^2/t)	0.071	0.088
Project investment under the same conditions (million USD)	733	895

the transport distance of materials, energy, and medium, which reduced operation costs and improved production efficiency, see Tables 6.7 and 6.8.

6.1.4 Discussions

Based on the issues mentioned above, the following conclusions can be made:

1. In large-scale modern strip mills with a production capacity of 8–9 Mt/a, with the premise of process structure optimization in steel plants, the process network optimization of ferrite mass flow, energy flow, and information flow should be considered. Compared with the construction of three 4000 m^3 level BFs, building two 5000 m^3 level BFs offers more advantages in energy saving and emission reduction, cleanness and environment protection, investment saving, economic efficiency, etc.

2. After comparison and study of technical proposals of three 4080 m^3 BFs and two 5576 m^3 BFs, it shows that the latter requires lower quantities and weight of equipment: the quantity of equipment is lowered by 15%, and the total equipment weight is lowered by 14,820 t (22.6%). With the reduction of equipment quantity, the equipment investment of operation, maintenance, and consumption of spare parts is also reduced. Power consumption, job-site manpower, pollutant emission, etc. are also lowered, reducing the cost of production and operation.
3. Under the same conditions, the construction of two 5576 m^3 BFs could reduce the investment by about 100 million USD, which means a decrement of 12–14%. Meanwhile, because of the application of large-scale equipment with compact and intensive process flow, land occupation of BF was reduced by 158,100 m^2, a decrement of 20–24%. Reduction of occupied land not only saved land resources, but also greatly shortened the transport distance of materials, energy, and medium, which in turn reduced operation cost and improved production efficiency.
4. BF enlargement and rationalization of BF quantity and position on the premise of process structure optimization in steel plants have become a trend. The trend also provides a guide for small and medium scale steel plants for the production of construction bar/wire rod products.

6.2 INTERFACE TECHNOLOGY BETWEEN BF–BOF AND MULTIFUNCTIONAL HOT METAL LADLE

Historically, BF has been operated to cooperate with Open Hearth Furnace (OHF) over a long period; therefore, the mixer was built to hold the surplus hot metal in most steel plants. The surplus hot metal is caused by the discordant operation schedule of BF and OHF, so that it has been difficult to coordinate the duration of the hot metal tapping with the tap-to-tap period of the OHF. The mixer has now been eliminated because of its high energy consumption, serious pollution, and high investment. It was replaced by a torpedo car to coordinate the BF operation with BOF. But the torpedo car still involves high investment, serious pollution, and a large temperature drop. Moreover, the torpedo car is not fit for hot metal pretreatment and deslagging. Therefore, the suitability of the torpedo car has been called into question. There were suggestions to develop a new interface technology between the BF–BOF sections, based on the theory guidance of analysis-optimization of procedure functions and synergetic-optimization of procedure's relations.

6.2.1 General Idea of Multifunction Hot Metal Ladle

As mentioned above there have been questions raised over the use of the mixer and torpedo car—can they be replaced by a hot metal ladle or not?

It should be answered based on the function analysis of the ladle: which function should be provided with it, and how to multifunctionalize it. After consideration and research, with the evolution of the steel manufacturing process, the hot metal ladle and its transport system in the steel plant with a long process should have the following functions:

1. The function of accepting hot metal in time and reliably;
2. The function of transporting hot metal stably, reliably, and quickly;
3. The function of storing (buffer) hot metal within a certain period of time;
4. The function of favorable deslagging and desulfurization of hot metal;
5. The function of fine heat preservation (thermal insulation);
6. The function of weighing hot metal accurately and reliably;
7. The function of precisely informing ladle location and rapid turnover of the empty ladle.

It can be seen clearly that the interface technology of the BF–BOF sections is close to the multifunctionality of the hot metal ladle (the so-called One Ladle Technology). Moreover, simply to replace the torpedo car with hot metal ladle is not enough, it requires a series of studies of technological development and engineering design.

The integrated technologies package of the multifunctional hot metal ladle should include the following technologies and should be designed from an engineering viewpoint.

1. The calculation for the gravity center and reasonable construction of large volume hot metal ladle;
2. Study on the stability and safety of the transportation equipment, transportation system, and the transporting process of the large volume hot metal ladle;
3. Compatible design of the height of the BF tapping groove and hot metal ladle;
4. Design of reasonable layout and study of the spatiotemporal operation program of hot metal between BF and desulfurization station;
5. Precise weight measurement and control technologies of hot metal and hot metal ladle;
6. Study on reasonable turnover frequency of hot metal ladle (time/day per ladle) and its reasonable number of online operations;
7. Measurement of the cooling of hot metal in the ladle;
8. Measurement of the cooling of the empty ladle in operation;
9. Study on high-efficiency desulfurization of hot metal with higher temperature and higher level of activity;
10. Study on the service life of refractory materials of the hot metal ladle (study of refractory material of hot metal ladle and its service life);
11. Energy saving and smoke/dust emission reduction;
12. Economic analysis of investment, operation cost.

During the research into the multifunction hot metal ladle, the first question encountered was to compare the heat preservation function of the torpedo car and hot metal ladle. Therefore, some measurements of temperature drop for the torpedo car and the hot metal ladle were carried out in Baosteel, Shougang Qian'an Steel and Shagang Group, the results are as follows:

The measured dropping rate of the hot metal temperature of the 320 t torpedo car at Baosteel, when motionless, is 0.2–0.23°C/min; when in motion, it is 0.27–0.4°C/min. The measured dropping rate of the hot metal temperature of the 300 t hot metal ladle at Baosteel, when motionless, is 0.1–0.2°C/min; when in motion, it is 0.2–0.4°C/min (estimated). Also, the temperature dropping rate of the 45 t hot metal ladle in Shagang Group has been measured, the results together with that in Baosteel are listed in Table 6.9; Figs. 6.14 and 6.15.

Thus, the temperature dropping rate of hot metal in the large ladle is lower than that in a torpedo car of a similar size. The main reason is that more heat is absorbed by the refractory lining of the torpedo car; in addition, there is more heat loss per unit of hot metal from the surface of the torpedo car.

TABLE 6.9 Measurement of the Temperature of the Hot Metal in the Torpedo Car and the Ladle

Equipment	Status	Measured Temperature Drop
Torpedo car of Baosteel (320 t)	Motionless	0.2–0.23°C/min
	In motion	0.27–0.4°C/min
Hot metal ladle of Baosteel (300 t)	Motionless	0.1–0.2°C/min
	In motion	0.2–0.4°C/min (estimated)
Hot metal ladle of Shagang Group (45 t)	Motionless	0.79°C/min
	In motion	0.89°C/min

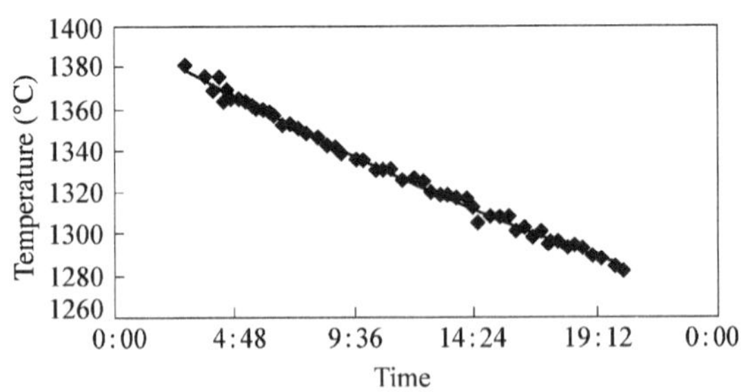

FIGURE 6.14 Dropping rate of the hot metal temperature in the motionless 300 t hot metal ladle.

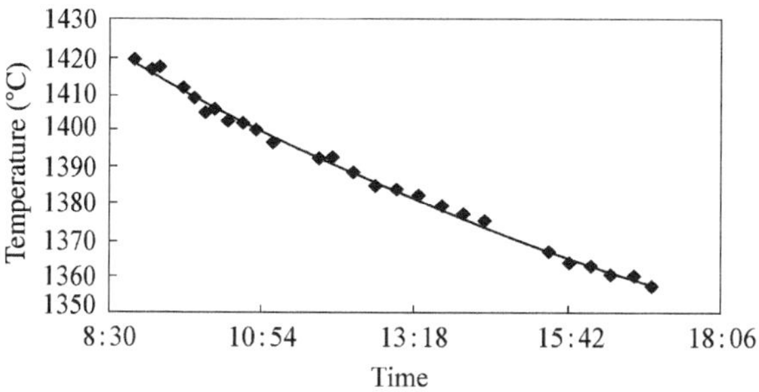

FIGURE 6.15 Dropping rate of the hot metal temperature in the motionless 320 t torpedo car.

As hot metal is transferred after the multifunction hot metal ladle, its temperature and [Si] content and [C] content are all higher than that with desiliconization pretreatment in the tapping groove. In addition, the activity coefficient of [S] in hot metal is larger, which is beneficial for improving the desulfurization efficiency of the pretreatment of hot metal.

Nowadays, the economic benefits of multifunction ladle technology (so-called One Ladle Technology) have become notable in some steel plants in China.

6.2.2 Multifunction Hot Metal Ladle and Its Practice at Shougang Jingtang Steel

There are two large BFs, 5576 m^3 each, at Shougang Jingtang Steel. The De [S] – De[Si]/[P] pretreatment has been adopted. There is a multifunction hot metal ladle and railway transportation system with standard rail gap of 1345 mm in dynamic-orderly operation between the BF–BOF sections.

In order to build the integrated operation system with two 5576 m^3 BFs and four KR De[S] stations (in fact, it is a dynamic operation system consisting of eight hot metal iron notches, four KRs, and two $BOF_{De[Si]/[P]}$), the multifunction ladle with capacity of 300 t has been adopted for accepting, transporting, buffering, heat preserving, De[S]/deslagging, and charging hot metal into BOF.

During the research of the multifunction ladle, the design of the ladle should be optimized, including its capacity, shape, and height at first, and especially its gravity center calculation. The height of the ladle should fit with the height of hot metal tapping groove of BF; the capacity of the ladle should be in accordance with the hot metal charging requirement of the BOF and there should be precise weight measurement to ensure an accurate charging; the gravity center position of ladle should be taken into account for the stability and safety of the transportation process, especially the transporting safety on the curving railway; and the heat preservation of the ladle should be taken into account, especially the heat preservation of empty ladle. Moreover, the dynamic operation on railways (or roads, rails) of the hot metal ladle and its transporting car should be carried out to measure the corresponding parameters and the stability and safety of these parameters.

The hot metal transporting vehicle with sixteen axles was specially developed to stably and safely transport a hot metal ladle with 300 t capacity on rail with the standard rail gap of 1345 mm, and the tracing and locating system of the ladle has been equipped throughout the whole process. A hot metal weight measurement system and detection system with higher precision have been developed to meet the requirements of accuracy control between ± 1 t of the 300 t target charging weight of the hot metal ladle.

The optimized design of the railway web system for the transportation of hot metal from eight iron notches of two BFs to four KR stations was very important to achieve the coordinated-compacted spatiotemporal relationship of this area. Therefore, the compacted-simplified transportation web system of the hot metal ladle was designed, with a distance of hot metal transportation between 1270 and 1840 m, and transportation time within 20 min. It has resulted in a higher turnover frequency of the hot metal ladle and a higher temperature of hot metal arriving at the KR stations, ensuring the De[S] temperature above 1380°C and thus helping to improve the De[S] efficiency.

The design and operation of the multifunction ladle and its transportation system has been remarkably beneficial at Shougang Jingtang Steel, the details are as follows:

1. The turnover frequency increased continually. The turnover system of the hot metal ladle starts with the ladle charged with hot metal from BF, then it is transported to KR station to Deslagging–De[S]–Deslagging, and transported by crane and charging hot metal to $BOF_{De[P]}$. After the cleaning maintenance, it is marshaled by the marshaling station, and then returned to BF to wait for the next charge. The reasonable organization of this system will greatly influence the temperature of hot metal tapping into the ladle and its De[S] pretreatment. With continual improvement, the turnover duration of the ladle has been shortened to within 360 min, and the ladle turnover frequency can be four times per day at Shougang Jingtang Steel. Figs. 6.16 and 6.17 show the turnover duration of its hot metal ladle.

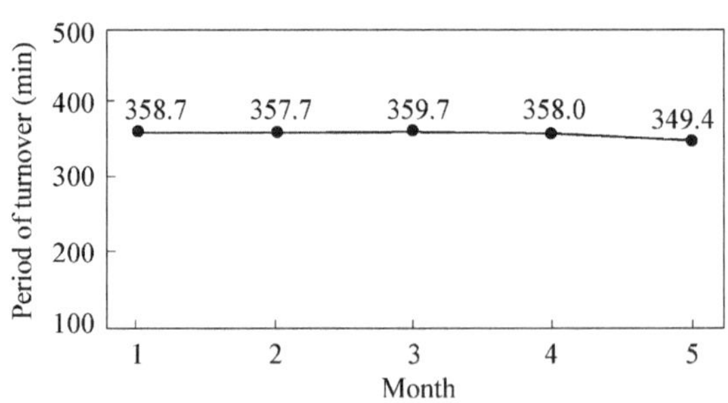

FIGURE 6.16 Average turnover period of hot metal ladle at Shougang Jingtang Steel from January to May in 2012.

2. The weight of hot metal in the ladle can be precisely controlled within a range of 288 t ± 1 t, as the hit rate of the weight of 288 t ± 0.5 t has been 95% since 2012 (Fig. 6.18), which is beneficial for the precise, stable operation of the subsequent processes. It will be good for speeding up the turnover frequency and improving the management of the hot metal ladle, since there is no half ladle of hot metal.
3. The rapid turnover of the ladle can decrease the temperature drop of the hot metal from BF tapping to the KR station (the temperature drop can be reduced by 30–50°C). At present, the turnover frequency of the hot metal ladle is only four times/day at Shougang Jingtang Steel, which can still be improved. Even so, the temperature of hot metal arriving at KR station is above 1380°C (Figs. 6.19 and 6.20). Compared with hot metal transported by torpedo car, the temperature drop is reduced by 30–50°C.

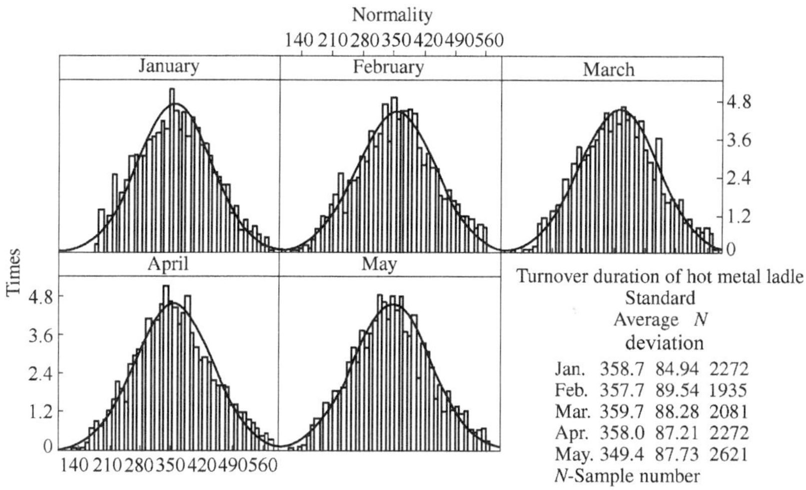

FIGURE 6.17 Probability distribution of turnover duration of hot metal ladle from January to May in 2012, Shougang Jingtang Steel.

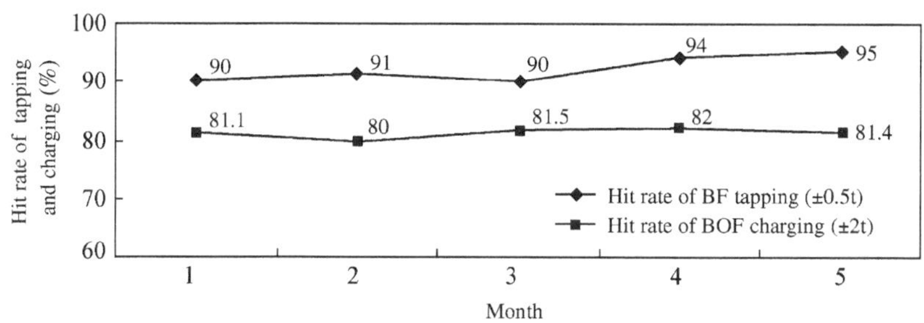

FIGURE 6.18 Accurate tapping ratio of BF tapping and BOF charging, from January to May in 2012 at Shougang Jingtang Steel.

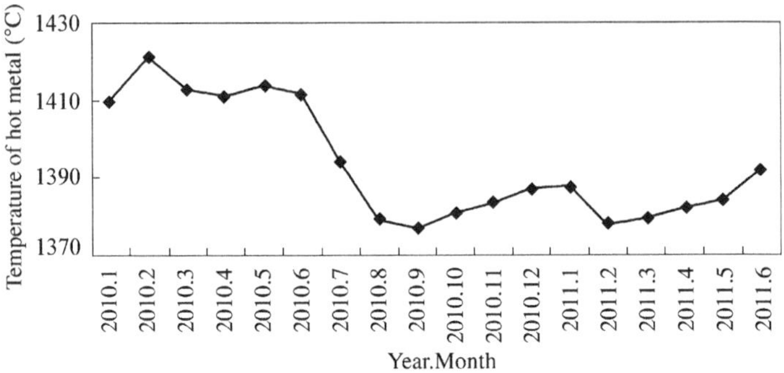

FIGURE 6.19 Temperature of hot metal at start of KR station since 2010, Shougang Jingtang Steel (Yang et al., 2012).

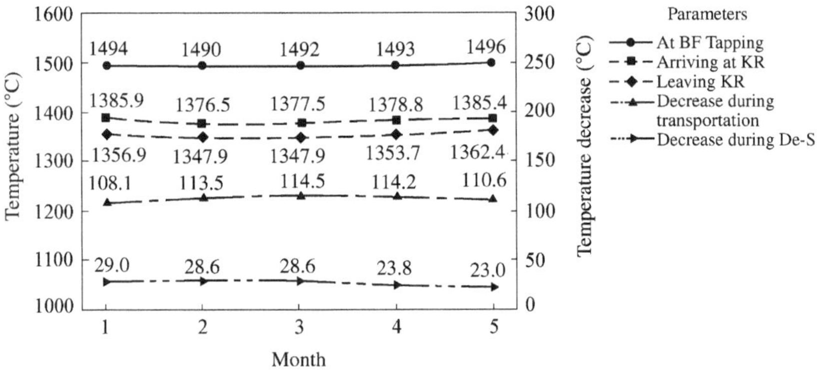

FIGURE 6.20 Temperature drop of hot metal during transportation and De[S] process, January–May 2012, Shougang Jingtang Steel.

4. The KR process has a favorable desulfurization effect and lower consumption of desulfurizing agent. The pretreatment period of De[S] has been shortened and the De[S] efficiency has been improved in the KR station. Moreover, it is easy to deslag in KR. There is a certain relationship between the pretreatment period of De[S] and the hot metal temperature. The higher hot metal temperature supports a higher efficiency of De[S] and a lower consumption of desulfurizing agent, and shortens the corresponding stirring time and deslagging time of the hot metal. Normally, the De[S] period of the KR is about 30–35 min (Fig. 6.21), and the stirring time of hot metal-slag is about 10 min (Fig. 6.22) at Shougang Jingtang Steel. The relationship between the desulfurizing agent consumption and the hot metal temperature is also obvious. Desulfurizing agent consumption is only about 7 kg/t-HM. The temperature drop of the hot metal during transportation and the De[S] process from January to May 2012 is shown in Fig. 6.20, indicating that the total

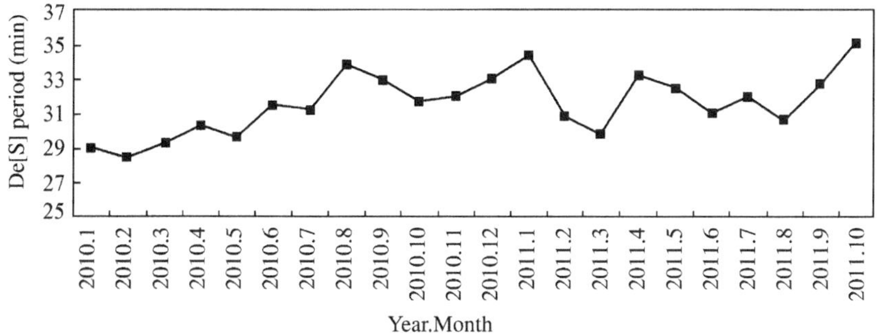

FIGURE 6.21 De[S] period monthly since 2010 at Shougang Jingtang Steel (Yang et al., 2012).

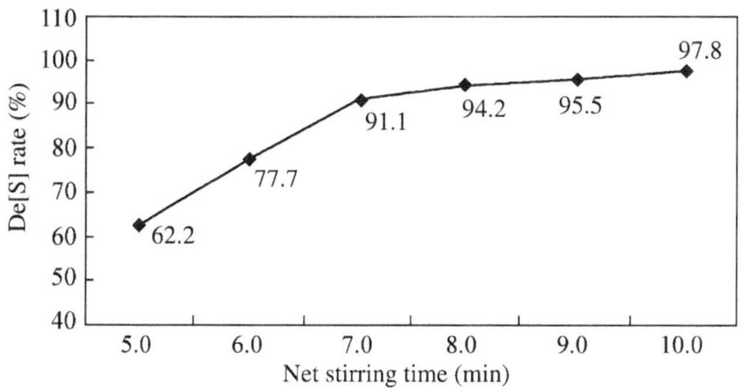

FIGURE 6.22 Relationship of De[S] efficiency with stirring time of KR in 300 t ladle at Shougang Jingtang Steel.

temperature drop of the hot metal is about 110°C; and the temperature drop caused by KR pretreatment is about 25–30°C.

The temperature of the hot metal at the KR station with the "One Ladle Technology" is normally above 1380°C. The process is De[S] first and then De[Si], and the activity of [S] in the hot metal is higher. Hence, over 50–60% heat achieves a [S] level of $\leq 5\times10^{-6}$ after KR hot metal treatment (see Fig. 6.23).

As shown in Fig. 6.23, from January to May, the ratio of heats with $[S]\leq 5\times10^{-6}$ and $[S]\leq 25\times10^{-6}$ after KR is about 59.1% and 97.06%, respectively. It shows that, the application of "One Ladle Technology" results in a higher temperature and activity of hot metal, therefore, this improves the efficiency and stability of De[S] in KR.

5. The investment benefit from the multifunction ladle is good. By adopting the multifunction ladle ("One Ladle Technology"), the torpedo car, transfer crane, transfer station, dedusting system of the station, and so on have been abolished. It has been calculated at Shougang Jingtang Steel that investment can be reduced by 6.93 million dollars; dust emission can be reduced by 37.1 kt/a; and electricity consumption can be reduced by 11.39 MWh/a.

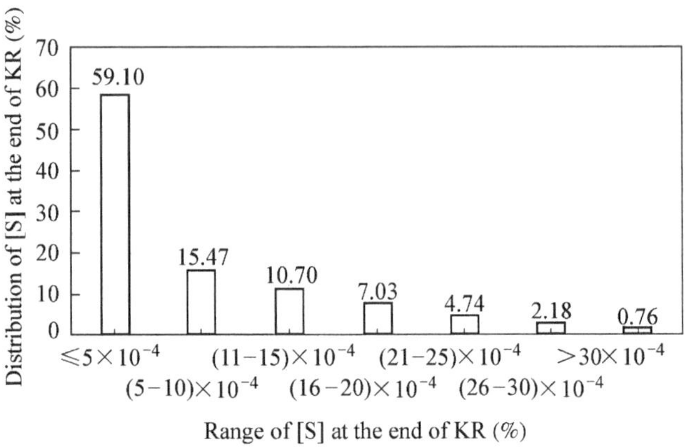

FIGURE 6.23 Distribution of [S] at the end of KR, January–May, 2012, Shougang Jingtang Steel.

6.2.3 Practice of Multifunction Hot Metal Ladle at Shagang Group

Shagang Group is one of the first plants to adopt the concept of the multifunction ladle, and it developed several different modes of "One Ladle Technology" in different processes. Some primary modes are as follows:

1. $1 \times 5800\ m^3$ BF (3 iron notches)−3 × KR station−3 × 180 t BOF (hot metal transported by standard rail gap with 1435 mm);
2. $3 \times 2500\ m^3$ BF (6 iron notches)−3 × Mg powder injection De[S] station−3 × 180 t BOF (hot metal transported by standard rail gap with 1435 mm);
3. $5 \times 380\ m^3$ BF (5 iron notches)−3 × 40 t BOF + 1 × 70 t EAF (hot metal transported by truck);
4. $4 \times 450\ m^3$ BF (4 iron notches)−4 × 100 t EAF (hot metal transported by truck).

The optimization of the interface technology and the mode of the multifunction ladle were developed based on the practice of the ladle being transported by truck, one truck for one ladle, between the blast furnace and steelmaking furnaces, such as the small BF−EAF process (4) and the small BF−small BOF process (3). And the multifunction ladle technology was further integrated and developed when it was applied to the large BF−large BOF process at Shagang Group. It has been innovated in the following aspects:

1. Innovation of high-efficiency interface technology connecting BF−BOF, leading to a simplified general layout, smooth operation of mass flow, prominent effect of energy saving and emission reduction; the

multifunction ladle, with functions of accepting hot metal, weight measurement, transportation, buffering, De[S] pretreatment, and accuracy charging to BOF was adopted. There is no transferring of hot metal during the whole process from the ladle accepting hot metal from BF to BOF, and the hot metal was directly charged into BOF after pretreatment in the same ladle. The torpedo car and marshaling station were abolished. The original hot metal weight measurement equipment between BF and BOF were replaced with an online weight measurement under BF. The hot metal transferring station and its dedusting system were abolished, as well. The compact arrangement during the section creates the shorter transportation route, smoother operation of mass flow, and less hot metal temperature drop.

2. Innovation of transportation of a super-height and super-width ladle by a standard-gage railway. Through the simulation of dynamics characteristics of a super-height and super-width ladle, "four consistencies" innovations were confirmed: (1) the ear axis of the heavy ladle and the vertical axis of the car are in the same direction; (2) the ear axis and the railway are in the same direction; (3) the moving direction of the ladle is parallel to the pillar column of the steel plant building; (4) the ear axis and the moving direction of the charging crane in the steelmaking workshop are the same.
3. Innovation on technology integration of accuracy control of hot metal weight tapping to the ladle. The weight of hot metal tapped into the ladle has been accurately controlled for every ladle to ensure stable charging to BOF. A management mechanism of hot metal weighting by the ironmaking plant has been established. Ten integrated technologies, such as weighing rail without foundation pit, online continuous weighing system, loop tracking weight of empty ladle, were developed to guarantee the precise measurement of hot metal weight and the high hit rate of the BOF charging weight.
4. Innovation on rapid turnover technology of the hot metal ladle. The marshaling station of the train was cancelled, through rapid marshaling of the hot metal ladle by the crane of the steelmaking plant. A management mechanism of hot metal ladle operation controlled at steelmaking plant was innovated, which is beneficial to the rapid turnover of the hot metal ladle, rapid online marshaling, rapid maintenance, and rapid operation of the hot metal ladle. And it guarantees the stability, high efficiency, and low cost of hot metal desulfurization. The online tracking system of hot metal information and the operation schedule of hot metal ladle were developed, which can guarantee that the waiting duration of an empty ladle is less than 3 h; therefore, the number of hot metal ladle online operations can be significantly decreased, and rapid turnover of the hot metal ladle was achieved. The average turnover frequency of the hot metal ladle is 5.6 times/day.

Based on these studies and analysis, the results are as follows:

1. Study on spatiotemporal relationship of BF–pretreatment of hot metal–BOF. Railway transportation of hot metal has been adopted from 5800 m^3 to 3 × 180 t BOFs; the layout of the railway at Shagang Group is shown in Fig. 6.24.

   The railway between BF and BOF is the route of hot metal transportation, with a distance of about 1500 m, and includes three transportation trains. It can be seen from Fig. 6.24 that the starting lines of the full hot metal ladle are S1–S6, the main lines are G1–G8, and the end lines are E1–E3.

   The layout of two BFs is shaped as a peninsula, and hot metal produced by BF is all transported by railway. Once the "One Ladle Technology" mode of transportation was adopted, the 180 t ladle full of hot metal was moved from the tapping area to the line G5–G6, then pushed by train to the steelmaking plant without a transfer operation, and after De[S] pretreatment at the KR station, the hot metal is charged into BOF rapidly; therefore, the temperature drop and heat loss are distinctly reduced, which is beneficial to improving the efficiency of De[S]. The railway line G5 of 5800 m^3 BF is linked up to the railway system of 2500 m^3 BF in the northern area, so that hot metal produced by 5800 m^3 BF can be sent to not only the No.2 steelmaking workshop, but also to the No.1 steelmaking workshop in the northern area. At the same time, hot metal from 2500 m^3 BF in the northern area can be sent to the No.2 steelmaking workshop in the southern area as well. Therefore, it is flexible in being able to allocate the hot metal between two systems.

   Fig. 6.24 shows that there are four transport paths of different distances for hot metal transportation and empty ladle transportation between the large BF and large BOF interface at Shagang Group (the real distance of ladle transportation is 1287 m, 1231 m, 1378 m, 1269 m, respectively), as listed in Tables 6.10 and 6.11.

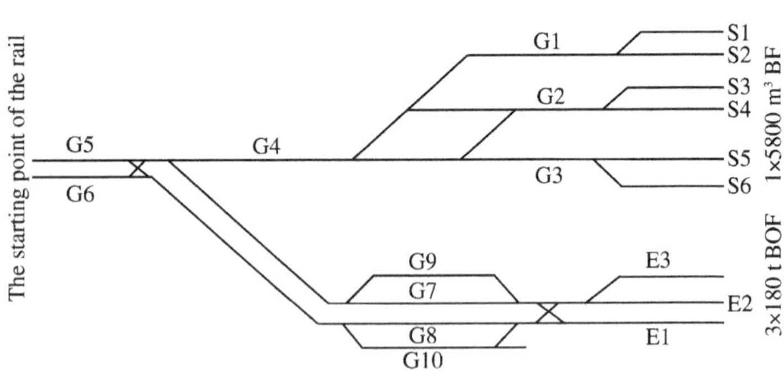

FIGURE 6.24 Schematic of hot metal transportation railway between 1 × 5800 m^3 BF and 3 × 180 t BOF section.

TABLE 6.10 Railway of Hot Metal Transportation by 180 t Ladle from 5800 m^3 BF at Shagang Group

Route 1	S1	G1	G1–G4	G4	G5	G7	E1
Route 2	S3	G2	G4	G6	G7	E2	
Route 3	S3	G2	G2–G4	G4	G5	G7	E2
Route 4	S5	G3	G4	G5	G8	E3	

TABLE 6.11 Railway of Empty 180 t Ladle Returning Back to 5800 m^3 BF at Shagang Group

Route 1	E1	G7	G5	G4	G1–G4	G1	S1
Route 2	E2	G7	G6	G4	G2	S3	
Route 3	E2	G7	G5	G4	G2-G4	G2	S3
Route 4	E3	G8	G5	G4	G3	S5	

The 180 t hot metal ladle is transported by a standard railway at Shagang Group, accepting hot metal from the three iron notches of the BF, along the railway line connecting with 3 × KR De[S] stations. Since the transporting speed of the train should be kept at less than 20 km/h, and the operation of the train is so busy, it can lead to a traffic jam during transportation, sometimes lengthening the transportation time. Since the ladle is open mouthed, the hot metal could possibly spatter out during the transverse deflection or longitudinal dashing of the ladle. Therefore, the area around the railway line should be kept clear.

Based on the conditions of the layout shown in Fig. 6.24, the ladle transportation mechanism of "one by one" was adopted. When the speed of the train with a full ladle and empty ladle are 10 km/h and 15 km/h, the occupation rate of every line of the railway system was calculated (Table 6.12).

In Table 6.12, which shows the occupation rate per day-night of every line, the maximum occupation rate is G4, so that the transportation capacity is determined by G4. The wait time of the lines and abnormal factors are not taken into account, so the results calculated are ideal occupation rates. Actually, under normal conditions, the occupation rate of every line is 10–20% more than the ideal (average wait duration occupies 50% of total transporting duration; average wait duration occupies

TABLE 6.12 Operation Time of Every Rail Line per Day-Night

Name f route	S1	S2	S3	S4	S5	S6	G1
Time (h)	2.63	2.63	2.37	2.37	3.30	3.30	5.14
Name f route	G2	G3	G4	G5	G6	G7	G8
Time (h)	4.72	7.44	14.49	8.01	8.01	11.33	11.33
Name f route	E1	E2	E3				
Time (h)	7.65	7.65	7.65				

TABLE 6.13 Hold Up Time and Occupation Rate of G4 Related to the Tapping Times of 5800 m^3 BF at Shagang Group

Tapping Times of BF Day and Night	G4 Section		
	Hold up duration (min)	Ideal occupancy rate (%)	Actual occupancy rate (%)
8	579.6	40.3	48.3
10	724.5	50.3	60.3
12	869.3	60.4	72.4
14	1014.2	70.5	84.5

40% of total duration needed by transporting one ladle). The hold-up duration and occupation rate of G4 related to tapping times of BF at Shagang Group are listed in Table 6.13, based on the layout shown in Fig. 6.24.

According to the results listed in Table 6.13, when the tapping times of BF is 10, the possible occupancy rate of the railway is 60.3% with the conditions of $1 \times 5800\ m^3$ BF-3×180 t BOF at Shagang Group.

2. The operation of the accurate tapping ratio of hot metal tapping. In order to ensure the operation effect of "One Ladle Technology," the hit rate of tapping should be kept at a higher level. Therefore, a management mechanism of the hot metal weighing measurement during charging by the ironmaking plant has been built at Shagang Group; and 10 integrated technologies, such as weighing rail without foundation pit, online continuous weighing system, loop tracking weight of empty ladle, were developed to guarantee the precise measurement of the hot metal weight and the high hit rate of the BOF charging weight. At present, the weight of

the tapping hot metal in the ladle was precisely controlled at 163 t ± 1 t at Shagang Group (Figs. 6.25 and 6.26).

From Fig. 6.25, the accurate tapping ratio of hot metal tapped to the 180 t hot metal ladle corresponding to the 5800 m^3 BF was controlled at 163 t ± 0.3 t by about 54%; and the hit rate of the hot metal tapping was controlled at 163 t ± 0.8 t by about 92%; while the hit rate of the hot metal tapping was controlled at 163 t ± 1 t by 100%.

From Fig. 6.26, the accurate tapping ratio of hot metal tapped to the 180 t hot metal ladle corresponding to the 2500 m^3 BF was controlled as 163 t ± 0.3 t, its share is about 50.2%; and the hit rate of hot metal tapping controlled as 163 t ± 1 t shares about 99%.

It shows that the accurate tapping ratio of hot metal weight tapped into the ladle from the large volume of BF at Shagang Group reaches a high level.

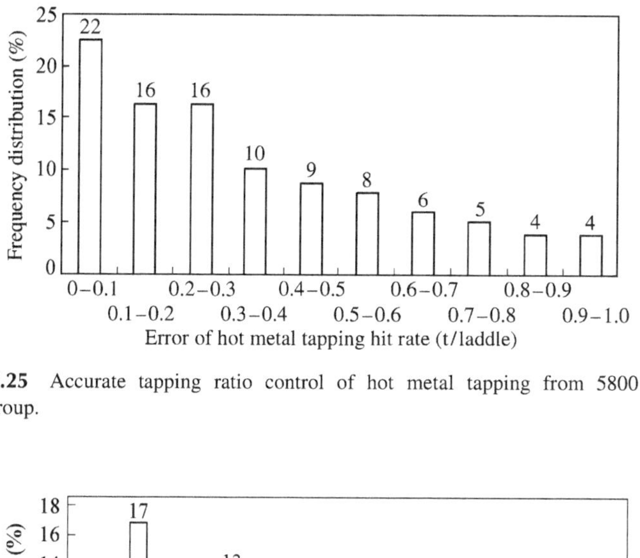

FIGURE 6.25 Accurate tapping ratio control of hot metal tapping from 5800 m^3 BF at Shagang Group.

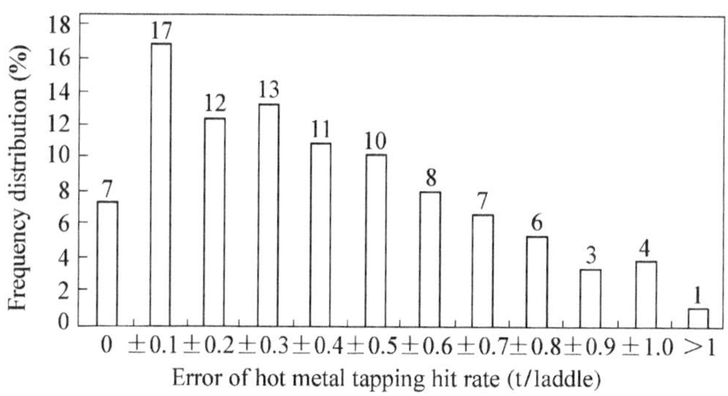

FIGURE 6.26 Accurate tapping ratio control of hot metal tapping from 2500 m^3BF at Shagang Group.

3. Study of the temperature drop of the 180 t ladle and its heat preservation measurements at Shagang Group. The temperature drop of the hot metal in the ladle is mainly caused by the thermal radiation from the hot metal surface and conductive heat dissipation by the liner and iron shell of the ladle. Conductive heat dissipation is mainly determined by the structure, material, and temperature of the liner. The liner of the ladle is usually constructed as a multilayer masonry refractory material, which favors heat preservation, so that the hot metal temperature drop was mainly caused by radiation heat dissipation from the hot metal surface when the ladle is full, and reduction of the heat dissipation from the surface is of great importance to decrease the temperature drop of hot metal in the ladle.

   In order to compare the heat preservation effect of different measurements of hot metal in the full ladle, the hot metal temperature drop in 180 t ladle was measured to three conditions: no top lid, covered with top lid, and covered with covering powder to the hot metal surface but no top lid, the results are shown as in Fig. 6.27.

   As shown in Fig. 6.27, there are remarkable effects on decreasing the hot metal temperature drop during the transportation and standing process when the full ladle is covered with the top lid or covering powder on the hot metal surface. The total temperature drop of hot metal during 5 h measured in the normal ladle with no heat preservation measurement is as much as 122°C, while they are 88°C and 95°C, respectively, in the ladle with top lid and covering powder on the surface of the hot metal, a decrement of 28% and 22% from the state. Therefore, if it is possible, the full ladle should be equipped with a top lid. It will potentially increase the hot metal temperature arriving at De[S] station. At the same time, the empty ladle should be covered with the top lid for heat preservation to decrease the temperature drop during the hot metal tapping process.

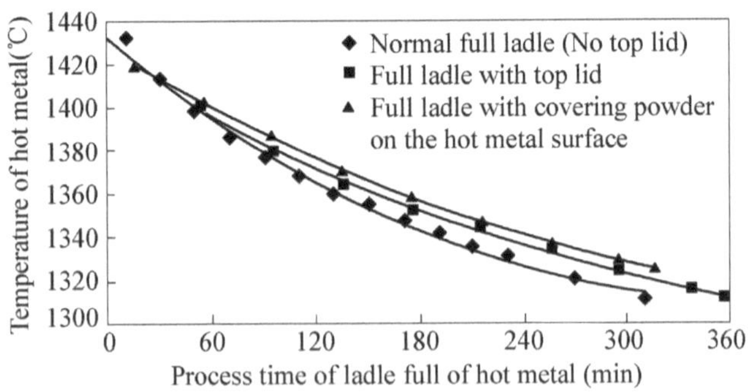

FIGURE 6.27 Comparison of temperature change in the 180 t ladle with different heat preservation measurements at Shagang Group.

Fig. 6.28 shows the hot metal temperature measured in 180 t ladle before deslagging at the KR station from the 5800 m^3 BF at Shagang Group.

Statistics of production samples show that, among the 327 ladles (randomly selected samples), the maximum temperature of hot metal in the 180 t ladle before De[S] is 1445°C, the minimum is 1326°C, the average is 1390.72°C, and the fluctuation coefficient is 0.02. The ladles with hot metal temperature higher than 1360°C occupy 84.4% of all of the random samples.

4. Effects of different De[S] pretreatment processes.

a. The relationship between De[S] effect and hot metal temperature of KR (desulfurization) process. The relationship between De[S] ratio of hot metal in 180 t ladle and hot metal temperature can be analyzed based on the 490 groups of data randomly selected from the records in the fourth quarter of 2011, such as [S] in hot metal is less than 0.01% after De[S] pretreatment by KR, hot metal stirring time of KR is fixed for 7–9 min, lime consumption is (6.6 ± 0.3) kg/t-HM, stirring intensity is 70–100 r/min, and hot metal temperature is 1280–1450°C, as shown in Fig. 6.29.

The analysis of practical production data shows that, the De[S] rate by KR process can be increased by 25.2% when the hot metal temperature in the 180 t ladle increased from the lower range of 1280–1300°C to the higher range of 1400–1430°C. Namely, higher hot metal temperature favors an increasing De[S] rate.

Impacts of different hot metal temperature, different initial [S] in hot metal, and different added amounts of De[S] powder on the De[S] rate were analyzed. The results are shown in Figs. 6.30 and 6.31.

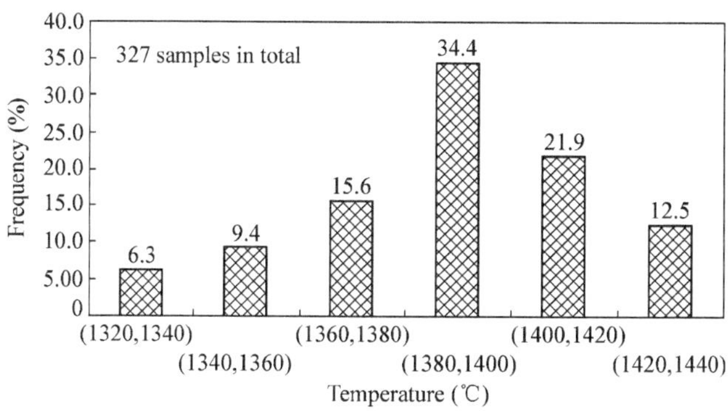

FIGURE 6.28 Distribution of hot metal temperature in 180 t ladle before deslagging at De[S] station, Shagang Group.

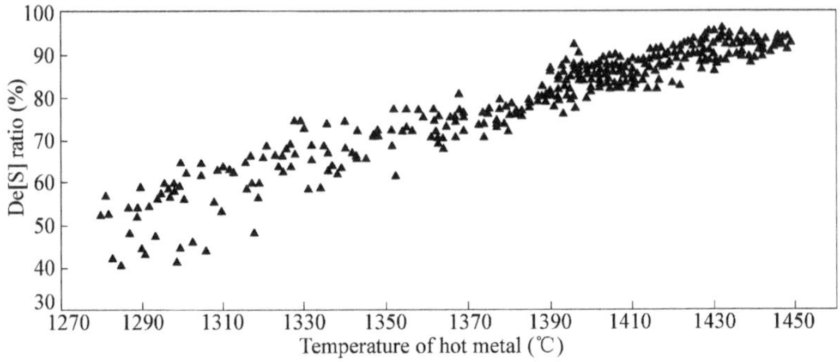

FIGURE 6.29 The relationship between hot metal ladle and hot metal temperature of the 180 t ladle at Shagang Group.

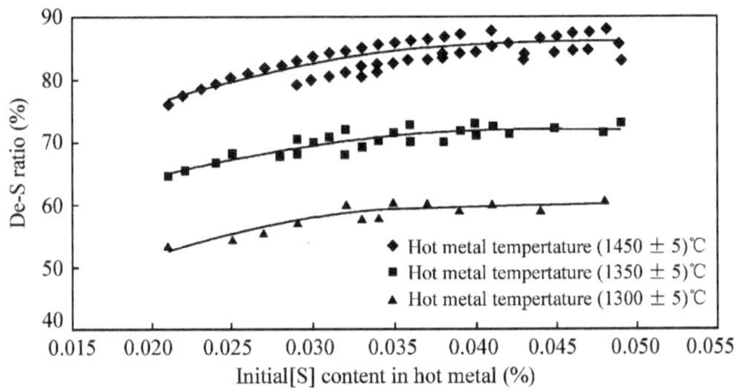

FIGURE 6.30 De[S] rate affected by initial [S] content in hot metal and hot metal temperature in the 180 t ladle at Shagang Group.

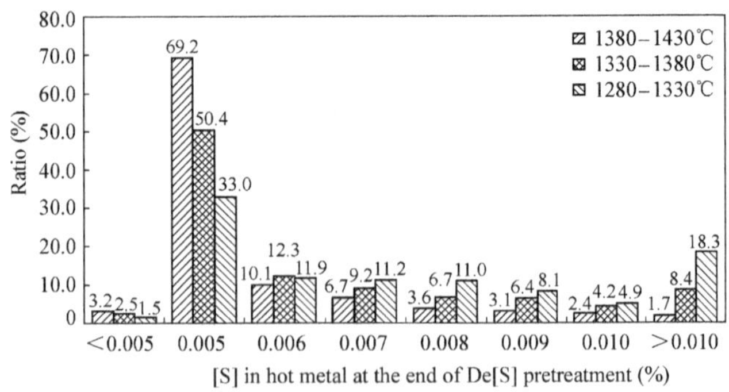

FIGURE 6.31 [S] in hot metal at the end of KR De[S] pretreatment in the 180 t ladle with different hot metal temperature at Shagang Group.

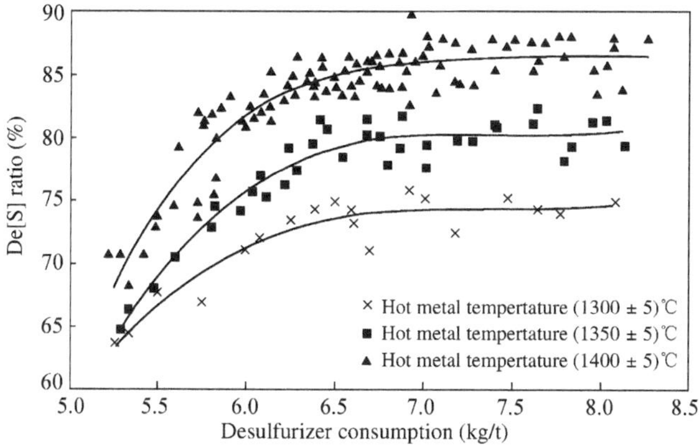

FIGURE 6.32 De[S] rate by KR De[S] process pretreatment affected by De[S] powder consumption and hot metal temperature in the 180 t ladle at Shagang Group.

As shown in Fig. 6.30:

- With the same initial conditions of sulfur content, the higher hot metal temperature is beneficial to a higher desulfurization efficiency.
- With different hot metal temperature, De[S] rate increases with the increasing initial [S] content in hot metal.

With different hot metal temperature, the De[S] effects affected by desulfurizer consumption have been analyzed. The results are shown in Fig. 6.32.

As shown in Fig. 6.32:

- Under the same condition of desulfurizer consumption, the higher the hot metal temperature, the better are the desulfurization efficiency yields.
- With different hot metal temperatures, the De[S] rate rises with the increase of desulfurization agent consumption; the increasing of the De[S] rate slows after desulfurizer consumption is more than 6.5 kg/t-HM. The slope of the De[S] rate of the desulfurization agent consumption in the range of 5.2–6.5 kg is greater than that in range of 6.5–8.2 kg.

De[S] agents used in the KR process are lime and fluorite, with a ratio of 9:1. Based on the analysis of practical production data, [S] in hot metal after De[S] pretreatment is shown in Fig. 6.33. Fig. 6.33 shows that the minimum [S] in hot metal after De[S] pretreatment is 20×10^{-6} at Shagang Group. After De[S] pretreatment by KR at Shagang Group, the hot metal shares about 60% in which [S] no more than 50×10^{-6}. It is probably related to the low quality of the lime used for De[S].

b. De[S] agent consumption and its effect on the Mg powder injection process. There are 3 × 180 t BOF corresponding to 3 × 2500 m^3 BF (9 iron notches) in Hongfa No.1 steelmaking workshop, the 180 t ladle was transported by standard gap railway. Mg powder injection process with three De[S] station is adopted by hot metal De[S] pretreatment. The Mg content of the Mg powder used as the De[S] agent is about 90%. The Mg powder is injected to stir hot metal by the carrier gas to De[S]. The effect of De[S] is better, and the minimum [S] content in hot metal after De[S] pretreatment is 20×10^{-6}. But due to easy resulfurization, the utilization efficiency of De[S] agent is decreased. The frequency distribution of the amount of Mg powder De[S] agent added by No.1 station in the month of 2011, at No.1 workshop of Hongfa steelmaking plant, Shagang Group is shown in Fig. 6.34. From Fig. 6.34, the Mg powder consumption is about 70–140 kg per heat, with much fluctuation.

The service life of the Mg powder injection gun is about 85–105 heats with an injection duration of about 11.7 min every heat; its service life is lower than that of the stirring head of KR. The [S] in hot metal at the end of the Mg powder injection De[S] pretreatment

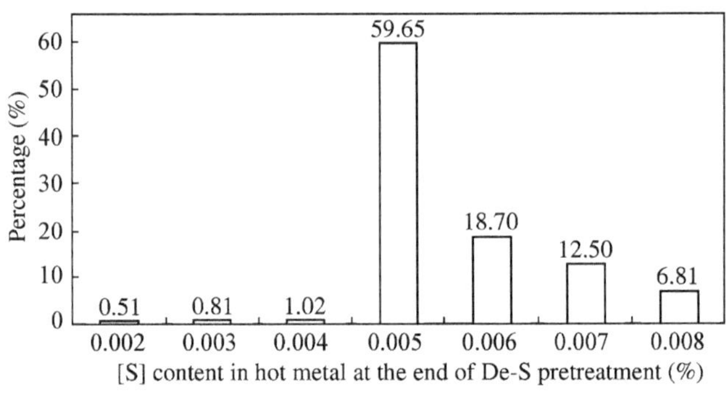

FIGURE 6.33 [S] content in hot metal at the end of De[S] pretreatment in the 180 t hot metal ladle at Shagang Group.

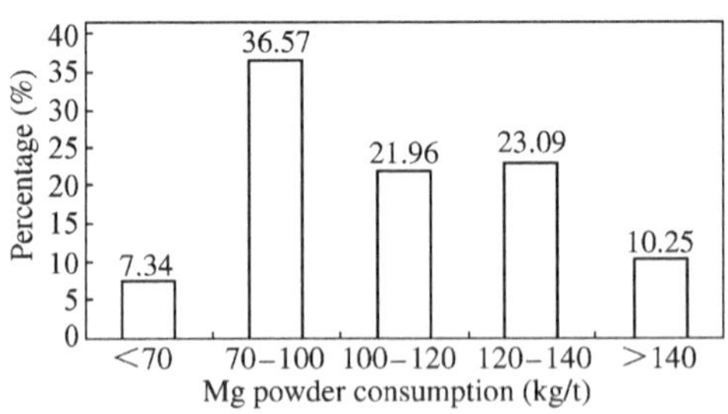

FIGURE 6.34 De[S] agent consumption by Mg powder injection pretreatment process in the 180 t ladle at No.1 workshop of Hongfa steelmaking plant, Shagang Group.

process can be up to $30 \times 10^{-4}\%$. The De[S] effect of the Mg powder injection De[S] pretreatment process at No.1 workshop of Hongfa steelmaking plant, Shagang Group is shown in Fig. 6.35.

In Fig. 6.35, [S] content in hot metal at the end point of the Mg powder injection De[S] process is 0.005–0.01%.

c. The comparison of temperature drop caused by deslagging of two De [S] processes. The comparison of temperature drop caused by deslagging of KR and Mg powder injection De[S] process at Shagang Group is shown in Figs. 6.36 and 6.37.

It can be seen from Figs. 6.36 and6.37 that:

- Temperature drop caused by deslagging after the KR process mainly ranged from 5°C to 9°C; and its average temperature drop in the 180 t ladle is 7.2°C based on the statistical analysis of samples randomly selected from deslagging after the KR process production at Shagang Group.
- Temperature drop caused by deslagging after the Mg powder injection process mainly ranged from 6°C to 12°C; and its average temperature drop is 9.8°C based on the statistical analysis of samples randomly selected from deslagging after the Mg powder injection process production at Shagang Group.

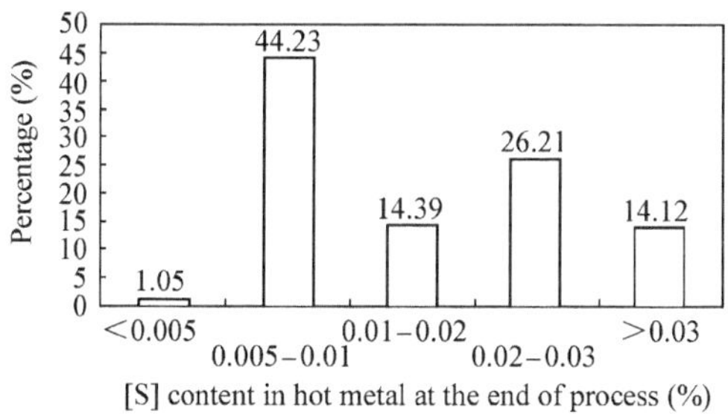

FIGURE 6.35 [S] of hot metal at the end of Mg powder injection De[S] pretreatment process at No.1 workshop in Hongfa steelmaking workshop of Shagang Group.

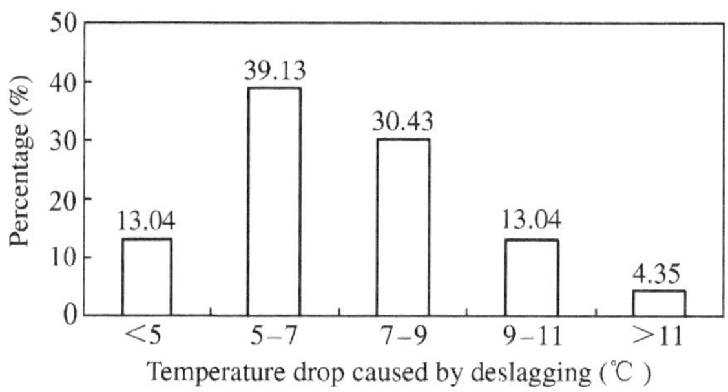

FIGURE 6.36 Temperature drop caused by deslagging after KR De[S] process in the 180 t ladle at Shagang Group.

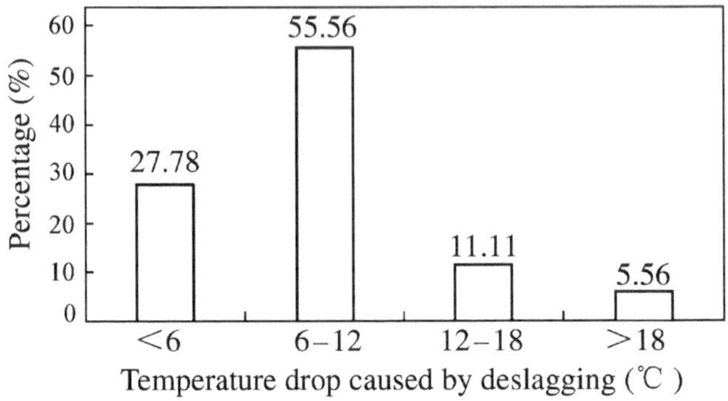

FIGURE 6.37 Temperature drop caused by deslagging after Mg powder injection De[S] process in the 180 t ladle at Shagang Group.

The results of the above analysis show that the temperature drop caused by deslagging after the Mg powder injection process is higher than that of the KR process. Because the De[S] slag produced by the Mg powder injection process is difficult to be grilled, and the deslagging duration is longer than that of the KR process, the temperature drop is higher than that of the KR process.

5. Management of turnover time and operation of hot metal ladle

   a. Ladle turnover duration of $1 \times 5800\ m^3$ BF-3×180 t BOF process at Shagang Group. Ladle turnover duration of the $5800\ m^3$ BF–180 t BOF section of No.2 workshop of Hongfa steelmaking plant has been measured, and the analysis of the ladle turnover duration is listed in Table 6.14.

   According to the data listed in Table 6.14, the average ladle turnover duration is 255 min, the turnover time is 5.65 times/day; among the turnover duration, the total waiting duration is 181 min, which is about 71% of the total duration of the ladle turnover, while the empty ladle duration is 149 min, which is an 82% share of total waiting duration. It shows that the turnover times of the 180 t ladle for $1 \times 5800\ m^3$BF–3×180 t BOF section at Shagang Group still can be improved.

   In order to investigate the potential of speeding up turnover of the ladle, the frequency distribution of turnover duration of the 180 t ladle was analyzed based on 300 samples randomly selected. The results are shown in Fig. 6.38.

   In Fig. 6.38, the turnover duration of 180 t for the $5800\ m^3$BF–180 t BOF (No.2 workshop of Hongfa steelmaking plant) section greatly fluctuates, ranging from 70 to 585 min, and the distribution of the total statistic samples basically show a normal distribution, most of which range from 180 to 300 min.

TABLE 6.14 Analysis of Turnover Duration of the 180 t Ladle for the 5800 m^3 BF at Shagang Group (min)

	Events	Average Time	Time Range	Total Number of Samples	Accumulation of Time
1	Hot metal tapping	18	10–40	300	18
2	Full ladle waiting for transportation	2	0–40	300	20
3	Full ladle transported to steelmaking plant	10	8–15	170	30
4	Full ladle craned to De[S] station	10	0–48	33	40
5	Full ladle waiting for crane	4	3–7	33	44
6	Full ladle craned to De[S] station	23	18–29	146	67
7	Full ladle after De [S] waiting for crane	6	0–84	146	73
8	Full ladle after De [S] craned to BOF	4	3–5	33	77
9	Full ladle waiting for charging into BOF	14	0–26	33	91
10	BOF charging	2	1–3	33	93
11	Empty ladle craned to ladle car	3	3–17	33	96
12	Empty ladle waiting for transportation	14	0–64	62	110
13	Empty ladle transported to BF	10	8–15	170	120
14	Empty ladle waiting for hot metal tapping	135	18–282	300	255

Note: Data listed in row 4, row 5, and rows 8–12 are surveyed directly, the other data are selected from the production records.

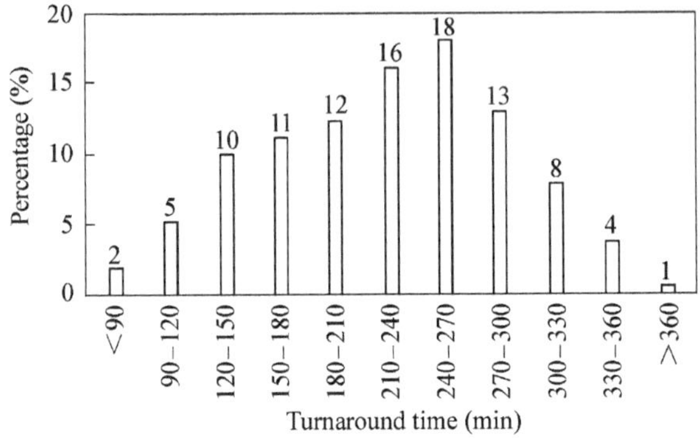

FIGURE 6.38 Frequency distribution of turnover duration of 180 t ladle for 1 × 5800 m^3 BF–3 × 180 t BOF section at Shagang Group.

TABLE 6.15 The Possible Turnover Duration of the 180 t Ladle for 1 × 5800 m^3 BF-3 × 180 t BOF Section at Shagang Group

Items	Turnover Parameters of 180 t Ladle				
Turnover duration of 180 t ladle (min)	255	240	225	210	180
Turnover times per day (time)	5.65	6	6.4	6.86	8.0
Hit rate of the statistic samples (%)	65	57	48	41	29

It has been ruled at No.2 workshop of Hongfa steelmaking plant, Shagang Group that the turnover duration of the ladle should not exceed 360 min. The analysis of the statistical data shows that all the turnover duration of the 180 t ladle for 5800 m^3 BF–180 t BOF (No.2 workshop of Hongfa steelmaking plant) section can almost fully meet the requirements. The samples with longer turnover duration than 360 min are mainly caused by the damping down of the BF, which is less than 1% of the total samples. The possible turnover duration of the ladles at Shagang Group can be estimated based on Fig. 6.38. The results are listed in Table 6.15.

If the management of ladle turnover between the 5800 m^3 BF and Hongfa No.2 steelmaking workshop was enhanced, the turnover times of the ladle could be more than 6 times per day, and even approach 7 times per day. At present, 41% of total ladles can achieve 6.86 times per day, and nearly 1/3 of the total ladles can achieve 8 times per day. Therefore, there is still great potential for the improvement of turnover times of the ladle used for the large BF–large BOF section at Shagang Group; it will be beneficial to improve the hot metal De[S] efficiency.

b. Turnover duration of ladles for the small BF–small BOF/EAF section. Small BFs are mainly located at Huasheng No.1 and No.2 ironmaking plants. Among them, there are five 380 m^3 BFs at No.1 plant, from which hot metal is mainly supplied to 3 × 40 t BOF and 1 × 70 t EAF in Yongxin steelmaking plant; there are four 502 m^3 BFs at No.2 plant, from which hot metal is mainly supplied to 2 × 100 t EAF in Runzhong steelmaking plant and 1 × 100 t EAF in Shajing steelmaking plant. There is only one iron notch for every BF. All the ladles are 40 t. There are so many users of the nine BFs in the Huasheng ironmaking plant. The user distribution is scattered, the transportation line is complex, and the local rivers limit the construction of transporting lines, so the 40 t ladles need to be transported by truck along the road.

A total of 297 and 258 samples were selected respectively from Huasheng No.1 and No.2 plants, Shagang Group. Their turnover duration distribution was analyzed and the results are shown in Figs. 6.39 and 6.40.

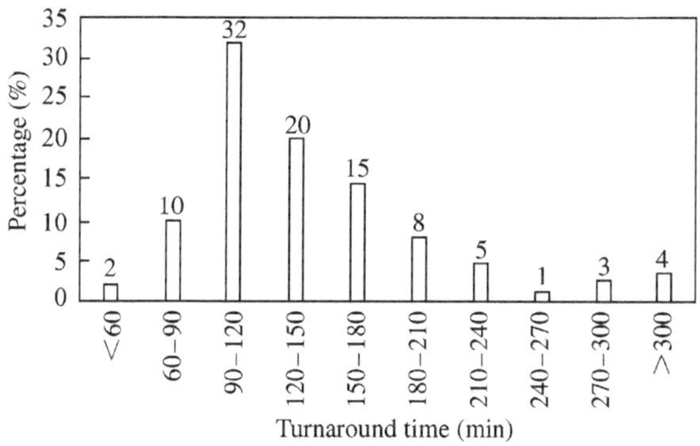

FIGURE 6.39 Turnover duration distribution of the 40 t ladle for Huasheng No.1 plant.

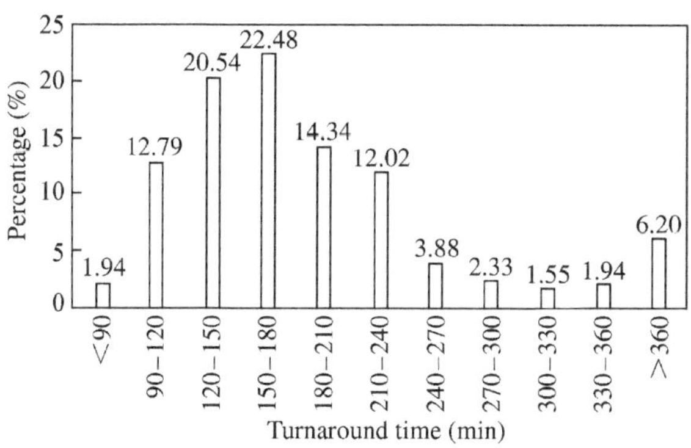

FIGURE 6.40 Turnover duration distribution of the 40 t ladle for Huasheng No.2 plant.

TABLE 6.16 The Expected Results of Turnover Optimization of the 40 t Hot Metal Ladle in Huasheng Ironmaking Plant

Items	No.1 Plant			No.2 Plant		
Turnover period of 40 t hot metal ladle (min)	145	130	115	195	175	155
Turnover times per day (time)	9.93	11.08	12.52	7.38	8.23	9.29
Percent of samples reaching the standard (%)	61	51	44	65	54	41

In Figs. 6.39 and 6.40, the turnover duration of ladles for the Huasheng ironmaking plant fluctuates greatly. The turnover duration of the ladle in the Huasheng No.1 and No.2 plants ranges from 40 to 420 min (mainly within 240 min, more than 90%) and from 75 to 747 min (mainly within 300 min, about 90%), respectively.

The average turnover duration of the ladle for the Huasheng No.1 plant is 145 min, calculated by statistical samples. It is possible to achieve a turnover 7.38 times a day. The expected turnover parameters of ladles for the Huasheng ironmaking plant after improvement of the ladle operation management are listed in Table 6.16.

6. Multifunction ladle and cleaner production. It is worth mentioning that there is still a mixer in operation online in the small BF–small BOF process due to historical reasons, and about 4% of hot metal needs to be treated in the mixer. With the mixer online operation, the loss of iron, dust emission, slag emission, etc. of the small BF–small BOF process are significant. The "One Ladle Technology" should be adopted to reduce the operation dependence on the mixer to reduce energy consumption and damage to the environment. The multifunction ladle of the large BF–large BOF process has a remarkable advantage and its experiences of rapid operation and cleaner production are worth learning. Based on calculations, the comparison of the environment load of the "One Ladle Technology" operation and a mixer for the "buffer" operation is listed in Table 6.17.

6.2.4 Discussions

Throughout the development of the steel production process, there have been great changes in the BF–BOF section; not only have the devices, processes, and functions been gradually optimized, but also the interface technologies of the connecting processes. The connecting interface and operation schedule of the ironmaking–steelmaking section is one of the key factors affecting

TABLE 6.17 Comparison of Different Mode of Interface Technology Between BF and BOF in Shagang Group

Items	Holder Changing of Hot Metal (time)	Environment Load of the Section		
		CO_2 emission (Nm^3/t)	Dust emission (g/t)	Slag emission (kg/t)
The mode of small BF–small BOF + mixer	2	68.78	240.6	105.70
The mode of small BF–small BOF with "one ladle"	0	65.23	130	92.12
The mode of large BF–large BOF with multifunction hot metal ladle	0	5	115	64.91

the stable operation of the whole process. It is of great importance to optimize the matching-collaborative model of the ironmaking–steelmaking interface to reduce the energy consumption, material consumption, and cost, and to promote environmental protection for the whole process. Interface technology of ironmaking–steelmaking section is integrated with a variety of heterogeneous technologies, including an integration package of the design and equipment, a technology package of the mass flow operation, and technology packages of mass flow and metallurgical effects.

1. Design is the foundation of the integrated design technology package of the process and equipment. In order to optimize the BF–BOF interface technology by the multifunction hot metal ladle, it is important to first understand this problem from the aspects of the concept and engineering design methods, as well as the level of process engineering. The problem of the BF–BOF interface technologies is not so simple as changing the hot metal container, instead it requires an integrated innovation in engineering design.
   a. Layout design. The volume, number, location, and the layout of BF significantly affect the tapping times, hot metal transportation duration and operation schedule to the pretreatment station; these factors are also decisive to the convergence time of process operation

schedule and mass flow connection. Therefore, the study of layout design is of great significance to determining the enterprise scale and a reasonable number of BF. Meanwhile, a reasonable layout design makes the regional layout compacted, transporting lines shortened, material flow smooth, and production highly efficient connected, which all could provide the hardware foundation for the rapid turnover of the ladle.

b. Administrative jurisdiction of the ladle. The key factor of determining the number of ladles online is that the hot metal ladle should be managed by steelmaking plants, which also provides the software foundation to guarantee the rapid turnover of the ladle and metallurgy effect. Traditionally, the ladle is not managed by the steelmaking plant; instead, the major factor considered is that the number of ladles online should yield the safety and convenience of BF tapping. Shagang Group is the first to develop the "life cycle management" technology of the ladle in being charged by steelmaking plants, it provides a useful reference for the operation management of the hot metal ladle.

c. Reasonable number of hot metal ladles and their turnover times. Under the established layout conditions, the number of ladles is the important factor affecting the hot metal transportation schedule, the temperature between BF and BOF interface, the connection of hot metal composition (mainly [S]), hot metal temperature between BF and BOF section, and also a key factor related to the high-efficiency De[S] of the multifunction ladle.

2. Operation technology package of mass flow. The details are as follows:

a. Rapid marshaling and management of the ladle. In order to cancel the locomotive marshaling yards, the ladle was marshaled by the crane of the steelmaking plant, the ladle operation management was based on the uniform principle of "first-in first-out" for the steelmaking plant and the ironmaking plant to achieve rapid turnover of the ladle.

b. The accurate tapping ratio of hot metal tapping is important to guarantee the stable production of steelmaking plants and to present the management level of the ironmaking plant. The technology of the multifunction ladle between BF and BOF cancels the molten iron pouring station; there is no other process to change the weight of hot metal except precisely controlling the hot metal weight originally in the BF tapping process. Therefore, precise weight measurement of hot metal at the iron notch area is another key factor to the successful operation of "One Ladle Technology." The hit rate of hot metal tapping must be charged by the ironmaking plant. This hit rate should also be considered as an indicator for BF performance assessment. High hit rate of hot metal tapping and a stable supply of hot metal also guarantee the stability of the steelmaking process.

c. For the flexibility of transportation with nonrailway-locomotive, technologies and equipment of ladle transportation by truck or electromobile should be developed, which will be beneficial to the rapid transportation of hot metal ladles between larger BF and larger BOF, as well as contributing to the temperature of hot metal pretreatment with hot metal ranging from 1400°C to 1450°C.

3. Technology package of mass flow operation and metallurgical effects. The details are as follows:
   a. The higher temperature of hot metal is beneficial to the desulfurization effect. The compact space layout of the interface between ironmaking and steelmaking plants leads to the optimization of the time schedule, effectively shortens the turnover duration of the ladle; and it can also cancel the transfer operation, so that the hot metal temperature drop decreases and the hot metal temperature at start of desulfurization station increases, the high-efficiency De[S] can be operated with higher hot metal temperature.
   b. Lowering investment and operation costs. The multifunction ladle is adopted for the interface between the BF and BOF section, which, in turn, eliminates the torpedo car and locomotive marshaling yards; the original weighing facilities between BF and BOF is replaced by the online weighing system under the BF iron notches; the transfer station and dedusting facilities are eliminated, as well as the mixers of the steelmaking plant, so that investment is reduced. The compact interface layout between BF and BOF has effectively shortened the ladle turnover duration and prolonged the service life of the ladle, which reduces the refractory consumption per ton of hot metal. The temperature drop of the full or empty ladle is reduced, and thus the temperature drop of hot metal. Therefore, the cost of operation is lowered.
   c. Better effects of energy saving, emission reduction, and cleaner production. Since the transfer operation is cancelled, the hot metal temperature drop is reduced, which saves energy. Accordingly, greenhouse gas emissions are reduced; and the problem of graphite dust pollution in steel plants could be solved.

6.3 De[S]–De[Si]/[P] PRETREATMENT AND HIGH-EFFICIENCY AND LOW-COST CLEAN STEEL PRODUCTION PLATFORM

The structure of the manufacturing process has been improved remarkably since the emergence of CC. The procedures before CC are now evolving in the direction of analysis-optimization: The number of procedures increases, the functions of procedures are simplified and integrated, procedures tend to be more coordinative-continuous and fast-paced. Meanwhile, the procedures after CC have taken more advantage of simplification-integration and compactness-continuation. The restructuring-optimization of the process structure resulting from the analysis-optimization of procedures' functions and the coordination-optimization of procedures' relationships are increasingly obvious.

Since the 1980s, the production process with basic elements of hot metal pretreatment–BOF–secondary refining–CC has evolved in the BOF steelmaking plant. Because the thermodynamics and kinetics requirements are different for desulfurization, dephosphorization, desiliconization, decarburization, deoxidation, alloying, etc., conflicts occur when different chemical metallurgical reactions take place in a reactor, which leads to a low production efficiency, low product quality, and high cost. According to principles of metallurgical physical chemistry, different types of metallurgical reactions should be carried out in different reactors to achieve procedure/device function analysis-optimization in the steelmaking plant. Based on the above, and through the optimization of the process network and dynamic operation program, it is possible to prompt the fluency of material flow and the high-efficiency operation of steelmaking plants, and thereby boost the optimization of element-structure-function-efficiency of steel production.

Since the 1980s, with the in-depth understanding and the creative development of hot metal dephosphorization technology, a high-efficiency and low-cost clean steel production platform has been formed which adopts De[S]–De[Si]/[P] pretreatment.

6.3.1 Why Adopt the De[S]–De[Si]/[P] Pretreatment

The original target of adopting De[S]–De[Si]/[P] pretreatment in BOF steelmaking plants to produce ultralow phosphorous steel has been diverted to establish the competitive new generation steel production process, namely, a high-efficiency and low-cost clean steel production platform with steady operation, in order to enhance the competitiveness of products. The advantages of the platform are as follows:

1. Through the analysis-optimization of steelmaking procedure functions, the high-efficiency fast-paced operation in the steelmaking plant and reliable molten steel quality can be achieved. The well-matched tap-to-tap time with advanced high-speed CC leads to the improvement of production efficiency in steelmaking plants.
2. The reduced slag smelting and low temperature taping (due to high-speed CC and fast-paced BOF steelmaking), recycling $BOF_{De[C]}$ slag as $BOF_{De[Si]/[P]}$ slag, etc., reduce the metal and lime consumption in the steelmaking process.
3. The desulfurization, dephosphorization, and desiliconization are almost completed in the hot metal pretreatment process, and no steel scrap and fewer auxiliary materials are needed in the $BOF_{De[C]}$, so the composition,

temperature, and weight of semisteel charged into the $BOF_{De[C]}$ can be properly measured, which is conducive to a better endpoint hitting rate, and meanwhile creates a better basis for the secondary refining process. The pretreatment not only promotes a fast-paced production process but also realizes the steady and mass-production of clean steel.

4. The possible utilization of higher [P] iron ore can reduce the purchase cost.
5. The possible utilization of manganese ore can lower the cost of alloy and avoid extra [C] and [P] being used in molten Fe–Mn.
6. For a new steel plant, the tonnage of the BOF can be reduced for the same productivity.
7. The $BOF_{De[Si]/[P]}$ slag has a lower basicity ($CaO/SiO_2 = 1.8-2$), and it can be used as a road-construction material without a hydration process.
8. The casting performance of ultralow carbon steel (namely the amount of casted steel within the life period of a submerged entry nozzle) can be improved and enhanced, which increases the number of CC heats.

This new process is useful for large BOF steelmaking plants producing high-quality sheet, high-quality seamless steel tube, plate, or rod, but it is not necessary for small ones producing long products for construction.

6.3.2 Analysis-Optimization of Procedure Functions and Coordination-Optimization of Procedure Relationships in the De[S]–De[Si]/[P] Pretreatment

The high/constant-speed sequence CC technology has led to the fast-paced and steady operation of BOF or EAF, which guides the analysis-optimization of the BOF procedure and the integration-optimization of upstream/downstream procedures in steelmaking plants. The steps of the analysis-integration are as follows:

1. Abiding by the rules of physical chemistry in metallurgy, distribute the functions of procedures reasonably and develop reactor devices conforming to the requirements of metallurgical kinetics.
2. Continuously improve and optimize the steel plant layout, try to build a coordinative and quasicontinuous/continuous running spatial-temporal network with smooth logistics;
3. Construct a dynamic operation system with rational element-structure-function-efficiency through the parameter optimization of the operation of procedures/devices and the information optimization of relationships among procedures.

The functions of BOF steelmaking are shown in Table 6.18 (Yin, 2009b).

Abiding by the rules of physical chemistry in metallurgy, the steelmaking tasks of desulfurization, dephosphorization, and desiliconization have been

TABLE 6.18 Procedure Function Analysis of BOF Steelmaking Process

Procedure Functions in BOF Steelmaking	Hot Metal Pretreatment	BOF	Secondary Refining
Desiliconization	⊙ ←	○	
Desulfurization	⊙ ←	○	→ ◎
Dephosphorization	⊙ ←	◎	
Decarburization	◎ ←	⊙	→ ⊙[①]
Heating		⊙	→ ◎
Degassing		○	→ ⊙
Including morphology control		○	→ ⊙
Deoxidation		○	→ ⊙
Alloying		◎	→ ⊙
Cleaning	⊙ ←	◎	→ ⊙

⊙: Main procedure for a function.
◎: Secondary procedures for a function.
○: Degradation procedures for a function.
①: Only in the condition of ultralow carbon steel, using vacuum decarburization.

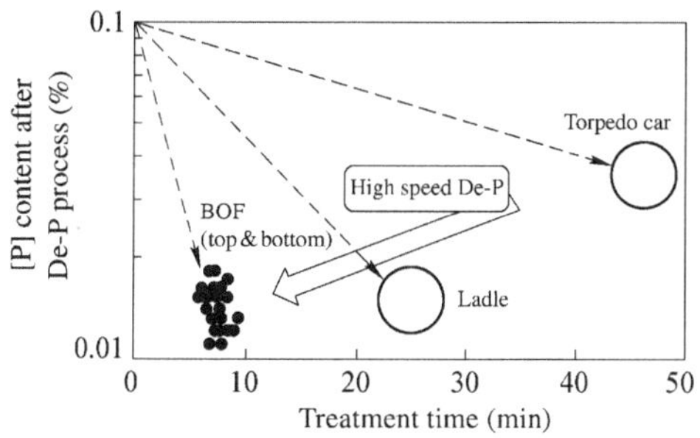

FIGURE 6.41 Effectiveness comparison for three dephosphorization methods (Kawamoto, 2011).

moved forward to a specialized procedure—the hot metal pretreatment. Further, the original BOF is intensively used to realize high-efficiency decarburization, rapid heating, secondary energy utilization, and moderate dephosphorization.

After continuous research and practice, the hot metal pretreatment sequence and process are improved and relatively perfect. That is, the poor thermodynamics and kinetics condition of hot metal pretreatment in the torpedo and the desiliconization pretreatment are no longer completed in the BF tapping ditch and nor is the hot metal pretreatment in the torpedo. Moreover, hot metal must undergo pouring and repouring if the torpedo is adopted, which leads to temperature drop, energy loss, and environmental pollution. Hot metal's direct charging into the iron ladle is beneficial to carrying out KR desulfurization under high temperature and activity after deslagging, which has a high desulfurization rate, low energy consumption, low cost, and low pollution. The KR desulfurization process, with the advantages of easy deslagging and the low resulfurization rate, has epitomized its reliability and stability in industrial practices. More discussions on KR can be found in Section 6.2.

After solving the problems of sequence of desulfurization and desiliconization (ie, taking desulfurization first), the coordination, devices and methods for desiliconization, dephosphorization, and decarburization should be discussed.

Since the 1980s, engineers in Japan have repeatedly studied and developed dephosphorization theories and technologies. They have compared the effectiveness of dephosphorization in the torpedo, iron ladle, and specialized converter, respectively. It has been gradually cognized that the thermodynamics and kinetics condition for dephosphorization in the specialized converter is superior to other reactors (Fig. 6.41).

6.3.3 A Case Study on Full Hot Metal Pretreatment—Steelmaking Plant in Wakayama Iron & Steel Works of Former Sumitomo Metal Industries

In Japan, a representative project was the restructuring of the steelmaking plant in Wakayama Iron & Steel Works of the former Sumitomo Metal

Industries. In Wakayama Iron & Steel Works there were two BOF steelmaking plants, each with 3 × 160 t BOF, which were merged into one plant in 1999. The merged plant had a high-efficiency clean steel production process with 2 × KR, 1 × 210 t $BOF_{De[Si]/[P]}$, 2 × 210 t $BOF_{De[C]}$, 2 × $RH_{De\text{-}gas/S}$, and 3 × CC, as shown in Figs. 6.42 and 6.43. The target production scale of the new plant was 4.0–4.5 million tons of steel per year. The products included high-quality tube, sheet, and bearing steel for high-speed trains.

During the restructuring, the Wakayama steelmaking plant had developed a series of hardware/software techniques including the specialized $BOF_{De[Si]/[P]}$, the high-efficiency $BOF_{De[C]}$, technological processes, plant layout, and the optimization of dynamic operation, which had been a success.

The profiles are different for the $BOF_{De[Si]/[P]}$ and $BOF_{De[C]}$ in Wakayama steelmaking plant, as shown in Table 6.19; Figs. 6.44 and 6.45. Since the intensive decarburization reaction is not allowed in $BOF_{De[Si]/[P]}$, its height and volume-capacity ratio should be smaller than those of $BOF_{De[C]}$. In the practical production, a respective 20 min of tap-to-tap time in the KR device (40 min/2), $BOF_{De[Si]/[P]}$, and $BOF_{De[C]}$ was adopted in the dynamic-orderly operation. Then the dynamic structure 2 × KR-1 × $BOF_{De[Si]/[P]}$ − 1 × $BOF_{De[C]}$, with a tap-to-tap time of 20 min, was formed. After this ran the 2 × $RH_{De\text{-}gas/S}$-3 × CC, as shown in Figs. 6.46 and 6.47.

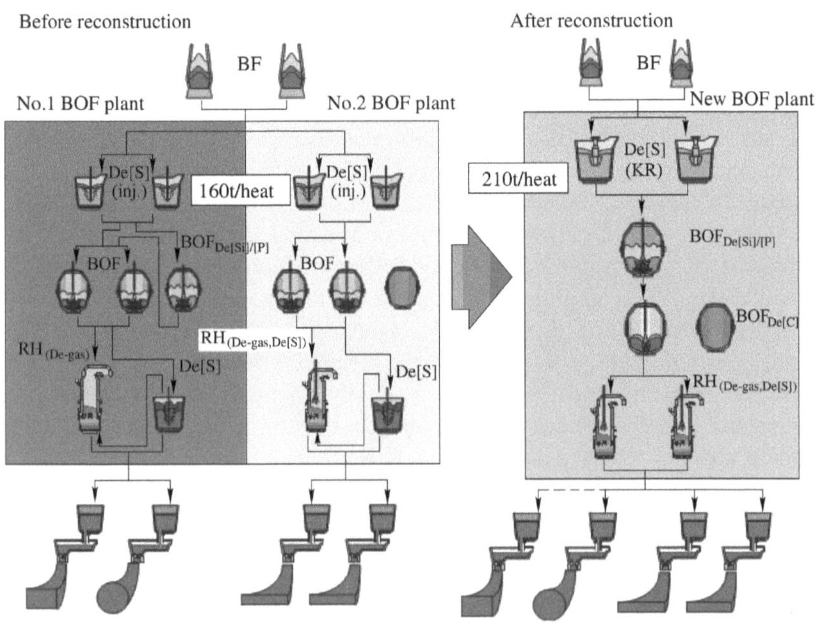

FIGURE 6.42 Structural change of Wakayama Iron & Steel Works (Kawamoto, 2011).

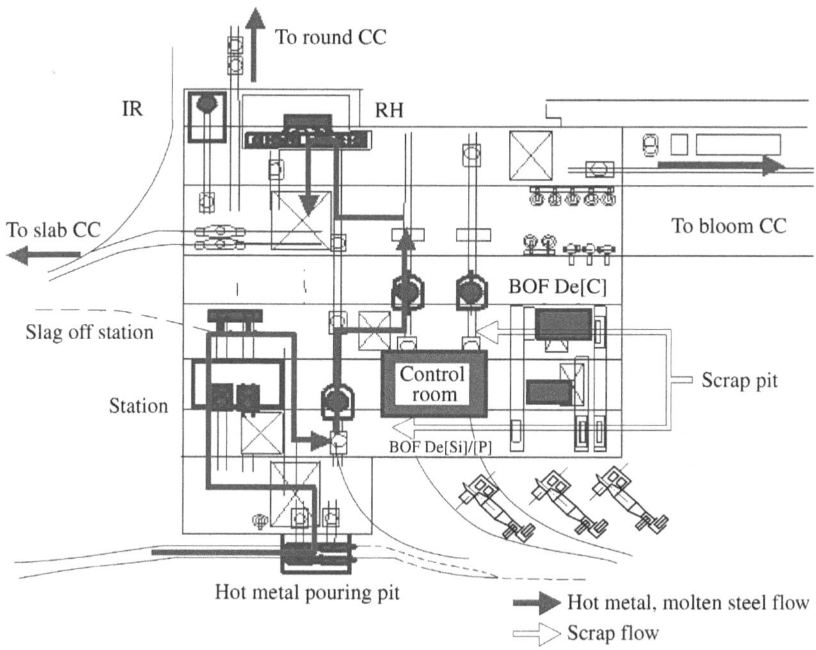

FIGURE 6.43 The layout of merged Wakayama steelmaking plant (Ueki et al., 2004).

TABLE 6.19 Parameter Comparison of $BOF_{De[Si]/[P]}$ and $BOF_{De[C]}$

Profiles	$BOF_{De[Si]/[P]}$ (210 t)	$BOF_{De[C]}$ (210 t)
Inner diameter, Φ (m)	7.7	8.0
Height (m)	10.5	11.5
Max. oxygen flow rate (Nm^3/h)	40,000	80,000
Bottom blowing	4 holes	4 holes
	CO_2, Ar, N_2: 5400 Nm^3/h	CO_2, Ar, N_2: 5400 Nm^3/h
Energy	OG + boiler gas recycle Auto blowing control system	OG + boiler gas recycle

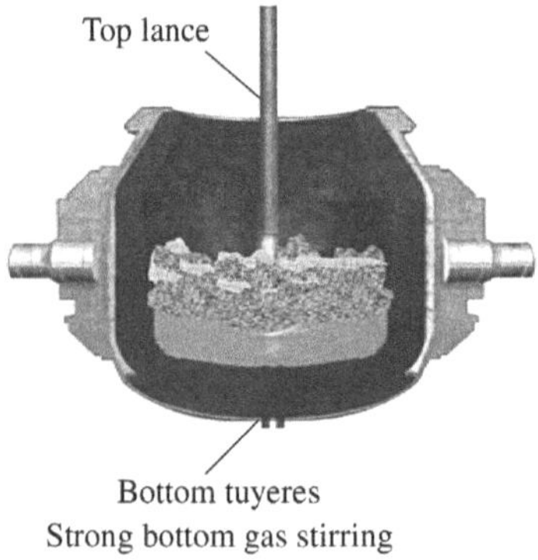

Type	Converter type
Heat size	210 t × 1
Furnace size	ID:7.7 m Height:10.5 m
Top blowing	O_2:Max. 40000 Nm^3/h
Bottom blowing	Tuyeres:4 CO_2, Ar, N_2:Max. 5400 Nm^3/h
Other	OG&boiler type gas recovery system Darts type slag stopper Automatic blowing control system

FIGURE 6.44 Profiles of 210 t $BOF_{De[Si]/[P]}$.

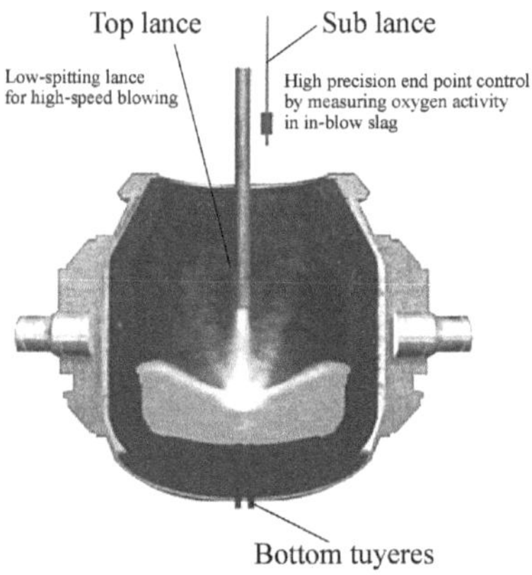

Heat size	210 t × 1/2
Furnace size	ID:8.0 m Height:11.5 m
Top blowing	O_2:Max. 80000 Nm^3/h
Bottom blowing	Tuyeres:4 CO_2, Ar, N_2: Max. 5400 Nm^3/h
Other	OG&boiler type gas recovery system Darts type slag stopper Automatic blowing control system

FIGURE 6.45 Profiles of 210 t $BOF_{De[C]}$.

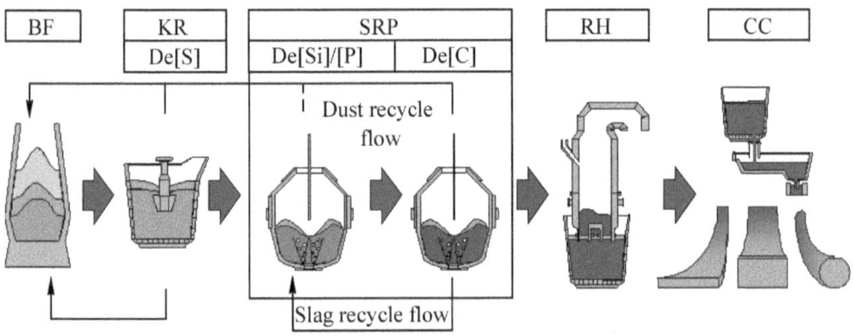

FIGURE 6.46 Process scheme of merged Wakayama steelmaking plant.

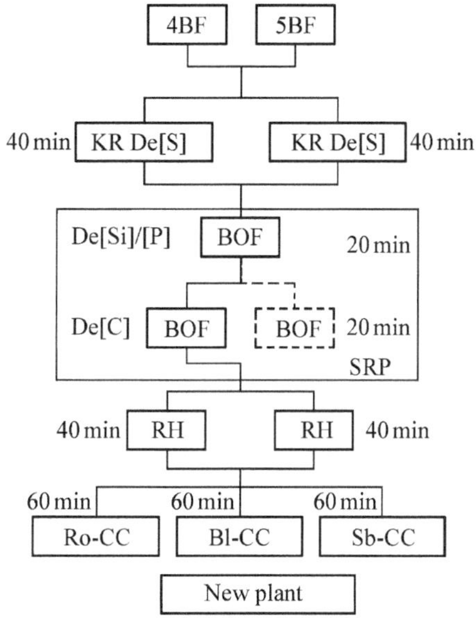

FIGURE 6.47 Technology process of merged Wakayama steelmaking plant.

From the innovative design and production operation of the merged Wakayama steelmaking plant, the following analyses could be made:

1. The metallurgical reactions in traditional BOF smelting process were analyzed and optimized. According to different technological operation requirements and different thermodynamics conditions of desulfurization, dephosphorization, and decarburization, their own reactors or devices were designed, especially the $BOF_{De[Si]/[P]}$ with intensive bottom blowing and an oxygen lance with excellent decarburization ability.
2. The reactors for desulfurization, dephosphorization, and desiliconization were coordinated and integrated, and the relationships among procedures were coordinated and optimized to increase the efficiency of material transformation and energy utilization.
3. Based on the technological operation parameter optimization for different reactors (procedures/devices), a new layout for the steelmaking plant was designed to rationalize the production material flow, reduce interference, improve the time/space structure of processes, and increase the material flow operation efficiency.
4. Aimed to match sequence CC and targeted at a 20 min tap-to-tap time of $BOF_{De[Si]/[P]}$ and $BOF_{De[C]}$, a time scheduling plan for the production operation process—the time-based steelmaking plant operation-state chart—was established. The operation time-rhythm was accelerated to a 20 min tap-to-tap time by all kinds of effective technological measures and information control.

Many years of production practice and continuous improvement had seen the following progress in the Wakayama steelmaking plant:

1. The total slag amount of $BOF_{De[Si]/[P]}$ and $BOF_{De[C]}$ had been reduced to 52 kg/t (steel) by adopting De[S]−De[Si]/[P] pretreatment, which was a marked decline comparing to the 97 kg/t of traditional BOF slag. At the same time, the BOF slag was fully utilized and the lime consumption declined significantly, as shown in Fig. 6.48.
2. Good dephosphorization results had been achieved in $BOF_{De[Si]/[P]}$. Under conditions of $T_{Molten\ pool} = 1300-1320°C$, $CaO/SiO_2 = 2.0-2.4$, and a $0.4\ Nm^3/(t \cdot min)$ intensive bottom-blowing stirring, the dephosphorization rate reached above 75% and the [P] content in semisteel fell below 0.025%, as shown in Figs. 6.49 and 6.50.
3. The $BOF_{De[Si]/[P]}$ had realized high-speed oxygen blowing. Owing to the ultrahigh-speed blowing technology, especially the improvement of

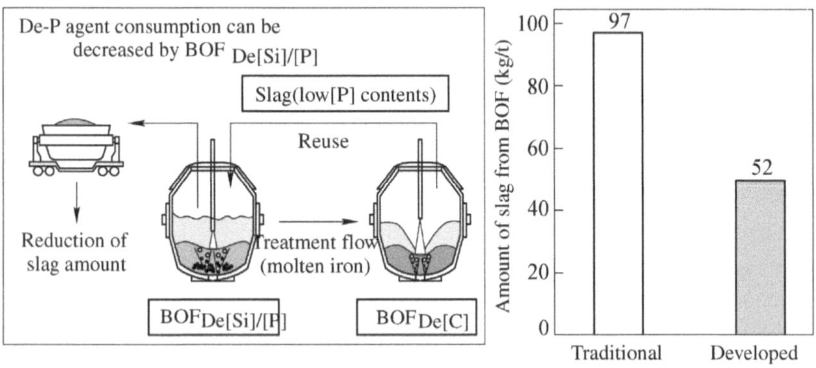

FIGURE 6.48 Lime consumption and total slag amount decreases obviously at Wakayama steelmaking plant.

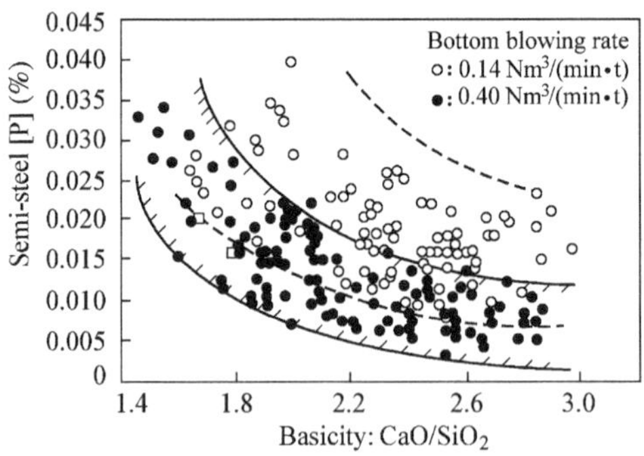

FIGURE 6.49 The relationship between basicity and [P] content after dephosphorization.

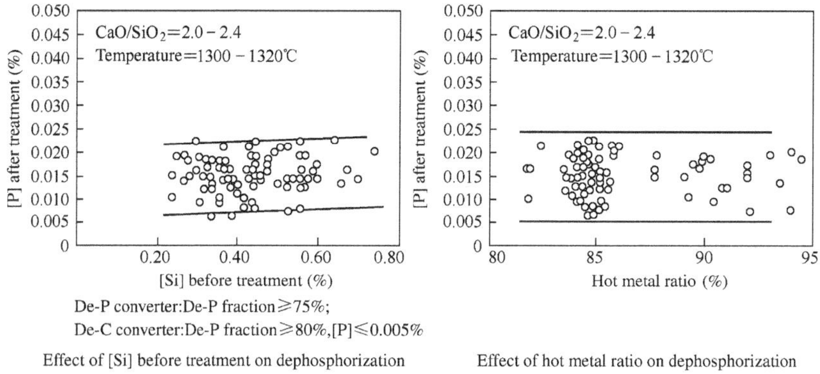

FIGURE 6.50 Improvement of the dephosphorization effect at Wakayama steelmaking plant.

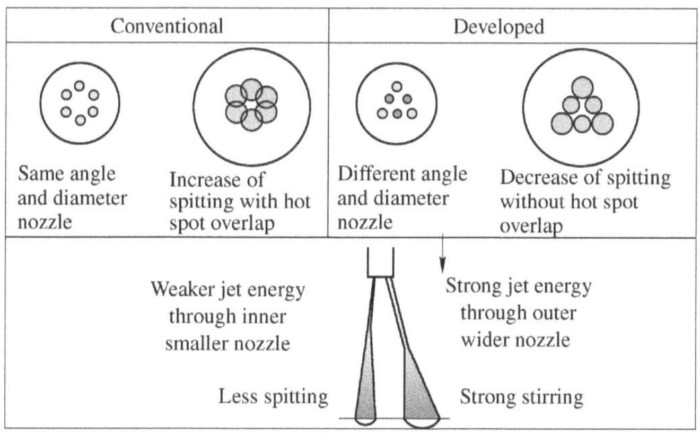

FIGURE 6.51 Improvement of the structure and parameters of oxygen lance nozzle (Ueki et al., 2004).

the oxygen nozzle structure and parameters (see Fig. 6.51), the oxygen blowing intensity reached above 4.5 $Nm^3/(t \cdot min)$, the blowing time lasted for about 9 min (see Fig. 6.52), the quick tapping rate increased significantly (see Fig. 6.53), and the tap-to-tap time reached about 20 min (see Fig. 6.54).

4. Owing to the dynamic-orderly and coordinative-continuous operation in the steelmaking plant, the entire steel plant realized a rapid, coordinative, and quasicontinuous operation and practically produced 45 heats per day. The general time-temperature parameters of the production process are shown in Fig. 6.55.

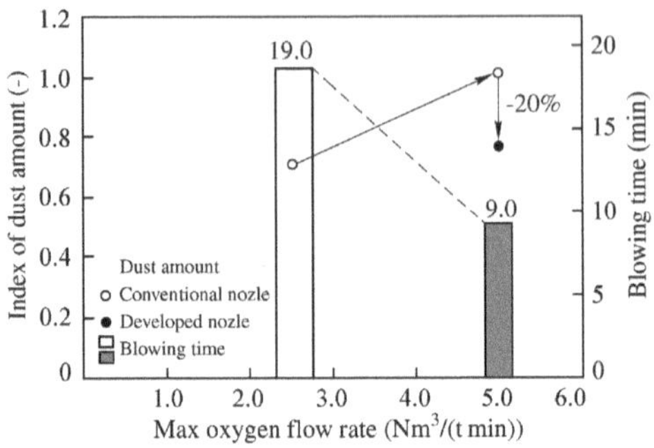

FIGURE 6.52 Effect of the improvement of oxygen lance (Ueki et al., 2004).

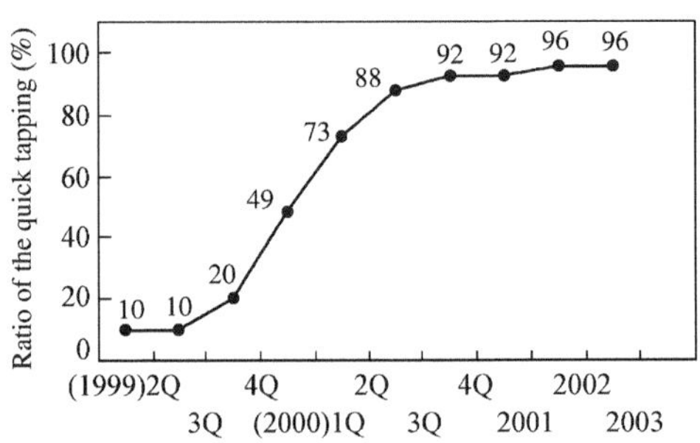

FIGURE 6.53 Changing of quick tapping rate of the end control of the decarburization converter (Kawamoto, 2011).

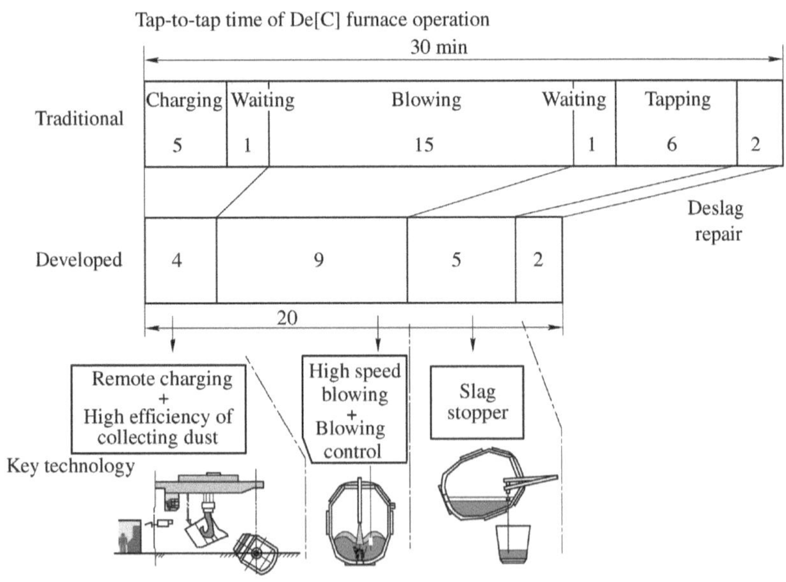

FIGURE 6.54 Reduction of tap-to-tap time of decarburization converter (Kawamoto, 2011).

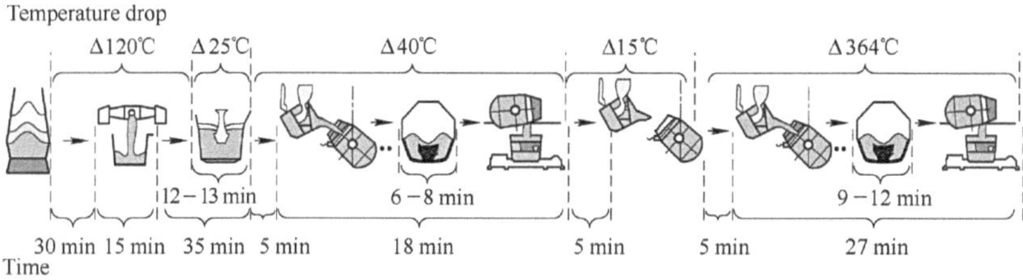

FIGURE 6.55 Schematic diagram of the time-temperature parameters of production process.

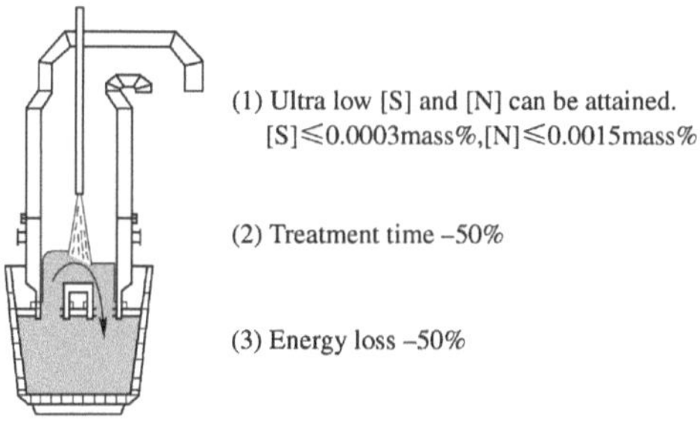

FIGURE 6.56 Schematic diagram of high-efficient production of RH−PB vacuum treatment devices.

5. $2\times$RH−PB, whose profile is shown in Fig. 6.56, could produce tubes, sheets, and high-quality bearing steel bar with ultralow [P] and [S]. The production scale was 4.0−4.5 million tons. Tubes were produced by the round billet caster. Slab caster supplied slabs for hot rolling mills to produce high-quality sheets. Billet casters supplied billets to produce high-quality long products. It is worthy of note that there were only two RH−PB devices, and no LF or VD devices in the plant. The production process is $2\times$KR$-1\times$BOF$_{De[Si]/[P]}-2\times$BOF$_{De[C]}-2\times$RH−PB and this can efficiently produce clean steel.

6.3.4 Different Types of Steel Plants with De[S]−De[Si]/[P] Pretreatment in Japan

Since 1987, many steelmaking plants in Japan had been concerned about the hot metal dephosphorization pretreatment in the converter and had made rapid progress on this technology. Different technological conditions and product requirements in steelmaking plants led to different dephosphorization pretreatment approaches, as shown in Fig. 6.57, from which we can see that the dephosphorization in the converter took a large proportion. Fig. 6.58 shows the ratio of dephosphorization pretreatment in a combined blowing converter in Japan from 1987 to 2005, from which we can see that in 2005 the converter dephosphorization rate was about 53%.

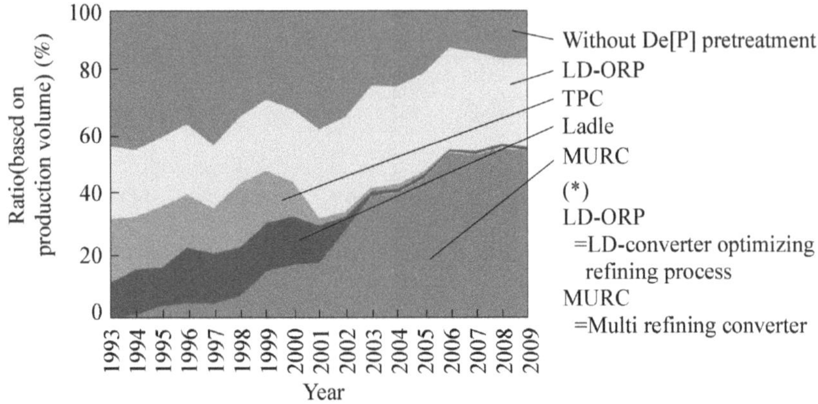

FIGURE 6.57 Development of hot metal dephosphorization process in Nippon Steel Corporation (Iwasaki and Matsuo, 2011).

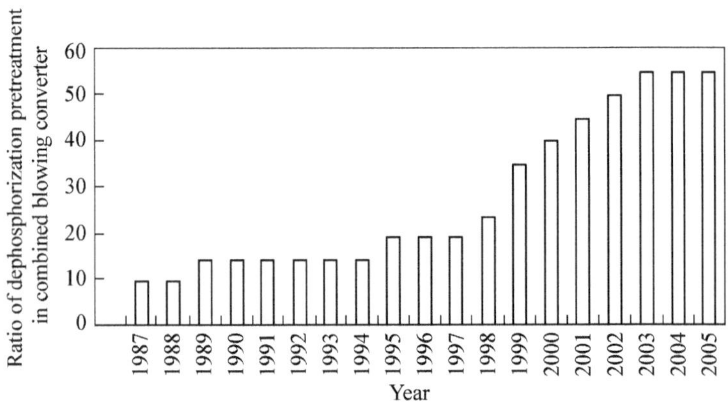

FIGURE 6.58 The ratio of dephosphorization pretreatment in combined blowing converter in Japan from 1987 to 2005.

From the viewpoint of technological process, there are three types of hot metal dephosphorization:

Type I: Dephosphorization pretreatment and decarburization are completed in two converters, respectively, as shown in Fig. 6.59, for example, the SRP in the Wakayama Iron & Steel Works and the LD-ORP of Nagoya Iron & Steel Works. The characteristics are as follows:

1. The BOF tap-to-tap time is reduced because of the simultaneous operation of $BOF_{De[Si]/[P]}$ and $BOF_{De[C]}$.
2. The BOF tonnage can be decreased if a newly built steel plant adopts this dephosphorization pretreatment method.

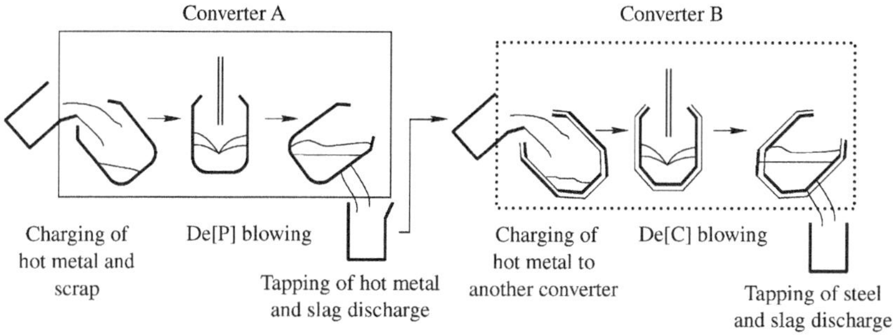

FIGURE 6.59 Dephosphorization pretreatment and decarburization are completed in two converters respectively (Kitamura and Ogawa, 2001).

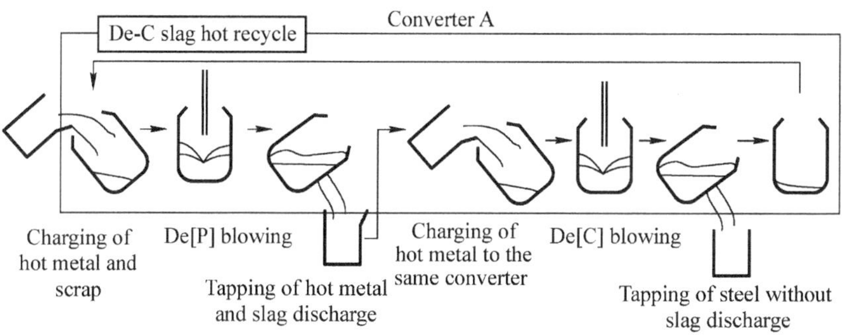

FIGURE 6.60 Dephosphorization pretreatment or decarburization and heating are completed in one converter.

3. The lime consumption obviously declines because of the metal-slag complete separation after dephosphorization pretreatment.
4. It is conducive to the $BOF_{De[C]}$ direct tapping with low [O] and [N] content in molten steel.
5. The products are of high, stable, and reliable cleanness.
6. The $BOF_{De[Si]/[P]}$ is a necessity.

Type II: Dephosphorization pretreatment or decarburization and heating are wholly completed in one converter, as shown in Fig. 6.60. The characteristics are as follows:

1. The metal-slag is completely separated after the dephosphorization pretreatment, which is helpful for producing ultralow [P] steel.
2. The high temperature decarburization slags can be directly used in dephosphorization.
3. The long BOF tap-to-tap time causes a lower production efficiency.

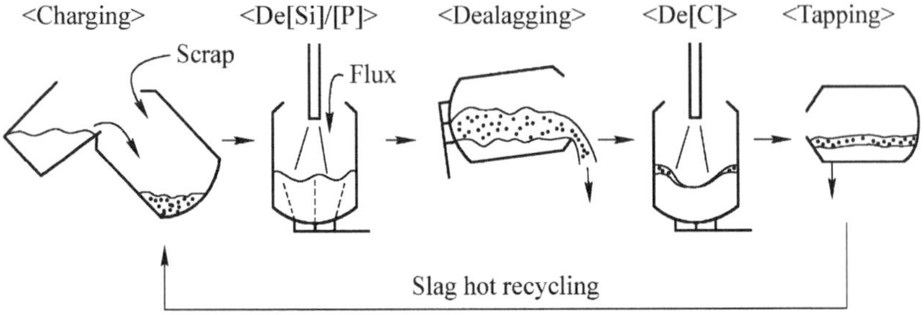

FIGURE 6.61 Slag-remaining and double slag steelmaking technology in one converter (MURC) (Ogawa et al., 2001).

4. The fluctuating BOF tap-to-tap time and imbalanced production rhythm will affect the sequence CC.
5. Can only be used temporarily in ultralow [P] steel production, rather than in a long-time, stable, and economical mass production.

Type III: Slag-remaining and double slag steelmaking technology in one converter, as shown in Fig. 6.61. The characteristics are as follows:

1. Slag from this heat can be used in the next heat to realize the early dephosphorization. The partial primary slag elimination can reduce the lime consumption.
2. The adequate heat can increase the coolant usage and decrease the hot metal charging ratio.
3. The deslagging process prolongs the operation time of BOF and may lead to the loss of hot metal.
4. The metal-slag separation cannot be fully completed in the deslagging process, which affects the [P] content in finished molten steel.

6.3.5 Development of De[S]–De[Si]/[P] Pretreatment in Korea

Since the start of the 21st century, De[S]–De[Si]/[P] pretreatment technology of hot metal has been developed in Korea and applied in Pohang Iron & Steel Works and Gwangyang Iron & Steel Works. This technology has also been used in the newly built Karatsu Iron & Steel Works.

6.3.5.1 New Dephosphorization Converter in the No.2 Steelmaking Plant of Pohang Iron & Steel Works

In 2008, the No.1 steelmaking plant of Pohang Iron & Steel Works was out of production so that there was excess hot metal being imposed on the No.2 steelmaking plant (3 × 300 t BOFs). A new 300 t $BOF_{De[Si]/[P]}$ was added to the No.2 steelmaking plant to alleviate the situation, as shown in Fig. 6.62.

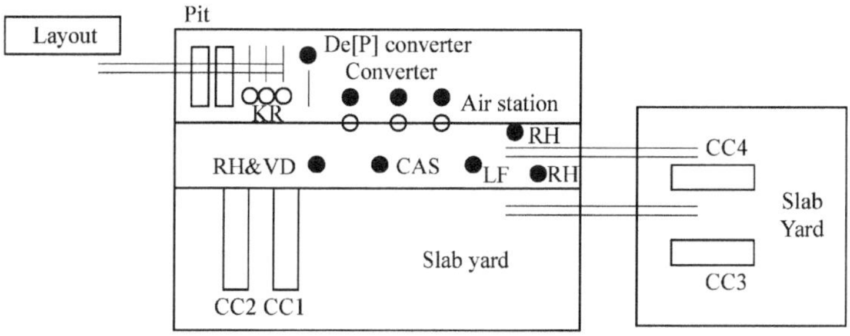

FIGURE 6.62 Sketch of De[S]–De[Si]/[P] pretreatment layout in No.2 steelmaking plant of Pohang Iron & Steel Works.

The $BOF_{De[Si]/[P]}$ was used to dispose of only 10% of the added hot metal in 2008.

The No.2 steelmaking plant was many years old. The location of the $BOF_{De[Si]/[P]}$ was limited by the layout of the plant and could only be installed in a crowded arrangement. Even so, the $BOF_{De[Si]/[P]}$ was not located in the same span as $BOF_{De[C]}$, which avoided interference between the two.

The No.2 steelmaking plant had adopted the KR desulfurization process so that the [S] content in hot metal declined from 0.025% to 0.0025%. The hot metal was then charged into the $BOF_{De[Si]/[P]}$ for the desilication-dephosphorization pretreatment with 10 minutes' oxygen-blowing and a blowing intensity half that in the $BOF_{De[C]}$. At the same time, slag from $BOF_{De[C]}$ was charged into $BOF_{De[Si]/[P]}$ in which the basicity was about 2. After this, the [P] content decreased from 0.12% to 0.02%.

After the technological process KR–$BOF_{De[Si]/[P]}$–$BOF_{De[C]}$ came the No.3 and No.4 slab CC with a respective 2.8 million and 3.4 million tons capacity. The casting time for a heat is 35–45 min.

6.3.5.2 New Dephosphorization Converter in No.2 Steelmaking Plant of Gwangyang Iron & Steel Works

In 2007, the No.2 steelmaking plant (3 × 250 t BOF) of Gwangyang Iron & Steel Works installed a new 250 t $BOF_{De[Si]/[P]}$ which was located at the same span as a desulfurization station and at a different span as the original $BOF_{De[C]}$, as shown in Fig. 6.63. According to the layout, the semisteel from $BOF_{De[Si]/[P]}$ could be directly charged into any of the original three $BOF_{De[C]}$, and the material flow from $BOF_{De[Si]/[P]}$ to three $BOF_{De[C]}$ was with little interference. However, the distance from the hot metal pretreatment area, where the $BOF_{De[Si]/[P]}$ was located, to the original BOF workshop was about 200 m, a long distance which caused a greater temperature drop of the semisteel.

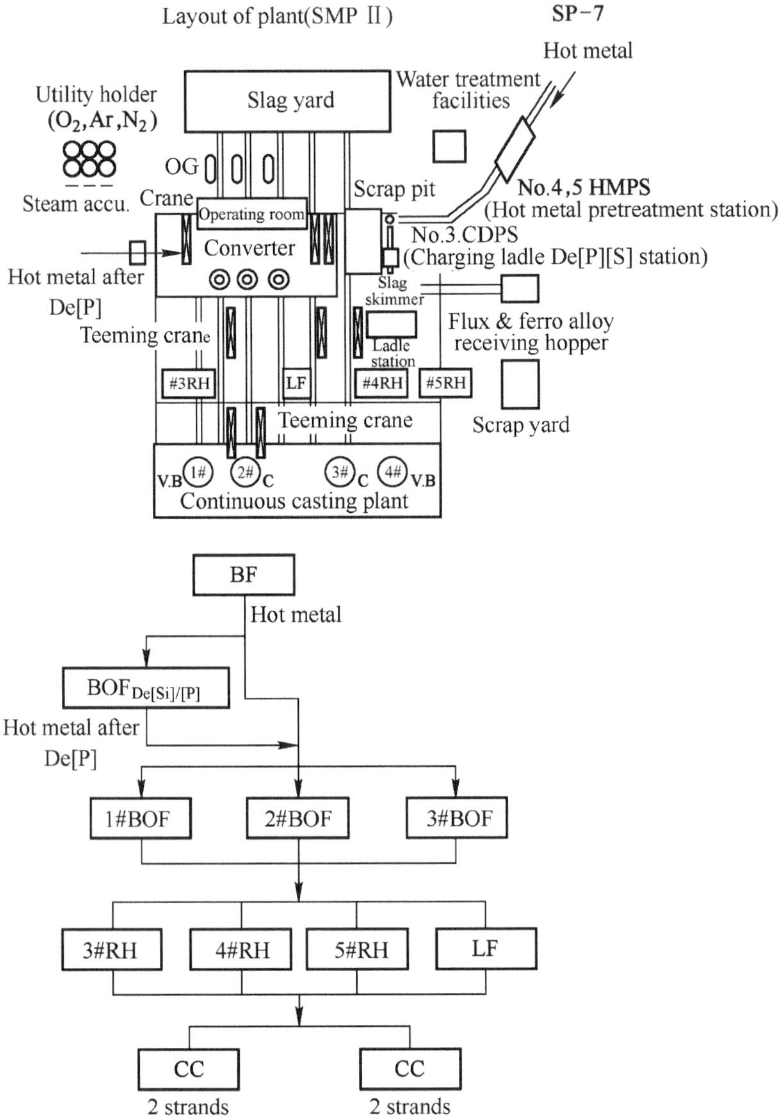

FIGURE 6.63 The layout of No.2 steelmaking plant of Gwangyang Iron & Steel Works.

Hot metal after KR desulfurization was charged into $BOF_{De[Si]/[P]}$ with a 7–10 minutes' oxygen-blowing and a blowing intensity of 1.5–2.0 $Nm^3/t \cdot min$ (a flow rate about 20,000–30,000 Nm^3/h), for the desilication-dephosphorization pretreatment. The basicity of slag was about 2.0, and its constitution included the lime of 10 kg/t (steel) and partial $BOF_{De[C]}$ returned slag.

After the $BOF_{De[Si]/[P]}$ blowing, the [C] in semisteel decreased from 4.3% to 3.3% and the semisteel temperature was about 1250–1400°C.

The BOFs in this plant could recycle gas; the $BOF_{De[Si]/[P]}$ with a flow rate of 35,000–40,000 Nm^3/h, and the $BOF_{De[C]}$ with 150,000 Nm^3/h. The productivity/heat of $BOF_{De[C]}$ was 275 t, the lime for slagging 15 kg/t (steel) and the tap-to-tap time 25–30 min.

The average productivity was about 105 heats per day in the $BOF_{De[Si]/[P]}$-installed No.2 steelmaking plant and 50% of total production underwent the full hot metal pretreatment.

In this plant, there were three RH devices and one LF device (see Fig. 6.63); and 70% of molten steel was processed by RH.

The cool rolled sheet occupied more than 50% of the total production in this plant. All the cold rolled sheet, the low carbon hot rolled sheet, and API oil tube were also processed by RH.

6.3.6 Design and Operation of De[S]–De[Si]/[P] Pretreatment at Shougang Jingtang Steel in China

The Shougang Jingtang Steel was put into production in 2009 and had a "2-1-2 structure," which is 2 × 5576 m^3 BF–1 × De[S]–De[Si]/[P] pretreatment steelmaking plant–2 × hot rolling mills.

The characteristics of the design and operation of the steelmaking plant were as follows:

1. The steelmaking plant covers the ranges from the BF hot metal charged into the ladle to the CC output. The dynamic-orderly and coordinative-continuous operation was adopted as the basic concept from the beginning of the design of the plant.
2. Procedures between BF and BOF were arranged as follows: hot metal dephosphorization pretreatment in KR device—desiliconization and dephosphorization in $BOF_{De[Si]/[P]}$—decarburization and heating in $BOF_{De[C]}$, as shown in Fig. 6.8, that is, an analysis-optimization on the dynamic-orderly restructuring of the metallurgical functions of traditional BOF.
3. The BF hot metal was charged by the hot metal ladle and then transported by a standard 1435 mm railway to the hot metal pretreatment station (4 × KR), a distance of approximately 800 m which took less than 20 min. The process of deslagging–desulfurization–deslagging was carried out in the hot metal ladle within 30–35 min, then the hot metal was charged into the $BOF_{De[Si]/[P]}$. Here the hot metal ladle replaced the torpedo or mixer and was changed into a multifunction hot metal ladle, which connected BF tapping–KR desulfurization–$BOF_{De[Si]/[P]}$ dephosphorization and therefore realized energy-savings, environmental protection, fast-paced turnover, and accurate weighing.

4. The $2 \times BOF_{De[Si]/[P]}$ were adopted to conduct the hot metal desiliconization/dephosphorization after the KR pretreatment. Then the semisteel was charged into $BOF_{De[C]}$ for decarburization, heating, gas recycle, and further moderate dephosphorization. The $2 \times BOF_{De[Si]/[P]}$ and $3 \times BOF_{De[C]}$ were arranged in different but adjacent spans to reduce interferences.
5. The composition, temperature, and weight of semisteel charged into the $BOF_{De[Si]/[P]}$ were steady and accurately determined, and the $BOF_{De[C]}$ needed no steel scrap, only 10–15 kg/t (steel) of lime and less than 30 kg/t (steel) of slag. So the hitting rate of the molten steel endpoint [C] and the temperature in $BOF_{De[C]}$ could be enhanced, which was conductive to the $BOF_{De[C]}$ direct tapping and its fast-paced production time-rhythm. Moreover, [N] increase due to a second or third blowing could be avoided, and the [O] content in molten steel was also relatively low.
6. High clean molten steel undergoing De[S]–De[Si]/[P] pretreatment and $BOF_{De[C]}$ direct tapping, went through CAS, RH, and LF, and then the high-speed slab CC with 230 mm thickness and different widths. Keeping laminar type running of $BOF_{De[C]}$–secondary refining–CC can realize the high-efficiency dynamic-orderly and coordinative-continuous operation.

The technology/devices were selected guided by the above design and operation principles. Based on the general layout, the steelmaking plant layout (see Fig. 6.64) and reasonable production operation arrangement, the structure optimization, and function expansion of the whole process could be achieved.

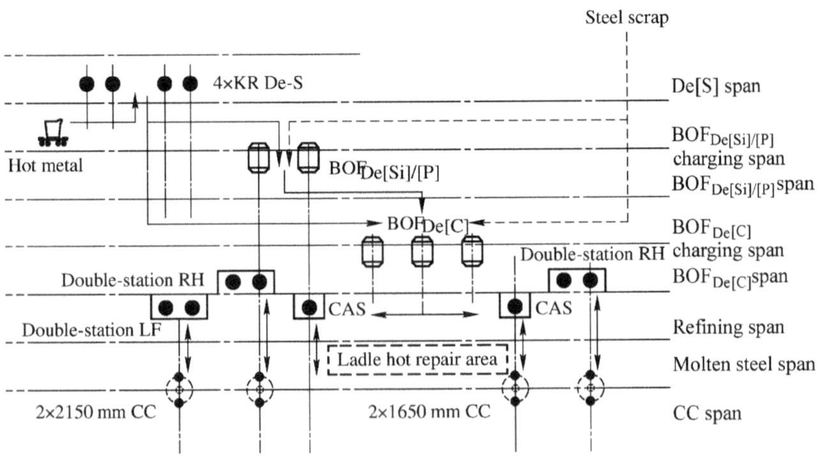

FIGURE 6.64 The layout of steelmaking plant of Shougang Jingtang Steel.

Two years of practices had seen the progress of the high-efficiency and low-cost clean steel production platform in Shougang Jingtang Steel, especially:

1. The advantages of interface technology between BF and BOF had been proved by the production practice, which was epitomized by the One Ladle Technology. The Jiangtang Steel abandoned the torpedo or mixer and the 300 t ladle was used to transport the hot metal form BF to KR desulfurization station via a standard 1435 mm railway within 20 min. Moreover, the installed accurate hot metal weighing system (see Fig. 6.65), with a 288 ± 0.5 t accuracy (the accurate tapping ratio of BF hot metal could reach 95%), provided advantages for the steady operation of the KR device, $BOF_{De[Si]/[P]}$, and $BOF_{De[C]}$. At the same time, there was no half ladle-heat of hot metal in the plant, which was beneficial to the management and fast turnover of hot metal ladle.

   Because of the One Ladle Technology, the temperature of hot metal arriving at the KR desulfurization station was generally above 1380°C or even 1440°C. Besides, due to the pretreatment order of desulfurization → desiliconization, The KR desulfurization process was conducted at a relatively high temperature and high [S] activity with a better desulfurization efficiency, after which the [S] contents in hot metal are almost less than 0.0025%; and 50–60% of those contents are less than 5 ppm.
2. The short tap-to-tap time of the 300 t $BOF_{De[Si]/[P]}$: After KR desulfurization, the 1350–1360°C hot metal was charged into the $BOF_{De[Si]/[P]}$ for blowing, so were the 14 kg/t (steel) lime and 12 kg/t (steel) $BOF_{De[C]}$ returned slag at the same time, as shown in Fig. 6.66. The blowing time

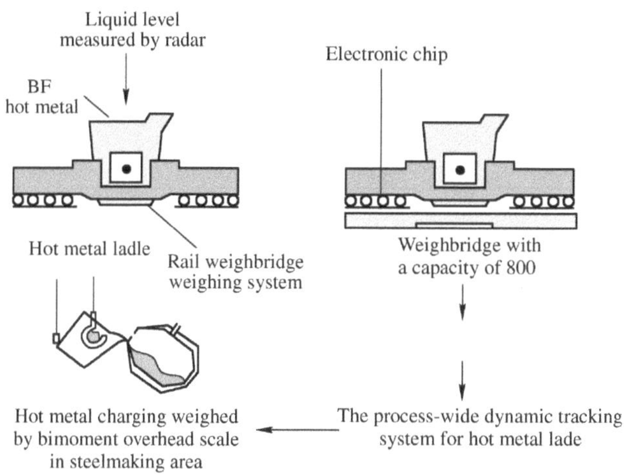

FIGURE 6.65 The hot metal weighing system in Shougang Jingtang Steel.

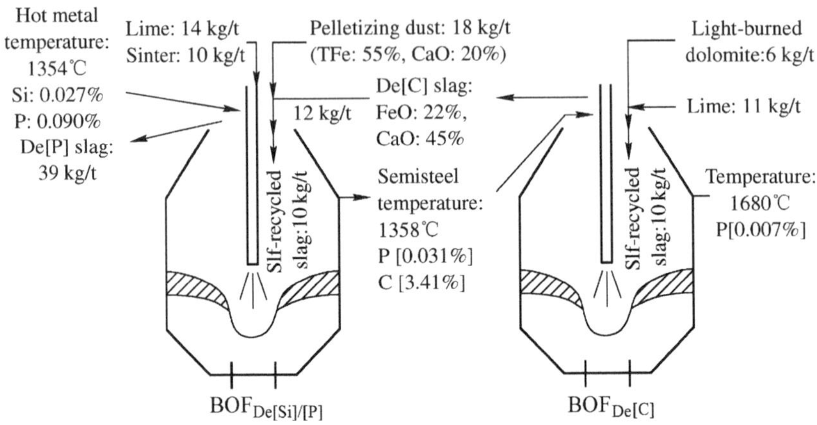

FIGURE 6.66 Relationship between slagging agent and slag in respective $BOF_{De[Si]/[P]}$ and $BOF_{De[C]}$.

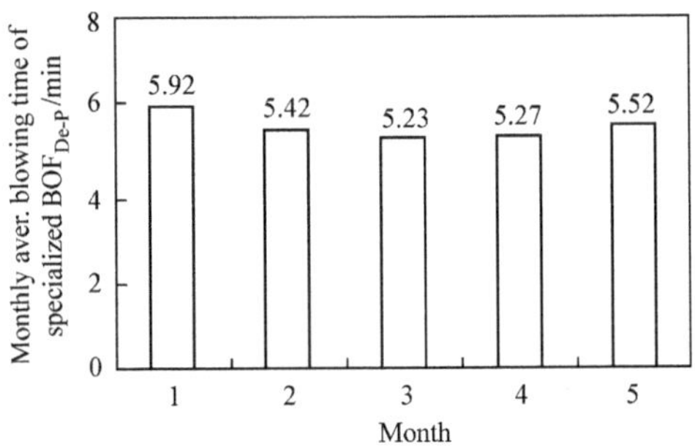

FIGURE 6.67 Monthly average blowing time of $BOF_{De[Si]/[P]}$ in 2010.

had declined from the initial 9 min to the then 6 min, as shown in Fig. 6.67.

Due to the optimization of the slagging process, increment of top oxygen-blowing intensity and enhancement of the maintenance of the bottom blowing system, etc., the slag basicity could be controlled at 1.8–2.0 and the FeO content at 12%. It should be pointed out that the designed bottom blowing intensity was relatively low at only 0.3 $m^3/t \cdot min$, while the practical intensity was even lower than the designed one, which had an adverse effect on dephosphorization rate. The average composition of the $BOF_{De[Si]/[P]}$ slag, and the average composition and temperature of $BOF_{De[Si]/[P]}$ semisteel at Shougang Jingtang Steel are shown in Fig. 6.68 and Table 6.20.

As shown in Table 6.20, it might be owing to the lower blowing intensity and higher endpoint temperature of molten steel in the $BOF_{De[Si]/[P]}$ that led to the endpoint [P] being above 0.030%.

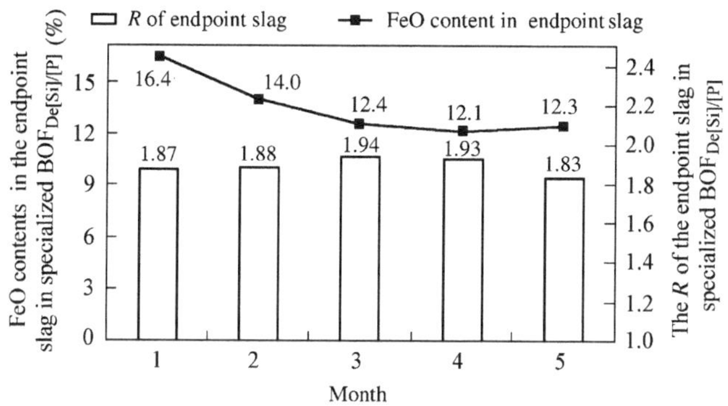

FIGURE 6.68 Monthly average index of final slag from $BOF_{De[Si]/[P]}$ in 2010.

TABLE 6.20 The Monthly Average Index of Endpoint Molten Steel in the $BOF_{De[Si]/[P]}$ in 2010

Item	Unit	January	February	March	April	May
Endpoint [C]	%	3.29	3.36	3.46	3.44	3.45
Endpoint [P]	%	0.034	0.035	0.033	0.031	0.032
Endpoint [S]	%	0.0087	0.0089	0.0074	0.0068	0.0065
Endpoint [Si]	%	0.020	0.022	0.021	0.020	0.020
Endpoint [Mn]	%	0.041	0.044	0.037	0.031	0.034
Endpoint T	°C	1339.3	1336.2	1335.4	1337.0	1332.9

3. Reduced slag smelting in the 300 t $BOF_{De[C]}$: After the KR process with a high temperature and a high [S] activity and the $BOF_{De[Si]/[P]}$ dephosphorization pretreatment, the metallurgical assignment on $BOF_{De[C]}$ could be largely reduced so that less slag smelting and high-speed blowing could be carried out. The lime consumption was reduced to 10–11 kg/t (steel); the hitting rate of endpoint [C] content and temperature of molten steel were both above 94%. The double hitting rate of [C] content and temperature was above 90%, as shown in Fig. 6.69. Therefore, the rate of direct tapping increased. Even if affected by the development of some new grade steels, the direct tapping rate reached about 50%.

The endpoint [P] content and the relationship between the endpoint [C] and [O] contents in $BOF_{De[C]}$ in Shougang Jingtang Steel are shown in Figs. 6.70 and 6.71, respectively.

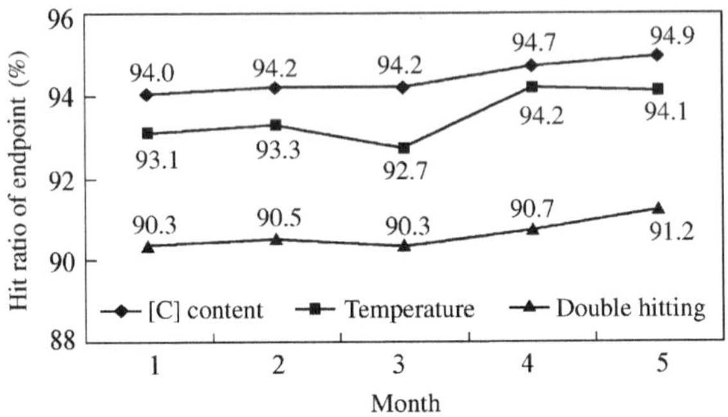

FIGURE 6.69 Hitting rate of endpoint [C] contents in the 300 t $BOF_{De[C]}$ from January 2012 to May 2012 in Shougang Jingtang Steel.

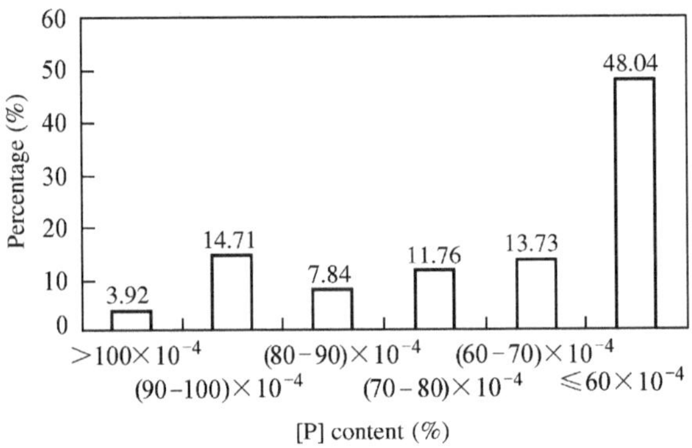

FIGURE 6.70 Endpoint [P] content distribution in the 300 t $BOF_{De[C]}$.

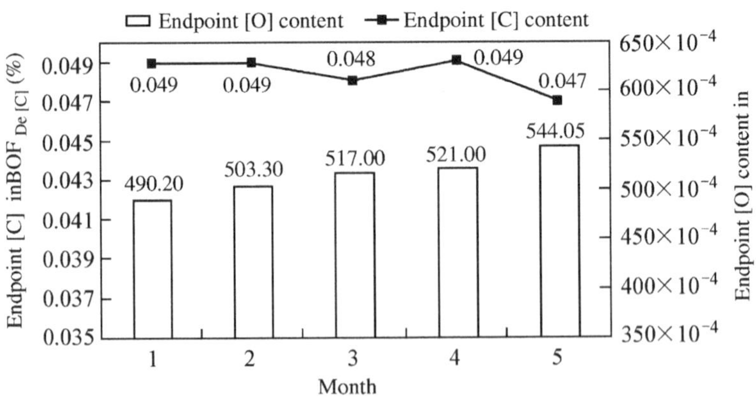

FIGURE 6.71 Relationship between endpoint [C] and [O] contents in the 300 t in $BOF_{De[C]}$.

6.3.7 A Conceived High-Efficiency and Low-Cost Clean Steel Production Platform (Large-Scale Full Sheet Production Steelmaking Plant)

In this conceived steelmaking plant, the products would be hot rolled sheet and corresponding deep-processed products. For the cold-rolled and deep-processed rate being above 60%, two hot strip rolling mills are adopted, including one 2050 mm hot strip rolling mill with an annual productivity of 5.0–5.3 million tons and another 1580 mm one with an annual productivity of 3.5 million tons. So, the total annual productivity of the steelmaking plant will be about 9.0 million tons. Targeting this and based on experiences of Wakayama Iron & Steel Works and Shougang Jingtang Steel, we could envisage the structures of the new plant as follows:

1. The high-efficiency and low-cost clean steel production process based on the De[S]–De[Si]/[P] pretreatment and its fundamental technological frames are:
   a. The One Ladle Technology should be taken as the interface technology between BF and BOF. The transport paths need to be adaptive, such as train-railway, flatcar-railway, or automobile-road, which should be selected by principles of accurate weighing (230 ± 0.3 t), concise and fast transportation (ladle turnover rate is 5 times/day per ladle), and low cost.
   b. A total of 2–3 KR devices are applied to first carry out the desulfurization at high temperature and [S] activity and the subsequent deslagging process. The KR device should be located as close as possible to BF in order to carry out the desulfurization process above 1400°C, increasing the desulfurization rate, and realizing both the [S] in hot metal below 0.001% and the after-processing temperature above 1370°C. The combination of pre-deslagging–stirring dephosphorization–post-deslagging in the KR device, with the tap-to-tap time of 20–24 min, can increase its utilization efficiency and decrease temperature drop owing to the long no-stirring interval.
   c. The 2 × 230–250 t $BOF_{De[Si]/[P]}$ are applied to carry out the desiliconization/dephosphorization process, in which the bottom blowing intensity should be highly valued and the steam and gas recycled. The profiles of $BOF_{De[Si]/[P]}$ should be smaller than that of $BOF_{De[C]}$ in the same tonnage to reduce the investment cost. The tap-to-tap time of $BOF_{De[Si]/[P]}$ should not be more than 22 min. The $BOF_{De[C]}$ returned slag should be recycled as part of $BOF_{De[Si]/[P]}$ slag and meanwhile moderate lime charged into the $BOF_{De[Si]/[P]}$ to keep the endpoint slag basicity at about 2.0. In this procedure, only 2 ladles are used for fast turnover and reducing the temperature drop in the tapping process.

d. The $3 \times 230-250$ t $BOF_{De[C]}$ are applied for rapid decarburization, temperature rise, gas and steam recycle, and auxiliary dephosphorization. The design of the oxygen lance of $BOF_{De[C]}$ should be promoted for rapid decarburization and reducing splashing at an oxygen blowing intensity of 4–4.5 $Nm^3/(t \cdot min)$. The BOF gas can be used to roast lime to reduce the sulfur content in lime.

By quick and direct tapping technology, the tap-to-tap time of $BOF_{De[C]}$ is 28–32 min.

e. The 3 × RH–PB and 1 × LF (1 × CAS if necessary) are adopted as the secondary refining devices to process the cold rolled sheet and hot rolled sheet, respectively. In this process, it is very important to fully utilize the RH devices so that the RH needs to be located on the orientation of the BOF tapping line to reduce the transport time (it is the same with the CAS). LF should be next to the 2050 mm hot rolling mill.

f. The 3 × slab CC, 230 mm in thickness, are adopted for the high/constant-speed casting and need to be acclimated by their metallurgical length. The basic operation widths of the three CC are 1650 mm, 1450 mm, and 1250 mm, respectively. The slab width is generally not changed during the CC procedure and the width adjustment function should be partially undertaken by the hot rolling mill, which will be helpful for the steady operation rhythm of the plant-wide production process, the high-efficiency and constant speed operation of CC, the stable product quality, and a better casting yield.

g. Specialized production based on the dynamic operation of three casters should be considered.

No.1 CC corresponds to the 1580 mm rolling mill which produces high-grade commercial steel products and the majority are cold rolled and with narrow gauge. The steel grades are low/ultralow carbon steel for cold rolling or steel grades fit for direct charging. The annual productivity is 2.7–2.8 million tons.

No.2 CC corresponds to the 2050 mm (or above) wide rolling mill which produces pipe line steel and cold-rolled and deep-processed products, such as automobile sheet. The annual productivity is 2.8–3.0 million tons.

No.3 CC corresponds to the 2050 mm (or above) wide rolling mill which produces hot rolled products. The No.3 CC mainly produces steel grades that are in mass production, in wider width and fit to the long-time casting, high-speed casting, and direct charging. The annual productivity is above 3.5 million tons.

The secondary refining devices for the three casters are as follows:

- No.1 CC corresponds to RH and, if necessary, CAS;
- No.2 CC corresponds to RH and sometimes LF;
- No.3 CC corresponds to RH and, if necessary, CAS.

2. The characteristics of the steelmaking plant layout, that is, process network optimization.

In the layout design of the steelmaking plant adopting De[S]−De[Si]/[P] pretreatment for clean steel production, the following characteristics should be noted:

a. If the train-railway is adopted as the mode of hot metal ladle transportation, the railway line should generally extend into the dephosphorization span along the length direction of the steel plant (the vertical direction if necessary). The transport distance should be as short as possible in order to speed up the ladle turnover. The KR station should be close to $BOF_{De[Si]/[P]}$ in order to reduce the ladle recharging times. If a flatcar-railway mode is adopted, it is also necessary to shorten the transport distance.

b. $BOF_{De[Si]/[P]}$ and $BOF_{De[C]}$ must be located at different spans, or the cranes-logistics will mutually interfere, which therefore affects the reasonable linking-matching of the plant-wide production time-rhythm.

c. The reasonable locations of secondary refining devices should be considered.

RH: RH is the main secondary refining device for top grade sheet (even tube and high-grade bars), especially the cold-rolled and deep-processed products. This device should be installed on the $BOF_{De[C]}$ tapping line so that the crane-transporting time from $BOF_{De[C]}$ to RH could be saved and the interferences between cranes at tapping span be avoided. This arrangement (especially if RH adopts two tapping lines and has three processing positions, that is, pretreatment position, vacuum-treatment position, and posttreatment position) is useful to realize sequence CC, to reduce the number of cranes, and to process a heat of molten metal every 30 min (not the RH processing time). It is also useful to realize continual operation of the hot vacuum tank, which reduces temperature drop, increases vacuum lifespan, and reduces operation costs.

CAS: CAS is the low-cost and high-efficiency production device for block hot rolled sheet. For the dynamic-orderly operation, the location of CAS is also very important and it should be installed next to $BOF_{De[C]}$, on the tapping line or close to the operation line, albeit its short processing time.

LF: LF is the primary secondary refining device for hot rolled coils with ultralow [S] and hot rolled products with medium/high [C] (eg, pipeline steel). Generally, LF is only applied in producing steel with ultralow [S] and [P] content or high-grade plate which are usually thick and wide. Therefore, LF should be installed close to the 2050 mm (or above) hot rolling mill. However, LF has a longer tap-to-tap time, and the hot rolled products with ultralow [S] content have only a small market share, so LF should be installed relatively far away from tapping operation line instead of being located next to the $BOF_{De[C]}$ tapping line. In order to keep the high-efficiency operation of the large BOF, LF should only be used to produce steels with ultra-low [S] and [O]. At the same time, it should be noted that LF may lead to [Si] and [N] increase in steel, and LF should not be taken as a general molten steel heating measure.

d. The output lines of the main secondary refining devices such as RH should be aimed at or next to the axis of the rotary table in order to reduce the transport time and realize the fast sequence CC.

e. The aligning operation time during the crane's traversing should be shortened. An interlock device could be installed at the $BOF_{De[Si]/[P]}$ and $BOF_{De[C]}$ charging position for the crane's fast aiming at the converter mouth and fast charging.

f. Attention should be paid on the looping path of the hot metal ladle, semisteel ladle, and molten steel ladle. Fast-paced turnover of ladles is conducive to reducing their number and the molten metal temperature drop.

g. The steelmaking plant layout should cater to the laminar operation of three specialized CCs with different widths and casting speeds, which is the core idea of the dynamic-tailored design of the plant. The layout is the static framework of the dynamic-orderly and coordinative-continuous operation in the steelmaking plant, as shown in Fig. 6.72, which must be discreetly and dynamically researched. Additionally, the slab hot charging path should also be carefully treated for the rational arrangement of the linkage of the steelmaking plant and the two hot rolling mills.

3. The dynamic-orderly and coordinative-continuous operation program in the steelmaking plan.

We should highly emphasize the multiprocedure dynamic operation program, the dynamic-tailored design, and the adjustment of time-rhythm for each procedure/device. Take the time parameters of the sequence CC procedure, such as time-process, time-rhythm, and time-cycle as the pull force of the dynamic-orderly operations of all the procedures/devices, and make out the time-scheduling plan of the entire steelmaking plant—a dynamic operation Gantt chart. All the arrangements of working times and running rhythms of KR, $BOF_{De[Si]/[P]}$, $BOF_{De[C]}$, RH, CAS, LF, ladles, cranes, etc., should be based on the dynamic Gantt chart. When casting different steel grades, billets with different transverse sections, all the three casters should keep a constant speed, which is the core of the time-rhythm of the dynamic operation from KR to CC. It is notable that the outspread of the Gantt chart dynamic operation should center on the CC. Typical examples have been shown in Figs. 6.73 and 6.74.

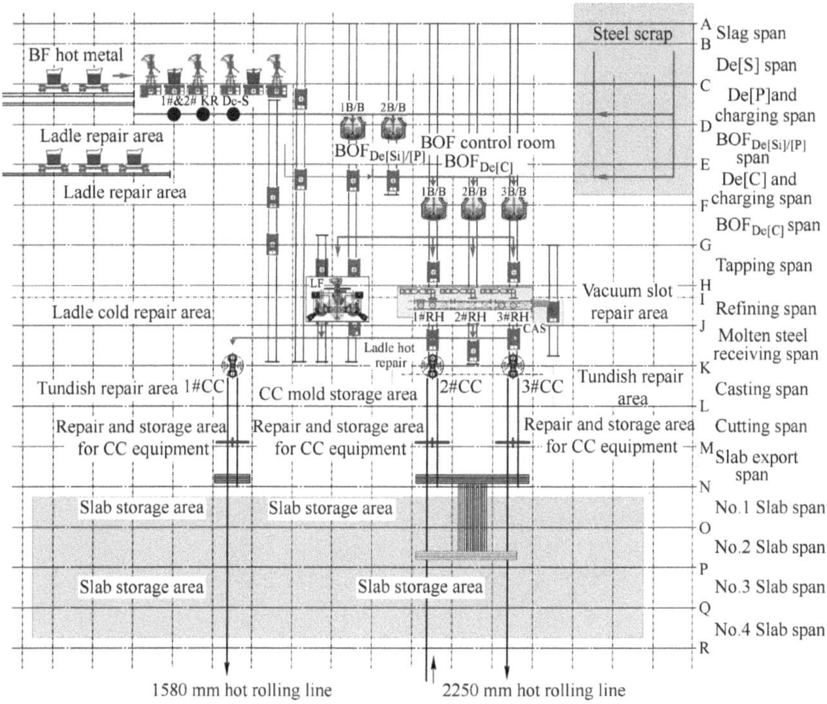

FIGURE 6.72 Layout of conceived high-efficiency and low-cost clean steel production platform (large-scale sheet production plant).

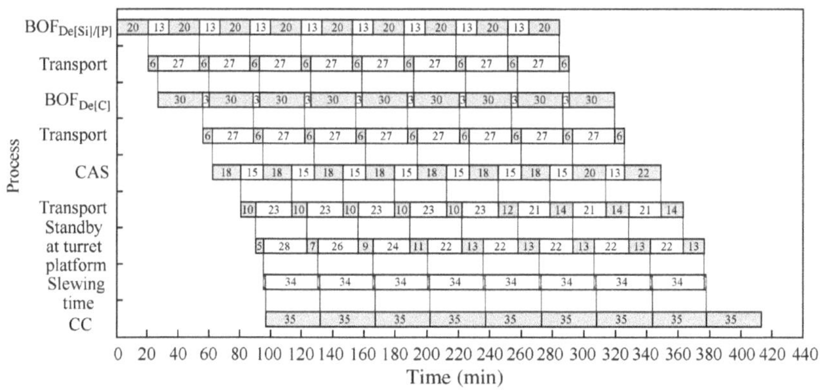

FIGURE 6.73 Gantt chart of 250 $BOF_{De[Si]/[P]}$–$BOF_{De[C]}$–CAS–CC.

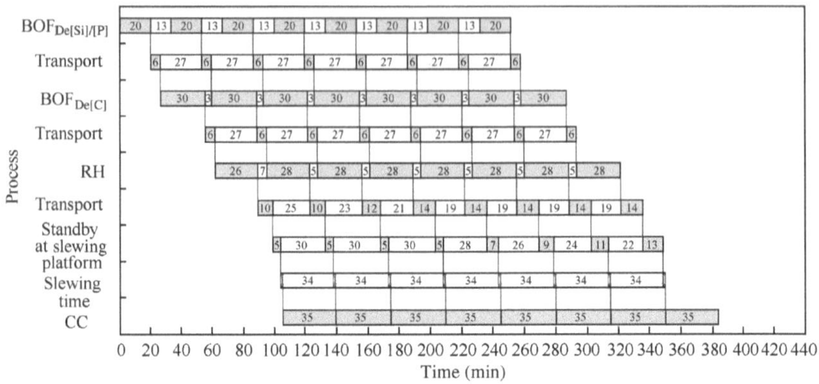

FIGURE 6.74 Gantt chart of 250 $BOF_{De[Si]/[P]}$−$BOF_{De[C]}$−RH−CC.

6.3.8 Theoretical Significance and Practical Value of De[S]−De[Si]/[P] Pretreatment

The theoretical significance of the De[S]−De[Si]/[P] pretreatment is as follows:

1. Present the analysis-optimization of the sets of procedure functions, the coordination-optimization of the sets of procedure relations, and the restructuring-optimization of the sets of process procedures.
2. Establish the basis for the steady running of the high-efficiency and low-cost clean steel production platform.
3. Be conducive to building computer models for the dynamic-orderly and continuous-compactness operation.

The embodiments of the practical value of De[S]−De[Si]/[P] pretreatment are as follows:

1. Build a stable and high-efficiency clean steel production platform, especially a platform for high-quality and high-speed casted sheets with ultra-low carbon contents.
2. The tap-to-tap time of the $BOF_{De[C]}$ is reduced by 8−10 min, therefore, the BOF tonnage, the crane tonnage, the load of steelmaking workshops, and corresponding investment cost could be reduced.
3. The production cost decrease. The reason is that metallic raw materials and lime consumption are reduced. If possible, the high [P] iron ore could be used; and the manganese ore could replace the majority or all the Fe-Mn alloy added in the $BOF_{De[C]}$.
4. The performances of high-grade products can be guaranteed, including machining property and utilization property.
5. Promote the sequence CC rate.
6. Promote the information management.

7. Be conducive to clean production and utilization of steelmaking slag.
8. Be conducive to the linkage between CC and hot strip mill and also to the billet hot charging rolling.
9. Be possible to moderately use the ore resources with high [P] content.

6.4 OPTIMIZATION OF INTERFACE TECHNOLOGY BETWEEN CC AND BAR ROLLING MILL

The No.2 steelmaking and rolling plant Tangshan Iron and Steel Corporation (hereafter referred to as Tangsteel) was originally built in 1958. Originally, it was just a side-blown converter workshop but gradually developed into a modern steelmaking and rolling plant after several technical revamps. The production line is composed of hot metal desulfurization pretreatment–55 t BOF–bottom argon blowing–165 mm × 165 mm billet casting. In addition, the No.1 and No.2 bar rolling lines were established in 1996 and 2003, respectively. The rebars in Φ12–Φ18 mm were produced with slitting rolling technology in No.1 bar rolling line, which consisted of 1 reheating furnace and 18 stands of bar rolling mills; while rebars in size larger than Φ20 mm were produced in the No.2 bar rolling line.

Tangsteel paid great attention to the plane layout relationship between the steelmaking plant and the bar rolling plant during the technical revamp, which, in fact, was the process network optimization. The compact and smooth transport line between No.6 caster and No.1 bar rolling line was 241.1 m in length, while the more compact and smooth one between No.5 caster and No.2 bar rolling line was merely 81.5 m. Billets from No.5 and No.6 CC were specially supplied for the two bar rolling lines, with a practical 220 million tons of bar steel output per year. The improved production technologies had played important roles, such as billet direct hot charging from casters to reheating furnace, fixed weight mode, and the related slitting rolling. Here the fixed weight mode was introduced later.

6.4.1 Technological Base of Billet Direct Hot Charging

The implementation of billet direct hot charging technology is based on a series of improvements in the producing techniques.

First, the compact and smooth plane layout (reasonable process network) is required so that the billets can be charged into the reheating furnace in the shortest distance, at the fastest rate, and in the relatively steady temperature range. The layout of the No.2 steelmaking and rolling plant in Tangsteel is shown in Fig. 6.75.

Second, the increase and stabilization of the casting speed, and the promotion of billets temperature after flame cutting must be paid attention to. The casting speed of 165 mm × 165 mm billet was stabilized at 2.15 m/min in Tangsteel.

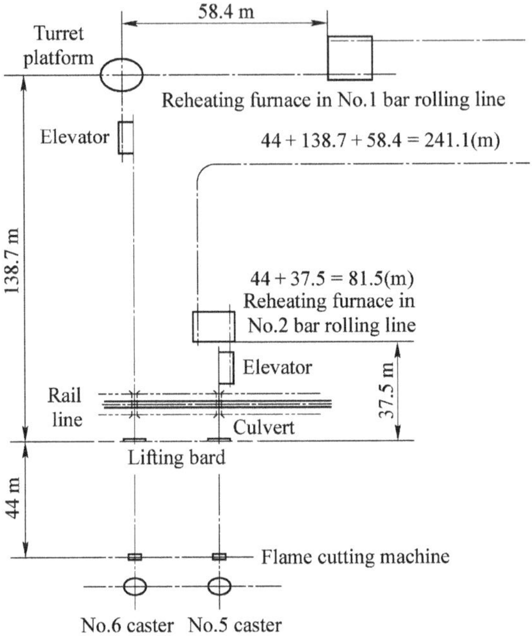

FIGURE 6.75 Layout of No.2 steelmaking and rolling plant in Tangsteel.

Third, it should emphasize the stabilization and decrease of the converter tapping temperature so as to increase and stabilize the casting speed. Through years of tackling key problems and taking multiple measures, the tapping temperature of 55 t converter had been stabilized around 1640°C in the No.2 steelmaking and rolling plant (detailed in the Appendix A of this chapter).

Fourth, it is necessary to pay more attention to the billet temperature after cutting. The billet temperature of No.5 and No.6 caster increased from 920°C to 970–980°C.

Fifth, it is necessary to keep the balance and continuity of the material flow between the billet caster and the bar rolling mill; and maintain the charging temperature of billet caster stabilized in a narrow range, which is good for energy savings for the heating furnace. Therefore, all the rebars below Φ18 mm were slit rolled in the No.1 rolling line with a then productivity of 100 million tons per year; while rebars above Φ20 mm are all rolled in the No.2 line with a then 120 million tons per year.

After all the work of detailed research, technical developments, and production information management related to the above aspects, obvious technical progresses, as well as energy saving and emission reduction effects, had been achieved in the No.2 steelmaking and rolling plant.

6.4.2 Practical Performance of Billet Direct Hot Charging Between No.6 Caster and No.1 Bar Rolling Mill

Corresponding to the production of No.1 converter (55 t), the 6-strand No.6 CC was a billet caster with billet section 165 mm × 165 mm. The tundish of the caster was 25 tons in capacity and the metallurgical length was 15 m. Through technical renovation, the casting speed of No.6 caster increased from 1.8 m/min to 2.1 m/min and the billet temperature after flame cutting increased from 940°C to above 975°C. The ratio of billet direct hot charging reached 84–90%. The billet was 11.5–12.0 m in length and more than 2.4 tons in weight. The billets were conveyed through rollers and hot charged into the reheating furnace in No.1 bar rolling mill. The conveying distance was about 241 m and it took about 10–17 min. The billet was charged one by one in heat number sequence, and the entrance temperature increased from the previous 690°C to 730°C or more. The gas consumption in reheating the furnace was reduced by an average of 26 Nm^3/t. The technical parameter and energy saving effects in different months are shown in Tables 6.21 and 6.22, and Fig. 6.76, respectively.

6.4.3 Practical Performance of Billet Direct Hot Charging Between No.5 Caster and No.2 Bar Mill

The strand number, section dimension, and metallurgical length of No.5 casting machine were similar to those of No.6 caster, and it matched the production of No.4 converter. The casting speed was stabilized at 2.14–2.15 m/min and the temperature after flame cutting was around 970–980°C. The ratio of billet direct hot charging is 100%, except for 1500–3000 billets per month which were transferred to No.1 bar rolling mill and added into the cold charged batches.

The billet was about 11.5–12.0 m in length and more than 2.4 tons in single weight. The billets were conveyed through rollers and hot charged into the reheating furnace in No.2 bar rolling mill. The distance of the rollers was around 81.5 m and it took only 3.4–4.4 min. The charging was one by one in sequence by following the heat number. Because of the short conveying distance, the charging temperature could be well controlled between 830°C and 850°C. Accordingly, the gas consumption was reduced to below 70 Nm^3/t. The slitting rolling was not adopted in No.2 rolling mill as all the products diameters were larger than 20 mm. The technical parameters and energy saving effects in different months are shown in Tables 6.23 and 6.24, and Fig. 6.77, respectively.

As shown in the tables and figures above, the complete successive direct hot charging connection had been realized between No.5 caster and No.2 bar rolling mill. Except for some billets (1500–3000 billets per month) that were transferred and incorporated into a cold charging batch in No.1 rolling

TABLE 6.21 Billet Direct Hot Charging Between No.6 Caster and No.1 Bar Rolling Line at Tangsteel

Item	Date	Casting Speed (m/min)	Billet Cut Temperature (°C) (Month average)	Billet Charging Temperature (°C) (Month average)	Gas Total Consumption (Nkm^3/Month)	Gas Specific Consumption[a] (Nm^3/t)	Billet Production (Number/Month)	Billet Direct Hot Charging Ratio (%)
Before optimization	July 2011	1.83	913.8	671.5	10,573	113	-	-
	Aug. 2011	1.85	922.2	676.3	10,253	105	-	-
	Sep. 2011	1.88	930.1	682.4	7713	128	-	-
	Oct. 2011	1.92	938.0	698.8	11,328	115	-	—
	Nov.2011	1.98	951.0	705.7	9745	107	-	-
After optimization	Dec. 2011	2.12	965.7	724.3	8608	90	36,909	88.38
	Jan. 2012	2.11	966.7	731.0	7869	85	36,455	80.82
	Feb. 2012	2.14	971.2	732.1	7426	81	38,045	91.81
	Mar. 2012	2.15	974.3	730.9	8217	91	37,283	84.61
	Apr. 2012	2.13	978.3	735.9	7447	90	32,513	83.47

[a]*Calorific value, 1700 kcal/Nm^3.*

TABLE 6.22 Distribution of Billet Direct Hot Charging Between No.6 Caster and No.1 Bar Rolling Line

Month	Jan. 2012		Feb. 2012		Mar. 2012		Apr. 2012	
Temperature range (°C)	Number	Ratio (%)	Number	Ratio (%)	Number	Ratio (%)	Number	Ratio (%)
Below 660	7973	20.14	0	0.00	0	0.00	0	0.00
660–690	7973	20.14	8638	23.19	4070	10.70	1016	3.17
690–730	8606	21.74	9752	26.18	10,130	26.63	9153	28.57
730–760	14,342	36.23	17,935	48.15	21,736	57.14	21,866	68.25
760–780	689	1.74	1296	3.48	1248	3.28	641	2.00
Above 810	0	0.00		0.00	0	0.00	0	0.00

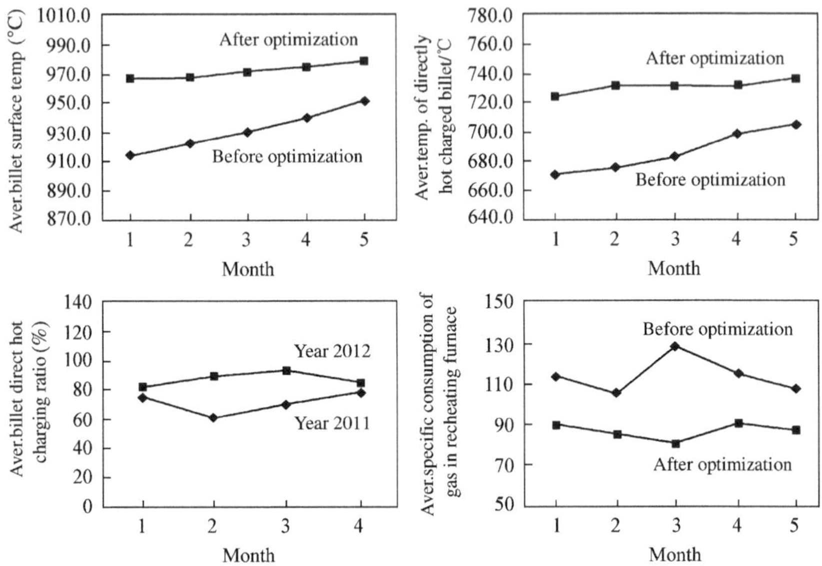

FIGURE 6.76 Comparison of relative technical index of billet direct hot charging between No.6 caster and No.1 rolling line.

mill, all the billets were in direct hot charging without intermediate storage. The surface temperatures of direct charged billets concentrated around 810–850°C had reached a ratio of up to 92–96% (see Table 6.24), which was especially beneficial to the improvement of heating quality in the reheating furnace and the further reduction of energy consumption. Consequently, importance should be attached not only to the promotion of the charging temperature, but also to the stabilization and the concentration of charging temperature coverage.

6.4.4 Progress on Fixed Weight Mode

Generally, the billet is supplied by length from the caster to the rolling mill, which is called the cut-to-length mode. The fixed weight mode is relative to the cut-to-length mode. When products in different sizes are rolled with billets in cut-to-length mode, a contradiction on fixed length ratio and rolling yield, etc., appears. In order to resolve these problems, fixed weight mode was developed in the No.2 steelmaking and rolling plant according to the requirements of diverse sections and different slitting rolling methods, and accordingly the rolling yield ratio increased.

The so-called fixed weight mode is that, under not only the guarantees of rational fixed product length ratio and negative allowance ratio of the rolled products in different section specifications, but also the consideration on the

TABLE 6.23 Billet Direct Hot Charging Condition Between No.5 Caster and No.2 Bar Rolling Line

Date	Casting Speed (m/min)	Billet Cutting Temperature (°C) (Month average)	Billet Charging Temperature (°C) (Month average)	Total Gas Consumption (Nkm^3/ Month)	Gas Specific Consumption (Nm^3/t)	Number of Billet per Month	Number of Billet Direct Hot Charging per Month
July 2011	2.13	966.3	835.5	6931	67	-	-
Aug. 2011	2.14	967.5	841.2	6356	65	-	-
Sep. 2011	2.14	970.1	846.6	6055	66	-	-
Oct. 2011	2.15	975.7	838.4	6697	67	-	-
Nov. 2011	2.14	981.6	842.3	7666	77	-	-
Dec. 2011	2.13	966.3	840.2	5044	65	34,969	32,267
Jan. 2012	2.14	967.5	844.3	7310	71	46,063	42,962
Feb. 2012	2.14	970.1	831.1	3175	70	21,259	18,987
Mar. 2012	2.15	975.7	828.1	6452	66	44,030	40,439
Apr. 2012	2.14	981.6	850.9	1608	62	39,676	37,544

TABLE 6.24 Temperature Distribution of Billet Direct Hot Charging Between No.5 Caster and No.2 Rolling Line

Month	January 2012		February 2012		March 2012		April 2012	
Surface temperature (°C)	Number	Ratio (%)	Number	Ratio (%)	Number	Ratio (%)	Number	Ratio (%)
Below 810	0	0.00	2628	13.84	8407	20.79	11,639	31.00
810–830	20,652	48.07	8453	44.52	7449	18.42	2433	6.48
830–850	18,843	43.86	6431	33.87	21,283	52.63	21,963	58.50
850–870	3467	8.07	1475	7.77	3300	8.16	1509	4.02
870–890	0	0.00	0	0.00	0	0.00	0	0.00
Above 890	0	0.00	0	0.00	0	0.00	0	0.00

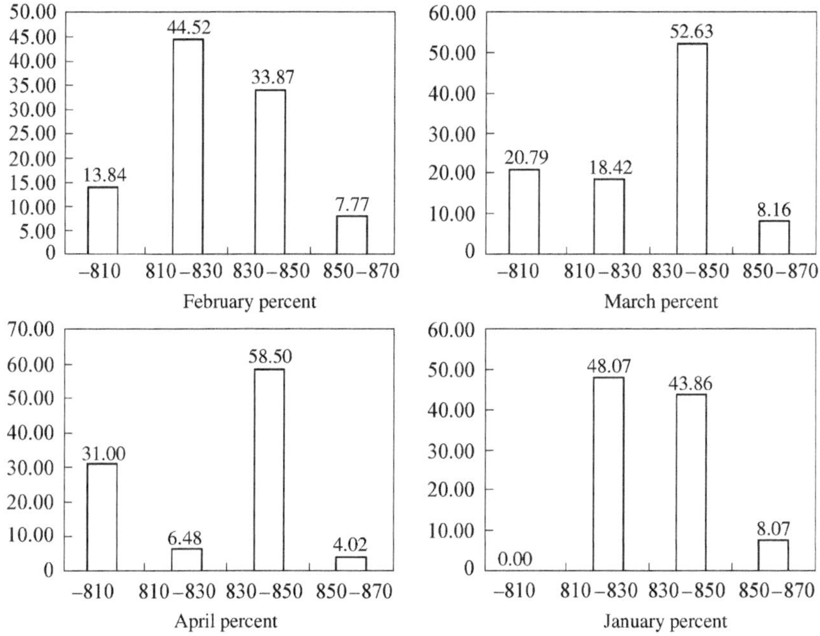

FIGURE 6.77 Progress of direct charging between No.5 caster and No.2 bar rolling line.

rolling yield (minimizing the cut tails shorter than 6 m in length) and the decrease of indefinite length ratio (the ratio and number of rolled products between 6 and 12 m in length), the billets for rolling are determined with rational but different weights and with accurate calculation and control. For instance, when the Φ12 mm rebar is produced, the theoretical weight of 12 m steel is 10.656 kg/m and the ratio of negative allowance is −3%. The billet weight should be 2441 kg in a condition of 8 times multiple lengths, when that of single multiple length is 340.992 kg. Consequently the billet should be cut by this weight. The technology of fixed weight mode is adopted instead of cut-to-length mode when producing different specifications of rebars or round steels, thus it is of benefit to the improvement of fixed length ratio and negative allowance ratio, as well as rolling yield ratio and indefinite length ratio. Finally, it achieves energy savings, emission reduction, lower production costs, and labor intensity reduction.

6.4.4.1 Fixed Weight Mode Between No.6 Caster and No.1 Rolling Line

The technology of fixed weight mode between No.6 caster and No.1 bar rolling line has been developed since September 2011 in Tangsteel, and obvious effects have been achieved. The results are shown in Table 6.25 and Fig. 6.78.

TABLE 6.25 Comparison of Fixed Weight Mode Index Between No.6 Caster and No.1 Bar Rolling Line

Month	Fixed Weight Ratio (%)	Fixed Product Length Ratio (%)	Production of Indefinite Length Steel (t/month)	Rolling Yield (%)	Production of Billet (Number/ Month)
Sep. 2011	43.53	99.22	478.51	96.83	-
Oct. 2011	47.83	99.31	689.46	96.98	-
Nov. 2011	45.94	99.30	643.66	96.80	-
Dec. 2011	47.63	99.41	558.25	96.90	36,909
Jan. 2012	49.53	99.47	488.39	97.00	36,455
Feb. 2012	48.74	99.51	437.59	97.11	38,045
Mar. 2012	47.93	99.47	489.43	97.22	37,283

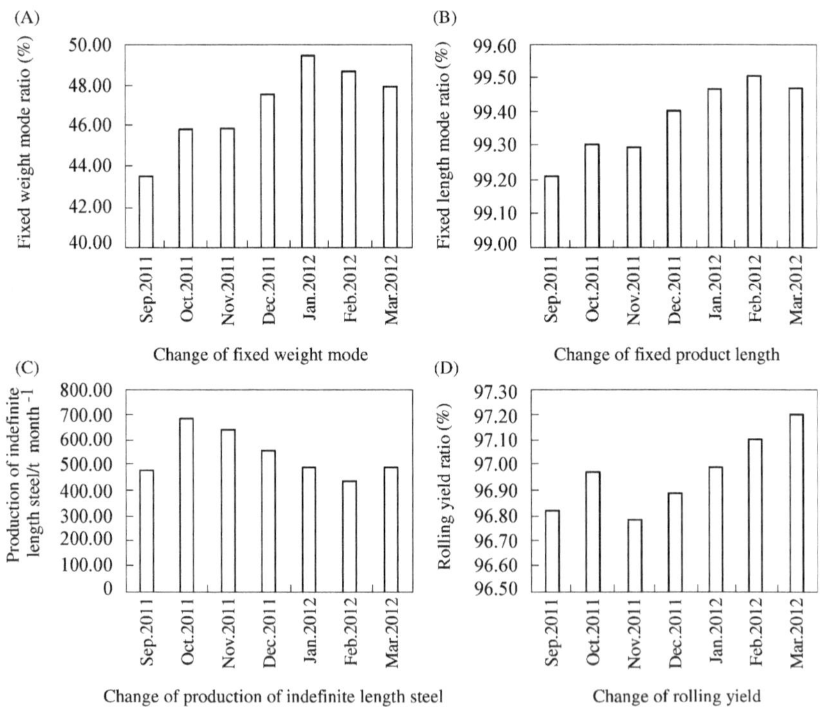

FIGURE 6.78 Effect of fixed weight mode in No.6 caster and No.1 rolling line.

The rebars of Φ12 mm, Φ14 mm, Φ16 mm, and Φ18 mm went through one of the 4-slitting, 3-slitting, and 2-slitting rolling in the No.1 bar rolling mill, thus a series of complex and extreme requirements were imposed on the fixed weight mode. For example, the relevant computer model, high-precision measure sensors (the C3 high-precision sensors with 1/3000 scale were adopted for weight measuring), were in demand. Meanwhile, as the shape and area of the billets section could be changed by sequence casting and crystallizer operation cycles, it was necessary to monitor, calculate, feedback, and adjust in real time. All these came with considerable technical difficulties and production management difficulties, which demanded further research and development.

By tackling the key technical problems, the fixed weight mode ratio between No.6 caster and No.1 bar rolling production line reached 48–50%. Meanwhile, the fixed length ratio 99.45%, and the rolling yield ratio 97.22% (the steel of shorter than 6 m is calculated as cutting tails). It could be seen that the production of indefinite length steel between 6 and 12 m was reduced to lower than 490 t (the total production per month was around 95,000 tons). The effect of technology improvement was obvious but there was still the potential of further increases.

6.4.4.2 Fixed Weight Mode Between No.5 Caster and No.2 Rolling Line

The technology of fixed weight mode was implemented earlier between No.5 cater and No.2 bar rolling line. All the billets were fed directly into the No.2 bar rolling line and not cut in rolling, except for the cold ones which were delivered to No.1 rolling line. The fixed weight mode ratio reached 67–70%, correspondingly the fixed length ratio reached 99.64% and the rolling yield ratio was as high as 97.4–97.5%. The production of indefinite length steel reduced to 321 t/month (the total production was 100,000 tons or so per month). The practical operation of fixed weight mode between No.5 caters and No.2 bar rolling line is shown in Table 6.26.

In conclusion, the fixed weight mode technology is a new proposition in the billet caster and bar rolling mill. It is necessary that numerous techniques are integrated together to form one new integration technology package, for instance, the billets accurate weighting technology, billets accurate cutting, billets section dimensions stable controlling (rhomboidity prevention), the final products accurate fixed-length cutting, the final products section dimensions accurate controlling, the stable slitting rolling, and the information feedback of technological parameters between rolling mill and caster.

6.4.5 Discussions

In the steel manufacturing process there has generally been a varying procedure of solid–liquid–solid, accompanied by a series of energy input/output

TABLE 6.26 Comparison of Fixed Weight Mode Index Between No.5 Caster and No.2 Bar Rolling Line

Month	Fixed Weight Mode Ratio (%)	Fixed Product Length Ratio (%)	Production of Indefinite Length Steel (t/month)	Rolling Yield (%)	Production of Billet (Number/ Month)
Sep. 2011	68.79	99.52	428.15	97.28	-
Oct. 2011	65.49	99.56	443.35	97.27	-
Nov. 2011	62.43	99.60	396.22	97.31	-
Dec. 2011	65.73	99.56	339.05	97.25	34,969
Jan. 2012	67.58	99.63	341.42	97.40	46,063
Feb. 2012	69.32	99.63	173.29	97.39	21,259
Mar. 2012	67.35	99.64	321.46	97.52	44,030

and heat absorption/emission. The highest energy consumption lies in the course of the solid–liquid phase transitions (ironmaking). Meanwhile, large amounts of excess heat still exist during the liquid–solid phase change. It is also an important issue of energy saving and emission reduction to utilize these excess heats of the solidified billet.

Starting from the process of solidification (continuous caster), the melted metal cools and solidifies to form billets from the state of high temperature (1540–1550°C), then the billets are charged into the reheating furnace at different temperatures and rolled after heating. The different processes mean different results of energy consumption and emission reduction, for example, direct rolling, direct hot charging into the reheating furnace, hot charging, as well as cold charging, etc., which also mean different technical levels.

The technology of direct charging one by one in a heat sequence is one technical integrated package that contains major technological difficulties but obvious economic benefits, and comprises multiple techniques and dynamic integration system. The essence of achieving this lies at the establishment of interface technology between the steelmaking plant and the rolling mill, especially the dynamic operation interface technology between the casting machine and the reheating furnace. These technologies include:

- Setting up a reasonable space and time relationship between the casting machine and the reheating furnace, which includes a reasonable and compact plane layout of the workshop, minimizing the billets running distance between the caster and the furnace, as well as minimizing the billets' conveying time and quasicontinuation.

- Setting up a production match between the CC and the rolling mill. The ratio of the correspondence should be set up as an integral number if possible. Therefore, for the bar rolling mills, it is necessary to apply the slitting rolling technology of two cut, three cut, or four cut, etc., in the production of small size bars, which helps to stabilize the flow rate of the high temperature material and hot charging temperature of billets as well as to realize the energy savings and emission reduction of reheating furnace, thus increasing the output of rolling mills.
- A series of basic technology is required to support the billet direct hot charging, including:

The technology of constant and high casting speed. In Tangsteel, the casting speed for billets with sections of 165 mm × 165 mm is 2.15 m/min, which corresponds to 2.60 m/min for the billets with 150 mm × 150 mm.

The technology of converter temperature tapping and the technology package of stable tapping. The 55 t converter tapping temperature is stabilized at 1664°C.

The technology of billets temperature enhancement after cutting (the surface temperature of billets after cutting is stabilized at about 970–980°C in Tangsteel).

The technology of high temperature billets production without defects.

- The billets supply changing from the defined length mode to defined weight mode is beneficial for increasing the fixed length ratio and production yield ratio, thus decreasing the amount of indefinite length products. It is favorable for the enhancement of the quality and economic benefits. There is also further research, for example, the slitting rolling technology, the monitoring of the billets shape and dimension, accurate weighting and information feedback, the design of the piece weight of unit billet with different sections, and the cutting adjustment.
- The interface technology between the casting machine and rolling mill exists not only in the billet caster and the bar rolling line, but also in the slab caster and thin slab rolling mills or the middle plate mills, the round billet caster and seamless steel tube rolling mills. The interface technology between the casting machine and rolling mill should emphasize the connection quantificationally and stably in a more compact and continuous state and higher temperature, which contains many more innovative techniques.

APPENDIX A TURNOVER TIME STATISTICS OF STEEL LADLE IN NO.2 STEELMAKING AND HOT ROLLING PLANT IN TANGSTEEL

1. Ladle turnover among procedures

   No.2 steelmaking and hot rolling plant has three 55 t BOFs, three 165 mm × 165 mm billet casters, and ten ladles between BOF and CC including three ladles in a hot maintenance state. Now, the turnover rate of the steel ladle is close to 5.5 times per heat. The lifespan of the ladle is about 70–75 heats per ladle. Through a survey of ladle turnover in the process of 14 heats BOF tapping–CC–maintenance, the average stay time of the ladle in every procedure is shown in Fig. A.1 and Table A.1.

   Fig. A.1 gives some conclusions as follows:

   a. Average turnover time, that is, the sum of each procedure operation time, is about 77.6 min. The longest time is 104.2 min and the shortest time is 63.2 min.

   b. During the whole turnover cycle of the ladle, the operation time increases step by step from CC, the empty ladle going back to the maintenance position, the empty ladle waiting for BOF tapping, the full ladle transported to the holding position to the ladle maintenance process, etc.

   c. The waiting time of each procedure accounting for the proportion of whole turnover time of ladle is about 31%. Among the waiting time, the time of the empty ladle accounting for the proportion of the whole waiting time of the ladle is about 76.5%.

2. The relationship between molten steel temperature and ladle turnover time

   In order to reduce molten steel temperature drop, No.2 steelmaking and hot rolling plant in Tangsteel has developed a technological measure of covering the whole ladle turnover process for heat insulation. In the combined fast-paced turnover process, the number of ladles has decreased and the molten steel temperature drop has also been reduced. Fig. A.2 shows the relationship of molten steel temperature drop, ladle turnover time, and empty ladle inner wall surface temperature. In the figure, the molten steel temperature from the end of tapping to the start of casting refers to the molten steel temperature in the ladle measured by a thermocouple. The temperature from the end of casting to waiting for tapping refers to ladle inner wall surface temperature measured by infrared temperature instrument.

   It can be seen from Fig. A.2:

   a. During the process of tapping–Argon blowing–casting, the temperature drops respectively are 45°C in tapping process, 22°C from starting argon blowing to CC table, and it adds up to 67°C from the start of tapping to waiting for casting.

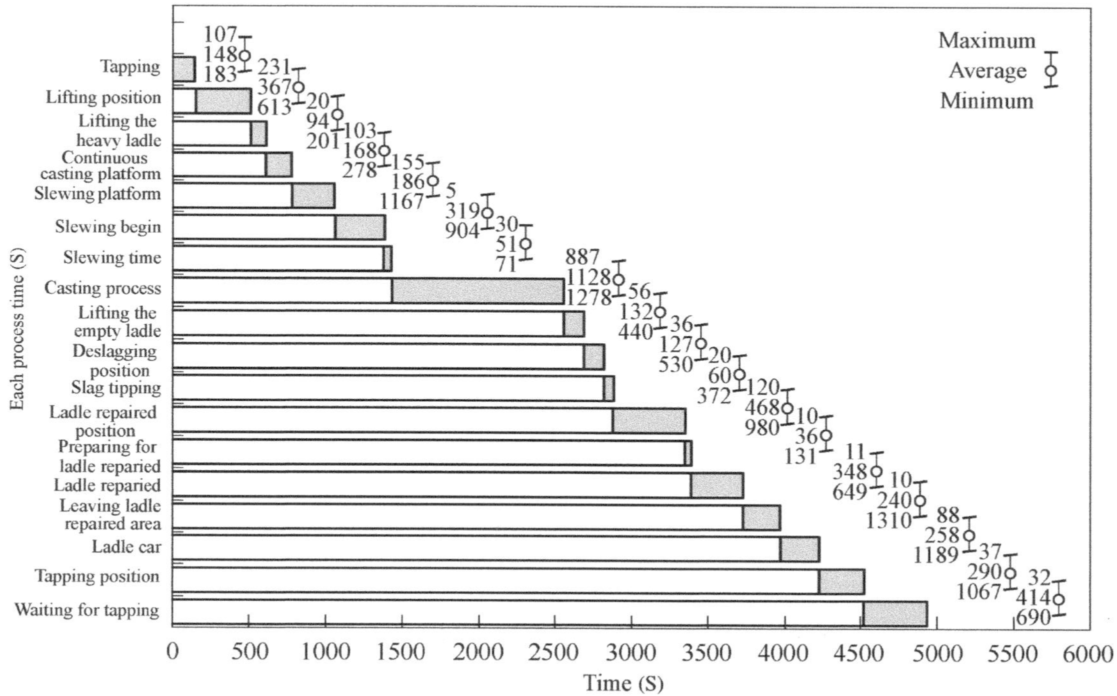

FIGURE A.1 Turnover time statistics of steel ladle in different positions.

TABLE A.1 Survey of Turnover Time of Different Ladles in Different Positions

Number Time Station	1	2	3	4	5	6	7	8	9	10	11	12	13	14
Tapping	172.2	172.2	172.2	183	166.2	139.8	126	106.8	142.2	111	160.2	135	130.2	150
Lifting position	460.8	231	460.8	346.2	307.8	373.2	295.8	360	268.2	613.2	480	405	265.2	270
Lift the heavy ladle	57	79.8	57	42	159	201	120	94.8	148.8	72	70.8	25.2	165	19.8
CC platform	204	277.8	204	123.6	247.8	103.2	139.8	180	133.2	135	121.8	154.8	138	195
Turret platform	214.2	184.2	214.2	160.8	169.8	291	286.2	189	373.2	160.2	1167	280.2	154.8	165
Slewing begin	36	480	36	15	216	340.2	369	751.8	394.8	4.8	903.6	630	274.8	10.2
Slewing time	37.8	34.2	37.8	46.8	46.8	58.2	69	70.8	55.8	64.8	43.2	30	64.8	60
casting process	1278	1252.8	1278	1237.8	1237.8	886.8	1119.6	1014	997.2	1101	973.2	1075.2	1170	1170
Lifting empty ladle	60	61.8	60	60	111	79.2	60	79.8	55.8	415.8	105	190.2	64.8	439.8
Deslagging position	81	81	81	82.2	36	109.8	75	57	270	117	75	64.8	120	529.8
Deslagging process	40.2	40.2	40.2	43.2	372	21	37.2	49.2	60	22.8	31.2	40.2	25.2	19.8
Ladle repair position	288	834	288	814.8	753	447	979.8	324	150	391.8	376.2	379.8	409.8	120
Repair Preparation	130.8	15	130.8	25.2	10.2	10.2	51	45	25.2	19.8	10.2	10.2	10.2	10.2
Repair process	649.2	364.8	649.2	10.8	150	183.6	316.8	465	469.8	424.8	294	315	76.2	499.8
Leave Repair position	208.2	15	208.2	45	325.2	249	565.8	40.8	10.2	34.8	169.8	165	1309.8	10.2
Ladle car	255	300	255	145.8	88.2	145.2	1189.2	211.8	165	220.2	120	220.2	127.2	169.8
Tapping position	159	712.2	159	379.2	286.8	535.8	67.2	1066.8	405	64.8	37.2	70.2	63	60
Waiting for tapping	352.2	426	352.2	31.8	180	75	390	579	675	675	228	640.2	690	499.8
Total (s)	4683.6	5562	4683.6	3793.2	4863.6	4249.2	6257.4	5685.6	4799.4	4648.8	5366.4	4831.2	5259	4399.2

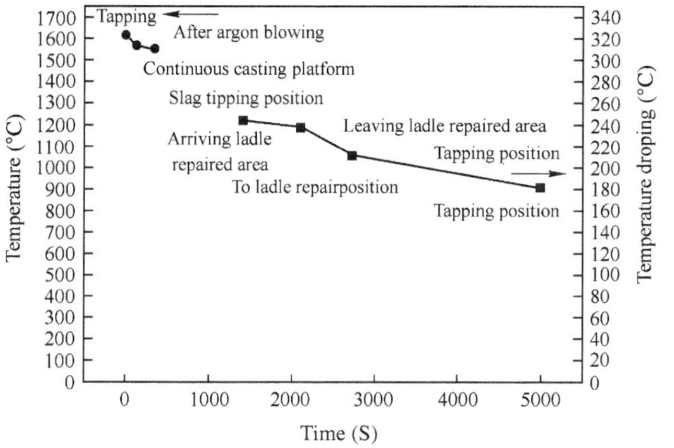

FIGURE A.2 Relationship between ladle turnover time and ladle inner wall surface temperature in No.2 steelmaking and hot rolling plant in Tangsteel.

b. In the process of casting stop, pouring slag, and waiting for tapping, the temperature of the empty ladle inner wall surface drops from 1220°C to 900°C. The average temperature drop rate is about 5°C per minute.
c. In the process of the empty ladle maintenance, waiting for tapping, and pouring out slag, the temperature drops are 12.6°C/min, 4.1°C/min, and 0.4°C/min, respectively. The maximum temperature drop is in the process of ladle maintenance.

3. Summary
   a. Under the condition of No.2 steelmaking and hot rolling plant in Tangsteel with three 55 t BOFs, three 165 mm × 165 mm billet casters, and ten ladles between the BOF and CC, via a survey of ladle turnover in the process of 14 heats BOF tapping–CC–maintenance, it is concluded that the turnover rate of the steel ladle is close to 5.5 times per heat,; the lifespan of the ladle is about 70–75 heats a ladle and the average turnover cycle is about 77.6 min. The waiting time of each procedure accounting for the proportion of the whole turnover time of the ladle is about 31% with fluctuation among different heats.
   b. When No.2 steelmaking and the hot rolling plant in Tangsteel developed a technological measure of whole turnover process ladle covering for heat insulation, the molten steel temperature drop from tapping end to casting starting is about 67°C, and the temperature drop of the empty ladle inner wall surface is about 291°C, mainly occurring in the ladle maintenance process.
   c. Some technological measures such as reducing the number of turnover ladles, fast-paced ladle turnover, and utilizing ladle covering are helpful in reducing the molten steel temperature drop during the tapping process. The average tapping molten steel temperature has decreased to 1640°C or less, which is conducive to the high-efficiency operation of billet CC, and the one-by-one direct hot charging rolling.

REFERENCES

Ariyama, T., Ueda, S., Natsui, S., et al., 2009. Current technology and future aspect on CO_2 mitigation in Japanese steel industry. J. Iron Steel Res. Int. 16 (S2), 55–62.

China Metallurgical Construction Association, 2008. Design Regulation for Blast Furnace Ironmaking Technology (GB50427-2008). China Planning Press, Beijing (in Chinese).

Iwasaki, M., Matsuo, M., 2011. Change and development of steelmaking technology. Nippon Steel Sci. Tech. Rep. 391, 88–93.

Kawamoto, M., 2011. Recent development of steelmaking process in Sumitomo Metals. The 2nd International Symposium on Clean Steel (ISCS 2011), Shenyang, China.

Kitamura, S., Ogawa, Y., 2001. Improvement of hot metal dephosphorization efficiency to decrease steelmaking slag generation. The 9th China–Japan Symposium on Science and Technology of Iron and Steel, Xian, China, 124–130.

Michael, P., Bodo, L.H., 2009. Iron making in Western Europe. J. Iron Steel Res. Int. 16 (S2), 20–26.

Miwa, T., 2009. Development of iron-making technologies in Japan. J. Iron Steel Res. Int. 16 (S2), 14–19.

Ogawa, Y., Yano, M., Kitamura, S., et al., 2001. Development of the continuous dephosphorization and decarburization process using BOF. Tetsu-To-Hagane/J. Iron Steel Inst. Japan 87 (1), 21–28.

Qian Shichong, Zhang Fuming, Li Xin, et al., 2011. Technical analysis on hot blast stove of large capacity blast furnace. Iron Steel 46 (10), 1–6 (in Chinese).

Tang Wenquan, 2005. Discussion on BF capacity factor. China Steel Focus 8, 52 (in Chinese).

Ueki, T., Fujiwara, K., Yamada, N., 2004. High productivity operation technologies of Wakayama Steelmaking Shop. Tenth Japan-China Symp. Sci. Technol. Iron Steel 11, 116–123.

Yang Chunzheng, Wei Gang, Liu Jianhua, et al., 2012. High efficiency low cost platform for clean steel production practice. Steelmaking 28 (3), 1–6 (in Chinese).

Yin Ruiyu, 2004. Metallurgical Process Engineering. Metallurgical Industry Press, Beijing, pp. 381–386 (in Chinese).

Yin Ruiyu, 2009a. Metallurgical Process Engineering, second ed. Metallurgical Industry Press, Beijing, pp. 276–277 (in Chinese).

Yin Ruiyu, 2009b. Metallurgical Process Engineering, second ed. Metallurgical Industry Press, Beijing, p. 140 (in Chinese).

Yin Ruiyu, 2010. Comment on behavior of energy flow and construction of energy flow network for steel manufacturing process. Iron Steel 45 (4), 1–9 (in Chinese).

Zhang Shourong, Yin Han, 2002. Evaluation of blast furnace intensified smelting. Ironmaking 2, 1–6 (in Chinese).

Chapter 7

Engineering Thinking and a New Generation of Steel Manufacturing Process

Chapter Outline

7.1 ENGINEERING THINKING

Engineering is considered as direct productivity and actual productivity. Engineering appears in history earlier than science. The relationship between technology, science, industry, economy, and engineering is close and complex. It is necessary to discuss the relationship between science, technology and engineering before introducing engineering thinking. The relationship among these three is studied in "Philosophy of engineering"

Theory and Methods of Metallurgical Process Integration.
DOI: http://dx.doi.org/10.1016/B978-0-12-809568-3.00017-6

(Yin et al., 2007) and "Theory of engineering evolution" (Yin et al., 2011). The outline is as follows:

Science: From the epistemic logic point of view, all those understanding and revealing the constitution, essence, and operational laws of the natural world and social objects belong to the category of science. In brief, the characteristics of scientific activities are the knowledge systems of studying the systematicness and regularity of the constitution, essence, and operational laws of the natural world and social matter. The key characteristics of scientific activity are exploration and discovery.

Technology: Technical activity is a specific knowledge system, reflecting an ingenious conception and experiential knowledge. In modern technologies, "instrumental" approaches, such as technique, instrument, equipment, and information processing-automatic control systems, are developed using ingenious conception and experience. The key characteristics of technology are invention and innovation.

Engineering: From the point of view of knowledge, engineering activity is an integrated knowledge system constituted by the combination of one or several core professional skills, the related professional skills, and nontechnological knowledge. The purpose of engineering activity is to establish a large-scale, professional, and sustainable system of productive relations or social services. The key characteristics of engineering are integration and construction.

From the discussions above, theory comes from a variety of practical activities. Human beings perceive and realize all kinds of experiences and problems in practice. The experiences and problems generate rational understanding of regularity through the thinking, discussion, comparison, induction, collating, and verification of human beings. Some of them are gradually transformed into theory through special experimental verification.

Therefore theory comes from practice, which is promoted into the former through thinking, inspiration, verification, and summarization. It is important to practice logical thinking and its transformation to theory.

Problems in the metallurgical manufacturing process, such as chemical reaction, physical phase transition, mass–energy transition, and metallurgical devices, include both basic science and technical science. However, during dynamic operation and engineering design, most come under the disciplines of engineering and engineering science. The thinking method of engineering and engineering science is open, dynamic, integrated-optimal, and complex, as engineering represents the integration of relevant and different functional heterogeneous technology, considering effective allocation of economic factors, such as resource, energy, land, fund, environment, market, and labor force.

Fig. 7.1 shows that engineering is a technology integrated system consisting of a series of related but different functional modules. The module group and the dynamic-orderly integrated system are basic elements of engineering. However, only with the reasonable allocation of technology integrated systems and economic factors, through the process

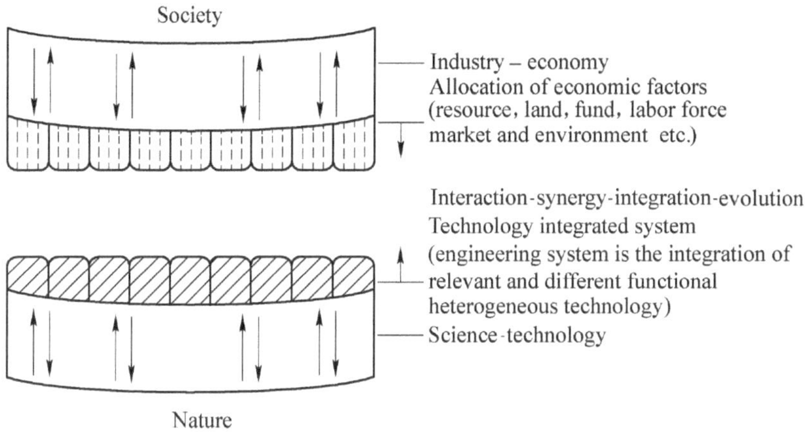

FIGURE 7.1 Connotation of engineering, elements of engineering and their integration (Yin, R., Li, B., Wang, Y., et al., 2011. Theory of Engineering Evolution. Higher Education Press, Beijing (in Chinese)).

of interaction-synergy-integration-evolution, can engineering be formed with functions, then generate values and thus have positive or negative effects on nature, economy, and society.

The logical thinking during the construction process of engineering model includes:

1. Establishing the correct engineering concepts, especially during the decision-making, planning, and designing process;
2. Establishing time-matching integration theories and methods, that is, paying attention to theories and methods of integration optimization and evolution innovation during the planning, designing, and production operation process;
3. Practicing and materializing engineering concepts and theories during the designing and constructing process, that is, obtaining an optimal engineering system of element-structure-function-efficiency;
4. Implementation during dynamic operation and management to obtain expected goals, that is, paying attention to multiobjective optimizations, as well as their selection and balance during practical dynamic operation of engineering systems;
5. Performing lifecycle evaluation, that is, evaluating the value and rationality of the process engineering system from natural resources, through operation, manufacturing, consumption, and landfilling process, to the digestion-treatment-recycling process of resource and energy;
6. Intensifying the recognition of adaptation, evolution, and evaluation of the engineering system on the natural environment and social environment, that is, expanding positive effects and avoiding negative effects.

7.1.1 Relationship Among Science, Technology, and Engineering

Science, technology, and engineering are involved, closely related but essentially different objects in engineering. For the three objects, only by distinguishing between the concept and method, combined with a clear understanding of their interrelation, can the correct engineering thinking be established.

7.1.1.1 Engineering and Technology

In the history of human society development, engineering has had an organized, designed, and creative artificial existence. Engineering activity is a practical activity depending on nature and reforms nature to serve as a benefit for human beings, which is a collective activity. In the process of human history, engineering has been always directing productivity and the material basis of advancing society. It's a technology flock (technology integrated system) gained by selecting and integrating relevant technology, analyzed from the characteristics of engineering. Technology is the core of engineering; meanwhile, engineering has to be a predictable and valuable artificial existence, interacting with economic elements (resource, land, capital, labor force, market, and environment etc.) under certain conditions.

Technology is the process operation methods and skills that have accumulated and developed gradually through the practice of understanding and transforming nature for human beings. In general, the corresponding product tools and equipment are included as well. Both technology and engineering come from the subjective initiative of human beings, while their objectives are different. The objective of technology is relatively simple, and this is emphasized by the accumulation of work experience and the understanding of nature and its rules for the aim and in the process of changing the world. In the development process of technology, the social economic factors are considered as external factors, while they are treated as inherent factors for engineering. For example, the technology of semimolten state steelmaking-Wrought Steel (Yang, 2004) in metallurgical history is discussed. Wrought steel is a metallurgical technology. Antique excavation and historical documents show that the wrought steel technology was very exquisite in ancient China, for example, the "Hundred Refining," "Fifty Refining," "Thirty Refining," and "Cofusion Process" (Yang, 2004). These, however, can just be classified as the category of technology since these techniques were only obtained by an excellent carpenter or family workshop. The puddling process in the 18th century in western Europe is wrought steel, adopting the flame reverberatory furnace technology, which combined the social ecological requirement of laying the railways and building the Eiffel Tower. It had reached the engineering scale, but technology and engineering can hardly be simply classified by scale. For example, to reform the Thomas steelmaking process, in western Europe countries hydrocarbon splitting chilling technology was adopted to restrain the burning jet of oxygen nozzle during steelmaking, forming bottom blowing basic oxygen furnace (BOF) (such as OBM); after the technology was introduced to the Americas, a series of pilot experiments were performed on the 30 t BOF to improve the dephosphorization after Q-BOP, and developed to the Q-BOP method. A 200 t Q-BOP furnace was built. (The capacity of OBM was only dozens of ton.) However, because of the limit of economic factors, Q-BOP was not a viable project, but only a new technology in metallurgy. Afterwards, bottom blowing furnace technology was collected into the top-bottom combined blowing furnace, which was an engineering innovation. Finally, it became one of the basic processes of the metallurgical manufacturing processes. Accordingly, the development of the technology is updated constantly, and advances are made not only by choosing new technologies, but also by considering economic factors.

7.1.1.2 Technical Invention and Engineering Innovation

Innovation is an economic concept, involving the development of new kinds of products, adopting a new mode of production, realizing a new organ of production, utilizing a new source of feed, and exploiting a new market. Engineering innovation can hardly be separated from the improvement of technology since engineering innovation lacks sufficient motivation without technical invention. After being effectively applied into engineering systems, a new technical invention can result in engineering innovation.

Invention and innovation are two different concepts. Technical invention and engineering innovation should be distinguished. Invention is unequal to innovation. In fact, many technical inventions can hardly be successfully marketed. Foreign scholars defined innovation as the first commercial application of the invention (Freeman and Soete, 2004). Such examples are usual in metallurgical engineering technologies. The WORCRA continuous steelmaking process was tested in Australia and Shanghai, China, respectively, in the 1960s. This is a continuous steelmaking technology with high smelting efficiency, using the countercurrent flow contact of slag and steel. But WORCRA can hardly work normally in production for various reasons. So it can only be treated as an exploratory technological invention but not an engineering innovation. A lot of methods have been investigated in smelting reduction ironmaking technologies. Most of them could hardly be treated as engineering innovation.

Engineering innovation is not the end of business application of technical invention. It can also occur in the process of the development of technological factors. For example, 30 t BOF is a mature steelmaking technique, combining continuous casting of square billet, sublating die casting, coordinating integration of continuous rolling of wire rod, and coordinating the market requirement of Chinese constructional engineering in the 1990s, thus, the quasicontinuous manufacturing process of wire rods was established. In the guidance of dynamic-orderly integration theory, the temperature of the entire transportation process is controlled above 800°C. As a result, a brand new continuous casting–rolling production line was established (see Section 6.4). It is a distinct engineering innovation.

7.1.1.3 Engineering, Technology, and Science

Both engineering and technology contain subjective initiatives, which can be presented as invention and innovation. Science explores and reveals the objective rules, and it can only result in a discovery but not an invention. There are only true–false instead of high–low in science, the so-called high technology customarily is an inapt presentation, which should be called "high technique" or "new technique" in its exact meaning. The mission of science is to reveal and illuminate the regularity of the constitution, essence, and operational laws of objective things. Scientific knowledge can be divided into many subknowledge with their own order, logic, and regularity.

Both engineering and technology serve human life and economic development, which have a strong value goal. However, the basic target of science is to explore, reveal, and discover the constitution, essence, and operational laws of nature, rather than aiming for the development of technology and engineering, and thus possibly has little direct relationship with advancing social economic activities. For science, being over utilitarian may prevent the development of science.

Engineering and technology have subjective choice factors, especially in engineering decision-making. Accordingly, it is an incorrect understanding that engineering is considered as the thing to reflect the will of people, which does not belong to the objective law. Engineering against scientific law can hardly reflect the aim and value of engineering reasonably and sustainably. Therefore it should be emphasized that engineering must conform to the objective law, and require scientific and democratic decision-making.

In the development and evolution of engineering, many experiments and improved technologies were performed continually and a wealth of knowledge and experience were accumulated. The validity of engineering scientific knowledge needs to be examined in engineering design, construction, and operational practice. The correct knowledge of engineering science can be reflected more in practice.

7.1.2 Characteristics of Thinking Mode in Chinese Culture

Traditional thinking mode in Chinese culture chronically emphasizes the observation and thinking of the whole, namely the holistic view; attention is paid to observing and thinking of problems based on the development evolution of things, analyzing the problem as time(s) changes, namely dynamic view; more attention is paid to exploring the nature of things through the dynamic change of entire phenomenon, and exploring and studying the root of the development of things deeply and discriminately.

Prigogine, putting emphasis on the opinion, said: "The Chinese traditional academic idea focuses on the integrity, naturalness, coordination and collaboration..., the development of modern science is more aligned with the thinking of Chinese philosophy" (Prigogine, 1980).

The most prominent characteristic of Chinese traditional academic idea is holism, which is to grasp the overall, observe and analyze everything integrally. China is known as the academic home of "oneness of man and nature," indicating that man is an integral part of nature. It is also inferred that the objective world is one, the subjective world is one, and the subjective world and the objective world are also one. Holism emphasizes things by cognition and problem handling, it is the key of understanding and solving the problem.

Holism is highlighted in Chinese traditional culture, which emphasizes a completely dynamic view at the same time. Attention should be paid to not only the overall and global, but also the constituent cells of the overall and global and the connection of cells, and their changes over time in the process of movement (adaptability only along with change). Attention should be paid not only to the entirety, but also to the connection of the constituent cells during observation and thought. The connection is sportive and metabolic, not static, even not ossified. Therefore the law of dynamic change must be investigated, indicating that a dynamic view is important.

Essence is also a distinct characteristic of Chinese academic idea. The process of research and analysis includes considering the entirety and the dynamic character, studying imagery through observation, and exploring deeply and revealing the internal essence. Everything is hierarchical, different hierarchical constituent elements and their movement have different rules. Because the constituent elements in various gradations are different, connections of elements are different, leading to form different constructions at different gradations and determine different essences. The study and discovery of the essence of things is to explore the composition, construction, nature, and law of motion.

Chinese traditional culture also contains mutual dialectics thoughts. It is considered that the way of change is the result of mutual transformation between positive and negative. Nature is highly respected in Chinese culture. People live depending on nature but hardly want to conquer nature.

It is noted that Chinese industrial society culture was not spontaneously generated since the development of the productive force was very slow over a long period of the feudal society of China. Some interrelated academic thought failed to achieve the systematic and quantitative level, which was hardly converted to theory. For example, there are only concepts of "stifled" or "dredge" in hydraulic engineering. The results of engineering vary with each individual, leading to river inundation occasionally. Water engineering, such as damming and hydraulic engineering, depends entirely on foreign mechanics and hydromechanics theory.

Although a large number of experiences are accumulated in the theory of meridians and collaterals, it carries somehow a mystery to explain using yin-yang and five elements. It is imagined that modern science and technology combined with the holistic view and dynamic view in Chinese traditional culture is beneficial to deepen the use of things. With the development of science and technology, some Western scientists have started to accept the thinking ways of revering nature; in oriental philosophy the harmonious coexistence with nature, and understanding rules as a whole are very important.

7.1.3 An Engineering Innovation Road in the "Reductionism" Deficiency

For a long time, the study of engineering problems usually starts from the discipline itself but not from a practical standpoint. Then, after continuous "reduction," "refinement," and "abstraction," an isolated, statistical, and broken pieces of recognition and concept is formed. Hence, the proposition is divided into many separate segments, and interface connections between segments are ignored, weakening the integrity of the proposition. Subsequently, a few key issues are refined from these "parts" (due to abandoning the other related questions, the structural and dynamic character is ignored), and then the key issues are concentrated on to become the "subject theory" (the technological realization and engineering feasibility are diluted). The methods of recognizing problems and solving problems means the way of thinking becomes more and more of "reduction," "separation," and "abstraction." Actually, a large amount of dynamic, connective, and synergistic information, especially time and space information, are usually lost in the process of "separation" and "reduction." The missed information contains many important issues like structure optimization and dynamic–synergistic operation, which deviates from the integrity and dynamic character nature of things. Furthermore, engineering problems are studied by "separation" and "abstraction." Then, it is generally considered that dynamic engineering systems could be formed by the stacked and jointed additive research results to solve problems (Fig. 7.2). However, the way of thinking that results in deficiency and disadvantage during the study process, even the engineering design process, appears frequently.

It should be noted that, no matter what the engineering design process or engineering production operational process, the essence of engineering problems is "everything flowed," "everything moved," and "everything changed." "Flow" is a dynamic operation, including dynamic integration

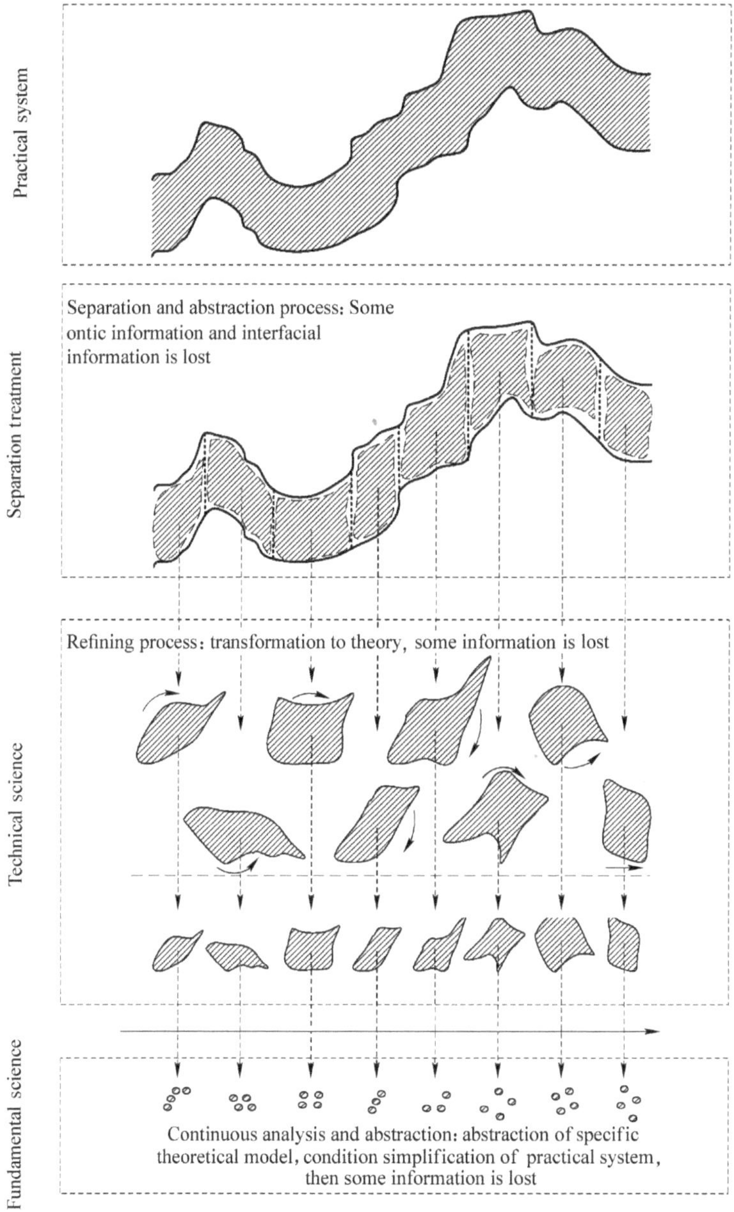

FIGURE 7.2 Schematic of "reductionism" method in engineering problems.

characters of multilevel, multiscale, multiple-factor. Multilevel means structural relevance; multiscale refers to the space–time concept, which means the directivity and time and spatial of "flow" in the dynamic operation process; multiple-factor changes with engineering type and industrial characteristic, and is related to the substance-energy-space-time-message. Actual problems can be solved fully by the concept and method of "reduction," "separation," and "abstraction."

Undeniably, the solution of crucial issues, key technologies, and key theories is vital. The revolution of key technologies may "detonate" the global evolution of engineering, belonging to the vital "tipping point" and be a catalyzer of the breakthrough engineering innovation process. However, it is worthy of note that the integration problems between related problems and crucial problems must be solved, forming a dynamic-orderly-synergistic operation structure and operating mechanism. Otherwise, the engineering problem can hardly effectively perform, operate, and exert comprehensive functions. Problems in different levels and different processes should be dynamic researched and orderly and effectively integrated, forming element-construction-function-efficiency optimized engineering systems to solve practical problems. Thus cross-disciplines are formed at the same time. In process manufacturing, the joint point of the process and process engineering system in different levels, different time and space scales is at the level of the manufacturing process, not the level of microscopic-mesoscopic-macroscopic, such as unit operation, the process/device.

Therefore, the advantage and contribution of reductionism, such as detailed analysis of problems, focusing on the basis of analytical and deepening cognition of every step, should not be neglected in the study of engineering problems. At the same time, the oversight of reductionism should also be noticed, for instance, various information losses, integration deficiency, and the distortion of dynamic operation caused by the constant reduction process.

The innovation of engineering theory and methods should get rid of the losses of information and the distortion of dynamic operation caused by the constant reduction process, for example, treating design as a limited way of providing solutions and engineering drawings. It has evolved to emphasize the deep understanding of the concept of holism and dynamic integration and to strengthen dynamic integration innovation, such as developing interface technology in technological process; emphasizing the process dynamic operation regulation, path and procedure, etc.; emphasizing the study of integration theory and elements-structure-function-efficiency relevance. These are the entry point, vantage point, and foothold of design philosophy and methods innovation in process engineering. In the academic study of engineering science, it is required to search the entry point from reductionism, expand new research frontiers and orientations, promote cross-disciplines, and even build a new discipline branch.

A part of spatiotemporal, dynamic, and interfacial information was lost in the analysis-reduction process of reductionism. The deficiency should be repaired and replenished in engineering design and the project dynamic operation process by holistic theory and dynamic research.

The engineering design in the analytical process of reductionism is integrated with integrality, dynamic character, and synergy. The lost spatiotemporal information, dynamic character information, interface information, and collaborative information should be integrated and restored to the essence of things. Under the guidance of top level design, advanced and reasonable elements should be selected to build reasonable structure, sophisticated function, and remarkable efficiency.

From conception, modern engineering always is associated with large systems, large production, and large operations. Large systems, large production, and large operations always contain related technology groups, various levels, and upstream or downstream interfaces, which are beyond the depth of cognition of people. In fact, the complexity of the process problem is always beyond the imagination of people.

Familiarity of the limited and partial knowledge in a certain level or fragment prevents study by people and leads to self-righteous. The comprehensive engineering problems are understood from a microcosmic point. For example, to understand a complex engineering problem from a micropoint is just like that using a microscope to measure distance between planets or using minutes to calculate the geological time, which are a typical fallacy of "lost in microcosmic." Engineering is complex, and involves multilevels, structures, and multifactors. Therefore, it is necessary and valuable to deepen the details of analysis. However, stubbornly focusing trivial matters may lead to lose in the "harmonious society," overgeneralization, and fallacy. For engineering linked with large systems, large production, and large operations, it needs to break through being addicted to microcosmic reductionism, and to pay attention to nature and disciplines in different levels and interfaces processes, and to be skilled in using different space–time scales and input–output concepts to unravel the essence and rules. Furthermore, it is necessary to perform the interaction function and dynamic nonlinearity of different process/devices and the microscopic processes therein based on the understanding of the essence and rules of things—all of which is the innovation of integrated engineering.

Therefore, to solve the engineering problem, conversion and operating programs are needed to work out how to achieve a process from a given initial state to a dynamic coordinating intermediate process, and then to reach a target. In the process of solving engineering problems, the following related problems need to be solved: technical parameter optimization selection problems, process/device optimization adoption problems, several interface cohesion and matching problems, dynamic operation rules formulating problems, structural optimization and function and efficiency optimization problems.

7.2 ENGINEERING EVOLUTION

7.2.1 Concept and Definition of Evolution

In the view of evolution, this world is not as a complex of ready-made things that can be grasped, but as a complex of processes. ("dass die Welt nicht als ein Komplex von fertigen Dingen zu fassen ist, sondern als ein Komplex von Prozessen" in German (Engels, 2009).) Evolution can hardly be separated from process. In fact, evolution means the running of process. Evolution process involves not only substance, energy, life and corresponding information, but also time–space and other factors. Furthermore, there are initial and time and spatial boundary conditions in the process of evolution. After initial and boundary conditions (external environment) change and reach a certain critical state, evolution will accelerate, slow down, or stop. Hence, evolution is certainly related to environment. Generally, when boundary conditions (external environment) are relatively stable, most of the evolution process forms are a gradual model. However, when boundary conditions (external environment) change and reach a critical value, a range of evolution process forms may occur.

The study of the concept of evolution certainly involves movement, elements, process, system, boundary conditions, function, effect and idea, and similar words. Evolution is considered as a conversion from one existing form to another. Evolution is a kind of movement, which derives from the innate quality of movement of everything on earth. Movement is a process related to many factors (essential parameters), particularly time–space parameters. Under specially promoted and restricted environment conditions, all relative factors are integrated through new ways in the movement process of objects (system), forming another well-aligned, effective system (objects) with specific structure, property, function, and effect. The feature that systems (objects) with different functions and effects are certainly artificial selected or natural selected determines their survival, development, or elimination, extinction.

From a historical and macroscopical perspective, the evolution process is continuous and nonexhaustive. Evolution processes are always continuing and extending with global macroscopic continuity, which is considered to be the principle of development. The characteristics of macroscopic continuity are continuity and the transition of the evolution process. External environmental conditions or internal factors and structure changes contribute to a singular point in the gradual, continuous evolution process, implying a continuous and discontinuous unification in the historical and microscopic development process.

From a concrete substance and logical perspective, the evolution process has characteristics of continuity and discontinuity in specific environments. Continuity, such as hereditary, succession, gradualness, is showed in the engineering tradition; discontinuity, such as variability, mutability, jumping, is showed in engineering innovation. The motivation of object development is the internal contradiction of movement and conflict (interaction, allelopathy) between objects (system) and the external environment conditions, which contribute to interconversion or advanced conversion to promote the development of objects (system). It is considered to be the uniform principle of the substance.

Unification of continuity and discontinuity under specific external environment conditions in the evolution process is determined by the internal cause (recombination) of objects (system) and the external cause (selection and adaption between environment and objects). In a sense, it is result of the combination of the development principle (competition, evolution) and the substance uniform principle (selection, adaption).

7.2.2 Technology Advancement and Engineering Evolution

Engineering activity is a practical activity in which human beings utilize all kinds of resources, knowledge, and related factors in a purposeful, planned, and organized way to create and build a new existence. Engineering activity reflects the comprehensive integration and process in nature and the configuration of artificial factors. Engineering, as an artificial system, has consisted of the evolution process in the historical process of development.

In human history, engineering is always a direct productive force. The development history of engineering is one of the direct productive force, and the evolution of the engineering can be considered as the evolution of the productive force.

In the evolution of material engineering, technologies and technological advances are fundamental factors and an important impetus. Technology is absolutely a necessary factor in material engineering. There is no material engineering that requires no technology or only one technology. Engineering reflects the integration of related and various technologies. Every material engineering needs basic conditions of technological means. It is difficult to develop engineering activities without technology.

In the discussion of technology and the technological advance of engineering evolution, the development process of technology should be distinguished. (It is a factory scale technology after the industrial revolution, instead of handicraft in primitive times or workshop skills in the agrarian age.) In general, according to the formation-development-process of technology, a technological state can be divided into two types: one is laboratory technology or research and developing technology, which is original but not immature and difficult to be directly "embedded" into engineering system. The other one is engineering technology, which is directly and smoothly embedded into the engineering system. Therefore engineering technology is the technology with directly significant function in engineering activities. However, laboratory technology, research, and developing technology are important due to their creativity and innovation, which are useless in related engineering systems until they have been through an "engineered" conversion process. Therefore, not only laboratory technique and research and developing techniques, but also deep thinking of the "engineered" conversion process under different conditions is very significant in technological advancement.

In the engineering systems of different industries, engineering technology can be divided into profession technology and supportive technology, according to their characteristics and functions. For example, in spinning industries, spinning, weaving, printing, and dyeing are professional technologies, and mechanical power, transportation technologies, civil engineering are supportive technologies. In steel industries, ironmaking, steelmaking, steel rolling are professional technologies, and blast, oxygen generation, hoisting, electric, transportation are indispensable supportive technologies. In its history, the engineering evolution of different industries has been influenced not only by gradualness and/or mutability of profession technology, but also by that of supportive technology. Because of the supporting integration and coadaptation among different technologies, technical advancement is not due to only one single technological advancement, but also due to commutative, networked integration.

In engineering practice, some technical advancement has indeterminacy between success and failure, which is induced by the adaption ability of the technology itself to the engineering system and the environment and by the mandatory adoption of the technical route and technical units focusing on the value and objectives of the engineering. Hence, the unit technology, especially the new module of the technology, should be selected based on the value and objectives of the engineering system. The unit technology will be embedded into the engineering system through the integration-interactive collaboration-integrated evolution. Therefore the evolution of engineering can be gradual and abrupt.

In the process of engineering evolution, innovation and elimination of technology are very important. The relationship of innovation and elimination is competitive. Under different market conditions, market selection profoundly influences the process of engineering evolution and technological innovation. In the history of technology development, technology advancement is the result of competition. In the selection of market competition, the main part of the market is human beings, which determines the role of subjective consciousness. In addition, the market mechanism has economical laws with characteristics of spontaneous force, which result in the objectivity of market selection. The selection mechanism with subjectivity and objectivity is not controlled by pure subjective forces or pure objective process without the influence of human will. In engineering evolution, the subjectivity of engineering evolution and the objectivity of engineering evolution are not the only points to be noticed. Engineering evolution and technological advancement are the result of interaction and the unity of the subjective and objective opposites. New technology includes not only new profession technology, but also new related supporting technology. This new supporting technology includes control-management, motivation, equipment, information, environmental protection, etc. In other words, the adoption of new technology and the elimination of outdated technology in an engineering system should be considered from the point of view of the entire procedure and the whole process.

In the adoption of technology or new technology, the philosophical proposition of "barriers" and "trap" should be emphasized (Xing, 2009). Generally, "barrier" is a visible barrier (such as the difficulty of technology, etc.), and "trap" is the invisible danger, for example, the application of new technologies may lead to structural imbalance of the entire process and reduce efficiency of engineering systems. The encountered "barriers" and "trap" during the adoption of new technologies are diverse and complex. The goal of selecting technology especially new technologies in engineering activities is overcoming technical "barriers" and preventing various "traps" to achieve the demands of higher, updated, and larger project objectives. For example, to prevent the "trap" of poor integration of new technologies, attention should be paid to "orphaned" leading of individual new technologies and the incongruity of related technology. Thus, in engineering activities, the blind pursuit of "orphans" leading local and individual technologies should be prevented, namely, the prevention of a blind pursuit across a "barrier" resulting in an unknown "trap" coming from technological advancement (new technology) itself (such as unstability and unreliability) or incompletion and noncollaboration of related technology.

Overall, there is a very close association between engineering evolution and technology advancement. Technology advancement is an important driver of the evolution of engineering. In turn, the target requirements of engineering systems (such as market demand, competitive needs, sustainable development needs) also have a driving or limiting effect on the invention, development, and application of technology since engineering is a comprehensive reflection of the direct productivity. Engineering directly reflects the value and associated markets.

7.2.3 Integration and Engineering Evolution

The emergence and evolution of engineering comes from the target value. Many related heterogeneous technology modules are integrated in an engineering system by choice, integration, interaction, collaboration, transformation, and other processes, aiming to operate and manage, plan, design, construct the engineering system through the comprehensive integration of economic elements. The adoption and integration of a technology module group is one of the important parts of engineering, which often appears in the dynamic process and management of engineering design and engineering systems.

Integration is not to simply stack and joint by technical modules. A simple sum-up of modules is not integration. The combination of modules should be after comprehensive selection, integration, interaction, collaboration, and optimization, by which an "organism" consisting of appropriate technical modules utilizing their advantages each other with functional optimization can be generated. The structural optimization, function optimization, and efficiency optimization of the engineering system are achieved by the whole process called integration.

In the history of engineering development, especially in the engineering evolution, it is found that the evolution of the engineering also reflects the related evolution of a heterogeneous technology integration module group.

The nature of the engineering activities is integration and structure (Yin et al., 2007). The following understanding of engineering integration is necessary:

From the nature of knowledge, especially the technical perspective, engineering is the dynamic integration of material-energy-time-space-information and other basic physical quantities. The integration includes judgment, balance, selection, optimization, formation of intersection, union set, and other connotations and forms. The purpose is to select and optimize the factors and to promote the evolution of the structure, function optimization and efficiency optimization, and is sometimes accompanied by the evolution and elimination of elements.

From the perspective of the logical understanding, engineering problems aim to first identify the entire situation, analyze it further to improve or innovate the selected weak link (elements), build newborn germination links (elements), eliminate backward links (elements), and finally achieve evolution of engineering integrity or structural upgrading through reconstructive integration optimization. The process is engineering evolution.

From the viewpoint of industry and specific projects, the integration of engineering is based on analyzing-operating-integrating the optimization of various elements and units, coordinating the integration with the basic economic factors (money, resources, land, labor, market, environment, etc.) to improve market competitiveness and the sustainable development capacity of project-enterprise, to promote industrial development, and serve the public.

From the point of view of philosophy, engineering is to exist, develop, or eliminate in a ternary system of "natural-person-community." Through the above integration and optimization, engineering promotes harmony and a win–win between engineering and natural, engineering and human, engineering and society. Engineering evolution should benefit mankind, serve the society, and benefit nature.

The integrated evolution of engineering should not be understood only from the viewpoint of technology integration, but also from the viewpoint of the integrated optimization perspective of technical modules and basic economic factors. In engineering activities, issues of people, mass flow, energy flow, information flow, capital flow, and ecological aspects are inevitably involved; therefore, understanding engineering from the overall perspective of technology, market, industry competitors, industry, capital, land, resources, labor, environment, culture, and the appropriate management, is necessary. In other words, engineering evolution reflects the integrated evolution process of factors under specific social and natural conditions. Specifically, there are a variety of technical routes, a number of options, a variety of implementation paths to choose in a project. Engineering innovation is integrated optimization of an engineering concept, development strategy, engineering decisions, engineering design, construction process, production running, and organizational management in the process (Yin et al., 2007). There is a strong relationship between engineering evolution (engineering innovation) and its integrated elements (integrated optimization).

7.3 THINKING AND STUDY OF A NEW GENERATION OF THE STEEL MANUFACTURING PROCESS

In China, the thinking about and study of a new generation of the steel manufacturing process started in 2003—the period of the national medium and long-term science and technology development plan. The important topic of the manufacturing process and science and technology development planning focused on the proposition, concepts, and specific connotation of "a new generation of the steel manufacturing process." The first problem that emerged in the thinking process was how to study process engineering. It had been noted that a new generation of the steel manufacturing process is a proposition of an integrated optimization of the engineering science and technology, rather than a simple group of several so-called frontier technologies uniting to form a new generation the steel manufacturing process. In other words, the combination of exploratory, frontier technologies such as the smelting reduction, direct reduction, and thin tape casting can hardly establish a manufacturing process with a stable operation and industrial production. Therefore thinking and study of a new generation of the steel manufacturing process should neither consider things as standard nor tend to the specific application of individual technology. The general theoretical study of integrity, openness, hierarchy, dynamic character in the steel manufacturing process is more important. The old method of figurative observation should be turned to the investigation of the physical nature of the dynamic running process. Based on this rational abstract, concept research, top level design, the dynamic tailored design study and dynamic operation rules, etc., were performed. Thus the concept, connotation, function, design method, and operation rules of a new generation of the steel manufacturing process were established.

7.3.1 Conception Study of Steel Manufacturing Process

In the thinking and study of a new generation of the steel manufacturing process, the concept study was the basic foothold and starting point. Since people are used to studying metallurgy and steel manufacturing process in an atomic/molecular scale or process/device scale, the entire manufacturing process is too complex and includes multiple subjects, so that it was difficult to find a common rule and a study mode from the superficial phenomena of a local process, which is lacking the rational and system studying of the dynamic operation of the entire process and dynamic integrated optimization. Many deficiencies were caused by the reductionism method. Therefore the study of a new generation of the steel manufacturing process should be started from the physical nature of the dynamic process of the steel manufacturing to perform the concept research at the entire dynamic process level.

From the process appearance of the steel manufacturing process, after its evolution over one and a half centuries, the manufacturing process of the modern steel plants has evolved into two basic processes:

1. The process, starting with natural resources, such as iron ore and coal, of "blast furnace → converter → hot rolling → deep processing process" or "smelting reduction → BOF → hot rolling → deep processing," which includes storing, processing, and sintering of raw materials and energy coking → ironmaking (smelting reduction) and steelmaking → steel refining → solidification and reheating → hot and cold rolling process → surface treatment (Fig. 2.5).
2. The process, with energy from renewable resources and electric power, of "electric furnace → steelrefining → continuous casting → hot rolling," which is the electric furnace process with resources of social recycling steel scraps, processing and manufacturing steel scraps, scraps self-production at steel plants, and electric power (secondary energy) (Fig. 2.5).

The method of observing the appearance of the manufacturing process can be described as follows: dividing and restoring of the manufacturing process, then, parsing, dividing and designing of the restored/divided process/device, and operation, respectively. For a long time, there were the following problems in the steel manufacturing process: individual and separate running of the processes/devices and upstream and downstream process/device, which led to waits for each other and random connections between these processes/devices. The results in the above production process were the frequently random pause, large middle inventories, long waiting time, repeatedly cooling/heating rate, low production efficiency, large energy consumption, unstable quality of products, etc. Hence, it was necessary to understand the problem via new thoughts by in-depth study of the entire dynamic operation of the manufacturing process based on its physical nature.

The method to rationally abstract the process of the dynamic operation can be described as follows: systematic thinking about the physical nature of the dynamic operation of the manufacturing process, studying the running rules and the rules of design and operation with the method of analysis and integration of the process dynamic operation.

The process of steel manufacturing is essentially the flow/evolution of mass flow, energy flow, and information flow within certain coverage. The essence of the dynamic running process is: driving the mass flow (mainly iron element flow) by the energy flow (mainly carboniferous flow), operating the specific "process networks" and realizing the multiobjective optimization according to the set "program." In the view of thermodynamics, the steel manufacturing process is an open, nonequilibrium, irreversible, and complex system formed by the nonlinearity interaction and dynamic coupling of different processes/units (Fig. 3.1).

After studying the nature of the dynamic operation in the physical process of the steel manufacturing, it was concluded that there could be a new concept where the dynamic operation system itself is a kind of dissipative structure and the process of the dynamic operation is a dissipation process. In order to minimize the dissipation in the process of the dynamic operation, an optimized dissipative structure should be generated first so that the mass flow can continuously run with a dynamic-orderly and coordinated-continuously mode.

In the study of the physical nature of the dynamic operation process of the steel manufacturing process, it was concluded that there was an understanding that there are three characteristic elements—flow, process network, and operational procedure—in the process of the dynamic operation process. The three elements are suitable for the ferruginous mass flow as well as the related energy flow and information flow.

In the study of the physical nature of the dynamic operation process in the steel manufacturing, a clear conclusion was reached that the steel plant, especially the integrated steel plant, should have the following three functions:

1. The function of the ferruginous mass flow operation, namely, the manufacturing function of the clean steel with high efficiency and low cost, which is the basic starting point to establish a steel plant. In the future development process, not only the market competitiveness but also the sustainable development ability should be considered. For the manufacturing function of the new generation steel plants, strategic issues of technological progress should be considered from the following perspectives:
   a. With the key objective of minimum process dissipation, the dynamic-orderly, coordinated-continuous manufacturing process in the new generation of steel plants is established.

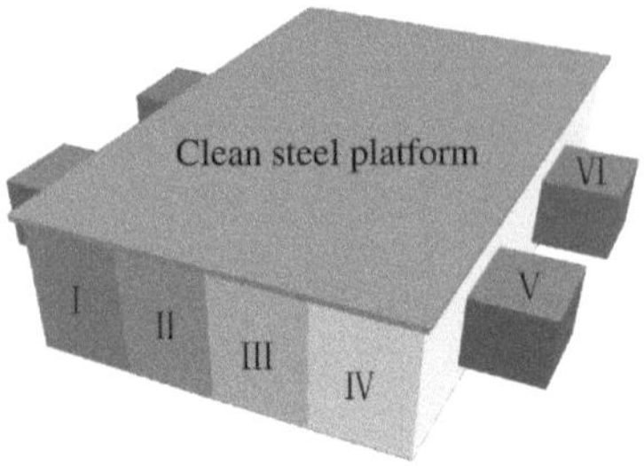

FIGURE 7.3 Clean steel platform with high efficiency and low cost. I, parsing-optimizing iron pretreatment; II, efficiently longevity steel furnace; III, quick-synergic secondary refining; IV, efficiently constant velocity continuous casting; V, updating-concise process network; VI, dynamic-ordering fulfillment technology.

b. Based on principles of parsing-integrated optimization in the metallurgical process, the "clean steel platform" (Fig. 7.3) with high efficiency and low cost is established for different product specialization and stable mass production.

c. Under the guidance of material engineering, reasonable parameters of different steel products are standardized to achieve product updating and environmental protection.

2. The function of the energy flow operation, namely, the function of reasonable energy utilization, high efficient conversion, and waste processing using residual energy. The physical nature and operating characteristics of the steel manufacturing process can be further expressed as: the mass flow consisting of each kind of material is driven by the input energy (primary energy/secondary energy) according to the set technological process and, after a series of variations in shape and properties, the ferruginous mass flow becomes the desired product. During this process, the mass flow and the energy flow sometimes accompany each other and sometimes separate. The accompanied ones interact and influence each other, paying great attention to their energy conversion efficiency and rationality. The separated ones show their individual behavior characteristics, respectively. The "secondary" and "residual" energy separated from the mass flow should be recycled reasonably in time and be efficiently utilized through establishing an energy flow network and a comprehensive control program of the energy flow, which more reasonably and efficiently utilizes energy (Figs. 3.2–3.4 and Figs. 7.4–7.7).

3. The function of the interaction process between the iron flow–energy flow, namely, the function to achieve process goals and recycle waste. From the viewpoint of mass flow and energy flow in the process of steel production, by-products, waste, remaining heat, and energy material are discharged to obtain the expected product. If the discharge process and emissions were mishandled, it will cause a negative impact on the ecological environment. In the steel manufacturing process, mass flow, energy flow, and the interaction between mass flow and energy flow actually provides the possibility of disposing of the given waste and recycling the energy. For example, steel scraps, processing dust, and iron scale can be recycled and metals such as

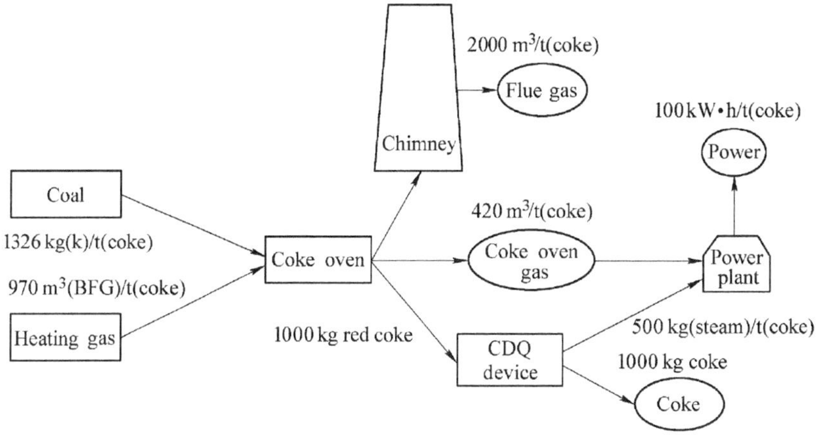

FIGURE 7.4 Block diagram of material and energy use in the coking system (Yin, R., 2008. The essence, function of steel manufacturing process and its future development model (translated). Sci. China Ser. E Technol. Sci. 38 (9), 1365–1377 (in Chinese) (Yin, 2008)). Description: Coking rate of boiling coal by 75.4%, or 1326 kg dry coal produce 1 t coke; calorific value of blast furnace gas by 1000 kcal/m^3 (1 cal = 4.1868 J); dry coal heat consumption of 728 kcal/kg.

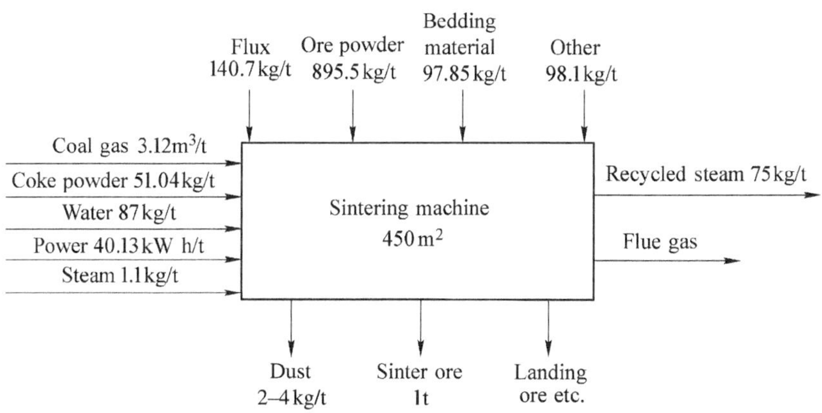

FIGURE 7.5 Schematic of material and energy utilization in sintering system (Yin, R., 2008. The essence, function of steel manufacturing process and its future development model (translated). Sci. China Ser. E Technol. Sci. 38 (9), 1365–1377 (in Chinese) (Yin, 2008)).

Cr, Ni, Cu, zinc, and other metals can be utilized in the mass flow. During the operation of the energy flow, waste plastics, waste tires, community sewage, and garbage can be disposed, and even a large amount of hydrogen can be produced, etc., which links with building materials, electricity, chemical industry and other industries in the process of the extending industrial chain (Fig. 5.7).

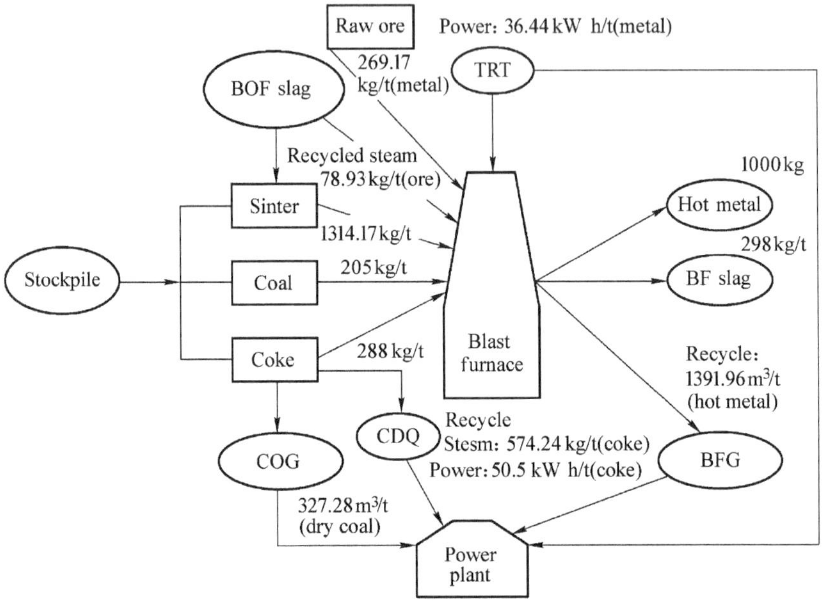

FIGURE 7.6 Schematic of material and energy utilization in blast furnace ironmaking system (Yin, R., 2008. The essence, function of steel manufacturing process and its future development model (translated). Sci. China Ser. E Technol. Sci. 38 (9), 1365–1377 (in Chinese) (Yin, 2008)).

In the future development of steel industries, based on the full understanding of the physical nature of the dynamic operation process in the steel manufacturing process, the function of the steel plants should be expanded to achieve the mode of ecological transformation and be integrated into the circulative economy society.

7.3.2 Study of Top Level Design in the Process

The concept study of the process is the foundation of top level design. The notion of the dynamic-orderly, coordinated-continuous operation and the design concept of the integrated accurate-dynamic operation are necessary. The dynamic concept of the "flow" in the manufacturing process is established, based on the dynamic-orderly, coordinated-continuous in the top level design. Entirety, openness, dynamic character, hierarchy, relevance, and environmental adaptability in the top level design need to stand out. The reasonable selection of elements in the process/device, process structure optimization, expanded function and efficiency are the guiding ideology and basic connotation for the top level design. Logical thinking emphasizes that the top level (process as an entire) determines, stipulates, and guides the bottom level (process/plant) on steps and routes.

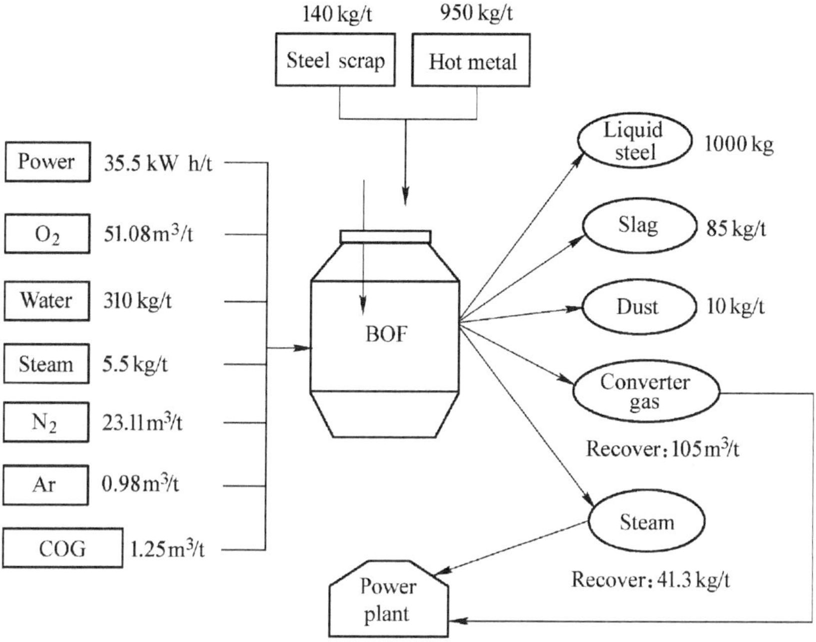

FIGURE 7.7 Schematic of material and energy utilization in steel furnace system (Yin, R., 2008. The essence, function of steel manufacturing process and its future development model (translated). Sci. China Ser. E Technol. Sci. 38 (9), 1365–1377 (in Chinese) (Yin, 2008)).

7.3.2.1 Elements Selection

In the top level design of a new generation of the steel manufacturing process, the choice and optimization of elements includes two main aspects: one is the choice and the optimization of technical factors; the other is the collaborative optimization of technology optimization and the basic economic factors.

In the design and construction of Shougang Jingtang at CaoFeiDian, for example, the selection and optimization of the technology elements mainly included the following points:

1. In the choice of rolling mill, there is a choice between sheet mill and plate mill, or two sheet mills. The reasonable production scale of the former is approximately 7 mt/year and the reasonable production scale of the latter is approximately 9 mt/year, affecting the process, installation, structure, and dynamic efficiency of the steel plant in the steel manufacturing process; and it will further affect the plane layout, volume, and number of blast furnaces (BFs). Two sheet mills were finally chosen in Shougang Jingtang.
2. For the BF, there is a choice between two sets of 5576 m^3 BFs or three sets of 4080 m^3 BFs. After detailed investigation, the plan of two sets of

5576 m^3 BFs was chosen to simplify the layout, reduce the investment, and improve the operation efficiency of graphite-ferruginous mass flow and energy flow (see Section 6.1).

3. For the interface technology between the steel factory and iron factory, there was a choice between the traditional torpedo ladle-transfer-hot metal desulfurization pretreatment process and the multiple functional hot metal ladle that unites the functions of BF tapping, hot metal weighing/transferring, hot metal desulfurization pretreatment and pouring quickly, and is known as "one-ladle technology." Finally, the "one-ladle technology" was chosen since it can reduce the engineering investment, accelerate delivery process, decrease the iron temperature drop, increase the efficiency of hot metal desulfurization, accurately weigh the hot metal, and promote the direct tapping ratio of BOF (see Section 6.2).
4. For the steel plant process/device design, in the sheet production, there was a selection between the traditional process and the process with De[S] − De[Si]/[P] pretreatment → BOF → secondary refining → high-speed and high-efficient continuous casting and low cost platform for clean steel production technology. The latter was chosen to achieve the production of clean steel with low cost and high efficiency. In fact, 2 × (230−250) t dephosphorization pretreatment furnaces and 3 × (230−250) t decarburization converters realize an annual production capacity of 9.2 mt in a steel plant, compared with the 300 t BOF schemes, meaning it is also beneficial to reduce engineering investment (see Section 6.3).
5. For the design of the energy flow network, according to the energy flow and different behavior of the energy medium running process, perfect energy supply system and energy conversion were designed. Moreover, the energy control center with functions of real-time monitoring, online scheduling, and process controlling was designed and constructed. Especially, the concept of "how much gas generates how much electricity" was broken. The pulverized coal is burned together with gas. Two 300 MW power units were constructed, which achieves a 30% gas mixing−burning ratio, increasing the generating efficiency and balancing the relationship between the power generation and the gas discharge in different seasons and different production conditions, and reducing electricity cost. On the premise of economic optimization, nearly zero emissions of gas in the whole flow is realized.
6. For specific energy saving techniques, a 7.63 m large coke oven and 260 t/h dry quenching were used, generating an electricity capacity of 112 kW h/t coke; a technology of 1300°C high temperature BF and 36. SMW dry TRT technology were chosen for the 5576 m^3 BF; the flameproof dry dust removal technology was chosen for the decarburization converter. It practically shows that all effective measures are beneficial to save water and recycle energy. The converter process can output energy rather than consume energy.

7.3.2.2 Structure Optimization

Through the optional selection of a series of process/device elements discussed above, a simple, smooth, and efficient process network, especially the "2-1-2" structure consisting 2 large BOF, 1 De[S] − De[Si]/[P] pretreatment steelmaking plant, two set of broadband hot rolling mill, was established. At the core was the establishment of an orderly, coordinated, continuous dynamic-orderly structure including the graphic mass flow network structure reflecting "minimum directed tree" and the energy flow characterized by the primary loop network structure. The optimization of structures of the mass flow and energy flow promoted simplicity, reliability, and optimization of the information flow structure.

7.3.2.3 Function Expansion and Efficiency Optimization

While selecting elements of the process/device and optimizing the structure of the process, the function expansion and the efficiency optimization are emphasized in the top level of a new generation steel manufacturing design. In other words, the function of a steel mill should be expanded from only manufacturing steel products to energy conversion and waste processing. The connotation of the function needed to be updated.

For example, the manufacturing function of steel products is divided into quality, variety, consumption, and other indicators, which should be integrated into the clean steel production platform with high efficiency and low cost. It is through the process of system integration that it is possible to achieve a clean steel production process with high efficiency and low cost, and guarantee production with high quality, large quantity, and stability.

The energy conversion function, for example, is beyond the original concept of mass and energy balance, and is not limited to the local energy saving of a monomer process/device. According to the optimization of the network structure of the entire steel plant, with characteristics of the input/output dynamic operation optimization, efficient energy conversion including production devices and energy efficient conversion devices, the utilization of by-product heat and recycling are performed in a timely manner and efficiently to achieve energy savings and emission reductions on a higher level.

As another example, the function of the classifying-treating-recycling of waste resources is to build eco-industrial parks related to the mass chain, energy chain, capital chain, and information chain of the steel plant, instead of being limited to recycling and processing (Zhang et al., 2011).

Through the process of the structure optimization and function explanation, the efficiency of mass flow, energy flow, and capital flow, environment friendliness and ecological optimization, which are related to the new generation steel manufacturing, will reach a higher level.

7.3.3 Process Dynamic Tailored Design

For a long time, the design method of a steel plant was to design different processes/devices separately by drawing the reactor diagram, and setting production efficiency and work rates of the reactor based on experience, estimating its annual production capacity, and keeping the surplus energy of reactors. Obviously, it is a kind of split and extensive design method. While in the real production process of a steel plant, different processes/devices have different functions, operation modes, and running rhythms. Therefore, with the concepts of "slow" and "active," the split-extensive design operation mode evolved into the dynamic tailored design method. The establishment of the dynamic tailored design method is the basis of the design concept. In particular, the following points should be emphasized:

1. Core concept. The dynamic tailored design method of the steel manufacturing process is based on the comprehension of the physical nature of dynamic operation. During the design process, three elements of "process network" and "operational procedure" were highlighted. Attention should be paid not only to the efficient conversion of the material and energy in the reactor, but also to the concept of time, space, vector, and network in the design of the dynamic-orderly and cooperative-continuously operational mass flow and energy flow. Under the principle of the consistency and irreversibility of time, the character of position, segment, rhythm, and periodicity should be designed during the dynamic operation of the flow. It benefits not only the dynamic precision of the design idea, but also the dynamic management of operational time plans during the real production process and thus benefits the organization and the execution of the manufacturing enterprise.
2. Establishment of the coordinated spatiotemporal relationship. For the dynamic tailored design, time reflects the continuity of flow, the coordination of procedures, the dynamic coupling of process factors between procedure/reactor on the timeline, as well as the energy loss caused by the temperature drop during the transport and waiting process. The determination of the main equipment adoption, the number of reactors, the process layout, and general drawing means that the static space structure of the steel plant has been already solidified, namely, the boundary of time–space has been fixed. Therefore the basic concepts of "flow," "process network," and "operational procedure" should be implemented into the design of the process layout and general drawing. It is also important to fully consider and optimize the line/arc connection distance and the connection path between "node," "node capabilities," "node capacity," "node quantity," and "node location."

   The method of expression and optimization design is to carefully calculate the coordination and matching of processes/devices in the upstream/downstream containing time process, time factors such as timing, time domain, time position, and time period.

3. Emphasis on the establishment and optimization of a "process network." The basic difference between conventional static capacity design and dynamic-orderly accuracy design is the design of a simple smooth "flow network," which is one of the time space synergy carriers and the framework of the time space synergy. The dynamic operation is specifically the reflection of the time–space relationship. In dynamic tailored design, dynamic operation and information regulation of the steel manufacturing process, it is necessary to establish the concept of "process network" to ensure the simple, compact, and smooth process layout and general drawing of the steel plant, which lets the "flow" run in a dynamic-orderly and cooperative-continuously state, leading to a minimum dissipation during the operation process. It should be noted that the "process network" is initially reflected as a "process network" of ferruginous mass flow. Meanwhile, it is very important to study and develop the process networks of energy flow and information flow.
4. Emphasis on the connective-matching relationship between processes/devices and the development and application of interface technology. One of the key ideas of dynamic tailored design is to optimize the ontology of each procedure/device, as well as to emphasize the connective-matching relationship between procedures/devices and the development and application of interface technology, such as the "one-ladle technology" between the ironmaking plant and steelmaking plant. Using tools like the dynamic Gantt graph, each device and its dynamic operation in the steel manufacturing process can be carefully designed in advance.
5. The integrative innovation of the top level design is highlighted. The design of the steel plant is the multispecialty crossing and coordinate-innovative integration process based on the procedure/device, and solving multiobjective optimization problems is the essence of the design. Every engineering design is different with a difference in locations, resources, environment, weather, terrain, transport conditions, and product markets. Meanwhile the designer will properly introduce new technologies in the design according to the relevant technical progress. The combination of the new technology and the existing advanced technology (or effective "embedded" technology) reflects the integrative innovation. Integrative innovation is the main content and method of the self-dependent innovation which requires not only the optimal innovation of the cell technology but also the organically and orderly combination of each optimized cell technologies. The integrated optimization in the top level design is highlighted, forming a dynamic-orderly, coordinative-continuously, and stable-efficiently flow system. Meanwhile, it should be noted that the "frontier" of individual exploration technology belongs to a partial test (probably not mature) which is not necessarily reflected in the top level design. Thus the application is based on the maturity and possibility of orderly and effectively embedding technology into the flow system.

Integrative innovation is not simply scraping various individual frontier explored technologies.

6. The stability, reliability, and efficiency of the whole process dynamic operation are highlighted. Dynamic-orderly and coordinative-continuously operational rules and procedures need to be determined for the dynamic tailored design method. Not only the dynamic operation of each procedure but also the whole process by the effect of the connective-matching and nonlinearity coupling-dynamic operation should be emphasized. Especially, the stability, reliability, and efficiency of the dynamic operation should be focused upon, which is the goal of the dynamic tailored design method.

7.3.4 Study of the Entire Process Dynamic Operation Rules

To design and establish a dissipative structure of the optimal flow operation and achieve the minimization of the dissipation of the dynamic operation process, procedures must be operated dynamically-orderly-coordinated-compactly. Therefore several operational rules for different procedures/devices or the entire flow operation are necessary, which come from and reflect the law of the dynamic operational process. The following rules are included:

1. Intermittently operated processes/devices should adapt to and obey the requirements of the quasicontinuous/continuous operated process/device dynamic operation. For example, the steelmaking furnace and refining furnace should adapt to and obey the parameter requirements, such as molten steel temperature, composition, and especially time rhythm proposed by sequence continuous caster.
2. Quasicontinuous/continuous operated procedures/devices should guide and standardize the batch operated procedure/device operation behavior. For example, an efficient constant speed continuous caster should match the molten steel temperature, molten steel cleanness, and time arrangement of the relevant iron pretreatment, steelmaking furnace and refining furnace.
3. Procedures/devices continuously operated at low temperature should obey the high temperature ones. For example, the manufacturing process and quality of the sintering and pellet should obey the dynamic operation requirements of the BF.
4. In the series–parallel flow structure, a laminar type running production line should be established as much as possible, for example, establishing a relatively fixed "casting machine-refining reactors-steelmaking furnace" professional production line in the steel plant.
5. Matching and using a compact layout of the capacity between the upstream and downstream procedure/device are the bases of the laminar

type running. For example, during the hot charging of billets and slabs, matching capability and fixing-coordinated operation between continuous caster and heating furnace and hot rolling mill are required.

6. In the operation of the overall manufacturing process, the kinetic mechanism of the "pushing source-buffer-pulling source" is necessary.

7.3.5 Some Recognization for the New Generation of the Steel Manufacturing Process

The flow manufacturing process can be abbreviated to the manufacturing process. The manufacturing process consists of the related, heterogeneous, and different structure-functional processes/devices, which form a dynamic integrated operation system. The manufacturing process must be a stably, efficiently, safely, and continuously operated production system. The manufacturing process reflects the constituent elements, overall structure, operation function, and operation efficiency of the enterprise, which is the basis of market competitiveness and sustainable development.

The new generation of the steel manufacturing process is different from the individual advanced technology exploratory study or pilot trials, such as strip continuous casting, nonblast furnace technology, etc., which are not a simple combination of all exploratory technologies. The new generation of the steel manufacturing process is based on its dynamic operation physical nature and is a new system consisting of the dynamic integration of mass flow, energy flow, and information flow, rather than on a figurative reconstruction of the existing procedure/device. Its core theory is to establish a "flow" concept (mass flow, energy flow, and information flow) by the synergistic consolidation, design, and establishment of a "process network" (including mass flow, energy flow, information flow network) and the corresponding dynamic operation program with advanced elements, synergy-continuous-efficient operation structure. It is also an engineering system with functions of high efficiency, low cost, high-quality steel products, efficient energy conversion, bulk waste treating, and recycling. The new generation of the steel manufacturing process is a stable, efficient, safe, environmentally friendly, and continuously operated production system, which reflects the advancement, the rationality of the whole structure, the efficiency of dynamic operation process, and the expansibility of the process function.

The conception study and top level design of the dynamic operation of the new generation of the steel manufacturing process indicated that metallurgy has transformed from an isolated theoretical segmentation study to an open dynamic system study; the operating characteristics of the manufacturing process are the transformation from "running alone-waiting for each other-random combination" of each procedure/device to "functional optimization-cooperated matching-nonlinearity dynamic coupling-ordered-continuous-compact process" of each procedure/device.

7.4 DEVELOPMENT DIRECTION OF METALLURGICAL ENGINEERING IN THE VIEW OF ENGINEERING PHILOSOPHY

From the perspective of engineering philosophy, for a long time, the relevant steel engineering design process and production operation process have focused attention on the local "substantiality," however, the global "flow" is overlooked. In the future, the engineering design, production operation and process management should not only address specific, local "substantiality," but focus on penetrating global "flow." If the dynamic concept is separated from "flow," it is just like losing its "soul"—the effect is bad. Judging from appearances, the production operation and engineering design of enterprises looks "substantial" (for process/design and operation of the device), but the "substantiality" is the very embodiment of dynamic running through the global "flow" in essence. Namely, the process/device of the "substantiality" is the dynamic form of operation and a local movement. The dynamic-orderliness, collaboration-continuous operation of "flow" (namely, self-organization optimization of dissipative structure makes process dissipative optimization) is the purpose of the production operation and the engineering design in the plant, which is the soul. Engineering design and factory production is the combination of "virtuality" and "substantiality." First of all, it must establish the concept—"virtuality," a dynamic-ordered, synergetic-stabile, continuous-compact open system-optimized dissipative structure, through the application of engineering design; and the dynamic production—"substantiality," is to pursue the dissipative processes "minimization" in the operation, and then to realize complex system multiobjective optimization. In other words, as a matter of the element-structure-function-efficiency integration optimization, it should reflect a dynamic-orderliness, collaboration-continuous manufacturing process running optimization in the engineering design, which is the theoretical core of the dynamic-precise design and the practical production operation process.

Dissipative structure and self-organization theory are physical propositions. In order to guide social production, the physical proposition needs to be transformed into an engineering proposition. In the production process, three elements of the process dynamic operation—process network structure, running program, throughput—should be analyzed to reduce the dissipation in the processes (to reduce energy consumption and material loss, etc.) as much as possible to improve the efficiency of the process operation.

1. About network integration. To build a relatively stable, dynamic-ordered process structure, it is very important to construct the network structure. Intuitively, a "network" is a graph composed of nodes and connected lines (arcs), which form a specific structure. Directly, the network structure in the dynamic-ordered-synergetic running in the process system is the optimized region of dynamic coupling of the "force" and "flow." The "force" and "flow" have nonlinearity interaction between the relevant nodes in the system. The nonlinear interaction and dynamic coupling of

the "force" and "flow" are the structural basis of the dynamic operation dissipation. The network integration means that the relationship between the nonlinear interaction of the relevant nodes in the system and dynamic coupling can be "frozen" in some optimized "resultant force" field, thus contributing to a dissipation structure system dynamic-ordered, synergetic-continuous operation.

In the steel manufacturing process, network integration means the rational choice among the process/device capacity (ability), function, number, position and the simplified, compact, smooth general layout; as well as, maintaining "laminar type" smooth running of the ferruginous mass flow, namely "resistance" minimization. "Resistance" minimization, basically, is embodied as decreasing the ferruginous flow motion process and shortening the process time. The network integration is also important to energy flow and operation optimization, which means that the static concepts of the material balance and thermal equilibrium are broken in order to study the energy flow with an open, input/output model view in the steel manufacturing process. The static computation of the local material balance and thermal balance in procedure/device should not be limited. The skeltering relationship between energy flow and ferruginous mass flow within a steel plant should be noticed. Therefore the energy flow is also necessary for network integration and programmatic synergy to improve energy conversion efficiency and utilize or recycle in a timely manner all types of secondary energy (such as waste heat, complementary energy).

On the contrary, if the "process network" is unreasonable (including the mass flow network, energy flow network, and information flow network) and the "flow" tends to space factor, time factors disorder or the chaotic state from time to time, this will lead to the increase of mass flow loss and energy flow dissipation. The dynamic operation ordering of the metallurgical manufacturing process depends on not only the order and stability of the respective operation of each producer/device, but also the promotion or restriction of the "process network" degree of integration.

2. About the programmatic synergy. Programmatic synergy of the steel manufacturing process is primarily programmatic synergy of various types of information. It is associated with analysis-optimization of the procedure characteristic set in the manufacturing process, synergy-optimization of the procedure relationship set, and the reconstitution-optimization manufacturing process procedure set. The space sequence should be compact, simplified, and laminar flow—these are closely related to network integration—determining the static structure of the manufacturing process. Programmatic synergy shows the rational function sequence design, and more importantly is reflected in the design and programming of the space program, time program, and time−space program. Design and programming of the time program requires a full understanding of the time factor's various manifestations in the steel manufacturing process dynamic running (such as time sequence, time point, time domain, time position, time cycle), and strives for coordination of the rapidity of the time program within the whole network and whole journey. The design of the time sequence is embodied in the optimized dynamic frame structure in high efficiency, coordination, stabilization of the mass flow, energy flow dynamic-ordered operation in order to realize throughput rationalization and material loss—energy dissipation "minimality." The programmatic synergy should prominently reflect the synergetic-optimized dynamic information, achieving a controllable nature.

3. About mass throughput. Reasonable mass throughput mainly depends on the product characteristics. For a steel plant, mass throughput of thin sheet production is relatively large, and the annual production capacity of operating line is between 2 and 5.5 mt. While the production of long wood products is relatively small, the annual production capacity of the operation line is between 0.6 and 1.2 mt, which determines mass throughput of various steel manufacturing process dynamic running (unit: t/min). Thus the so-called equipment "enlargement" is not a true reflection, and the key is to determine a reasonable mass throughput, rather than blindly emphasize "bigger is better" for equipment in the steel manufacturing process. Only the pursuit of equipment "enlargement" during production is uncoordinated and unreasonable. For example, the mass throughput of the production of long products using a 3 mt converter is uncoordinated and unreasonable.

The steel manufacturing process can be separated into ironmaking, steelmaking, and rolling process, which run independently. There may be the "best" solutions of their own independent operation. However, with the current development, "isolation" and "best" should be investigated, the optimization of the entire dynamic running process is more important in the manufacturing process of dynamic running. Thus the split method in the mechanism can hardly be used to solve problems of relevant and heterogeneous function and asynchronous procedure/device integrated running. Hence, it is important to study the process engineering of the multifactor, multiscale, multiprocedure, multilevel open system dynamic operation. The exterior–interior relationship, causal relationship, nonlinearity interaction, and dynamic coupling relationship between representation and physical essence must be distinguished to discover their inherent laws, which is significant for the steel design and optimization and operation regulation.

Metallurgical engineering is the basic component of the development of the steel industry (including science, technology, engineering and design, management etc.). The following problems should be investigated to drive the development of the discipline and its branches, and to promote interdisciplinary approaches:

1. The physical essence of the dynamic operation of the steel manufacturing process should be studied to explore the integration theory of mass flow, energy flow, and information flow in the dynamic process;
2. The "flow," "dynamic operation," and "running program" in the steel dynamic operation process, as well as their influence on the factor structure, function, efficiency, and impact on the steel business, should be studied;
3. The design level and running efficiency of the steel manufacturing process flow should be improved by the means of network integration, programmatic synergy (measures of process structure integration innovation);
4. The multiprocess, multilevel, multiscale, multifactor integrated optimization of development, design, and manufacture of new technology and new equipment is important, and should be dynamically "embedded" into the steel manufacturing process;
5. Energy flow and network integration in the entire process level are very important to improve the efficiency of energy utilization and further promote energy conservation and emission reductions from the entire level of system;
6. Studies on the effective regulation of the information optimization combination of the mass flow, energy flow in the dynamic operation process should be emphasized to promote the perforation of the information flow in the steel manufacturing process;
7. The cultivation, training, and utilization of elite talents with comprehensive knowledge—excellent engineers and strategic scientists—are very important;
8. The exploration of strategic countermeasures for the responsibility of the times and social–ethical propositions, such as environmental protection, ecology, climate change, and social ethics, are to be emphasized.

Through the studies related to metallurgical engineering discussed above, will be made important contributions towards the direction of development and source exploration of the competitive dynamics of the iron and steel industry in the future.

REFERENCES

Engels, F., 2009. Ludwig Feuerbach und der Ausgang der klassischen deutschen Philosophie. Bernd Müller Verlag, Zittau, p. 46.

Freeman, C., Soete, L., 2004. The Economics of Industrial Innovation. (H. Hua, H. Hua, Trans.). Peking University Press, Beijing (in Chinese).

Prigogine, I., 1980. From being to becoming. Chinese J. Nat. 3 (6), 11–14 (in Chinese).

Xing Huaibin, 2009. Analysis on the barriers and traps in engineering innovation. J. Eng. Studies 1 (1), 58–65 (in Chinese).

Yang Kuan, 2004. The Development History of Iron-Smelting Tech in Ancient China. Shanghai People's Publishing House, Shanghai, pp. 232–260 (in Chinese).

Yin Ruiyu, 2008. The essence, function of steel manufacturing process and its future development model (translated). Sci. China Ser. E Technol. Sci. 38 (9), 1365–1377 (in Chinese).

Yin Ruiyu, Wang Yingluo, Li Bocong, 2007. Philosophy of Engineering. Higher Education Press, Beijing (in Chinese).

Yin Ruiyu, Li Bocong, Wang Yingluo, et al., 2011. Theory of Engineering Evolution. Higher Education Press, Beijing (in Chinese).

Zhang Chunxia, Yin Ruiyu, Qin Song, et al., 2011. Steel plants in a circular economy society in China. Iron and Steel 46 (7), 1–6 (in Chinese).

Index

Note: Page numbers followed by "*f*" and "*t*" refer to figures and tables, respectively.

F

G

H

I

J

K

L

T

U

V

W

FSC
www.fsc.org
MIX
Papier | Fördert gute Waldnutzung
FSC® C083411

Kolibri 360 GmbH
Ferdinand-Jühlke-Straße 7
99095 Erfurt, Deutschland
produktsicherheit@kolibri360.de